Volume 71

Epigenetics and Cancer, Part B

Advances in Genetics, Volume 71

Serial Editors

Theodore Friedmann
University of California at San Diego, School of Medicine, USA
Jay C. Dunlap
Dartmouth Medical School, Hanover, NH, USA
Stephen F. Goodwin
University of Oxford, Oxford, UK

Volume 71

Epigenetics and Cancer, Part B

Edited by

Zdenko Herceg

Epigenetics Group
International Agency for Research on Cancer (IARC)
Lyon, France

Toshikazu Ushijima

Carcinogenesis Division
National Cancer Centre Research Institute
Tokyo, Japan

AMSTERDAM • BOSTON • HEIDELBERG • LONDON
NEW YORK • OXFORD • PARIS • SAN DIEGO
SAN FRANCISCO • SINGAPORE • SYDNEY • TOKYO
Academic Press is an imprint of Elsevier

Academic Press is an imprint of Elsevier

525 B Street, Suite 1900, San Diego, CA 92101-4495, USA
30 Corporate Drive, Suite 400, Burlington, MA 01803, USA
32 Jamestown Road, London, NW1 7BY, UK
Radarweg 29, POBox 211, 1000 AE Amsterdam, The Netherlands

First edition 2010

Notice

No responsibility is assumed by the publisher for any injury and/or damage to persons or
property as a matter of products liability, negligence or otherwise, or from any use or operation
of any methods, products, instructions or ideas contained in the material herein. Because of
rapid advances in the medical sciences, in particular, independent verification of diagnoses and
drug dosages should be made.

ISBN: 978-0-12-380864-6
ISSN: 0065-2660

For information on all Academic Press publications
visit our website at elsevierdirect.com

Printed and bound in USA

10 11 12   10 9 8 7 6 5 4 3 2 1

Working together to grow
libraries in developing countries

www.elsevier.com | www.bookaid.org | www.sabre.org

ELSEVIER   BOOK AID
            International   Sabre Foundation

Contents

SECTION III   **APPLICATION OF EPIGENETICS IN**
MOLECULAR EPIDEMIOLOGY
AND EPIGENETIC CANCER
PREVENTION   209

Contributors

Numbers in parentheses indicate the pages on which the authors' contributions begin.

Jia Chen (237) Department of Preventive Medicine, Mount Sinai School of Medicine, New York, USA

Dajun Deng (125) Key Laboratory of Carcinogenesis and Translational Research (Ministry of Education), Peking University School of Oncology, Beijing Cancer Hospital and Institute, Fu-Cheng-Lu, Haidian District, Beijing, 100142, PR China

Yantao Du (121) Peking University School of Oncology, Beijing Cancer Hospital and Institute, Fu-Cheng-Lu, Haidian District, Beijing, 100142, PR China

John K. Field (177) University of Liverpool Cancer Research Centre, Liverpool, United Kingdom

Triantafillos Liloglou (177) University of Liverpool Cancer Research Centre, Liverpool, United Kingdom

Jia (Jenny) Liu (79) Lowy Cancer Research Centre, and Prince of Wales Clinical School, University of New South Wales, Kensington, New South Wales, Australia

Zhaojun Liu (125) Key Laboratory of Carcinogenesis and Translational Research (Ministry of Education), Peking University School of Oncology, Beijing Cancer Hospital and Institute, Fu-Cheng-Lu, Haidian District, Beijing, 100142, PR China

John C. Mathers (1) Human Nutrition Research Centre, Institute for Ageing and Health, Newcastle University, Newcastle upon Tyne, United Kingdom

Tohru Niwa (41) Carcinogenesis Division, National Cancer Center Research Institute, Chuo-ku, Tokyo, Japan

Maté Ongenaert (259) OncoMethylome Sciences, Liege, Belgium

Caroline L. Relton (1) Human Nutrition Research Centre, Institute for Ageing and Health, and Sir James Spence Institute, Newcastle University, Royal Victoria Infirmary, Newcastle upon Tyne, United Kingdom

Jackilen Shannon (57) Department of Public Health and Preventative Medicine, and Center for Research on Occupational and Environmental Toxicology, Oregon Health & Science University, Portland, Oregon, USA

Gordon Strathdee (1) Crucible Laboratory, Institute for Ageing and Health, Newcastle University, Newcastle upon Tyne, United Kingdom

Kent L. Thornburg (57) Department of Medicine, Division of Cardiovascular Medicine, and Heart Research Center, Oregon Health & Science University, Portland, Oregon, USA

Philippe Thuillier (57) Department of Public Health and Preventative Medicine, and Center for Research on Occupational and Environmental Toxicology; OHSU Knight Cancer Institute, Oregon Health & Science University, Portland, Oregon, USA

Mitchell S. Turker (57) Center for Research on Occupational and Environmental Toxicology, and Department of Molecular & Medical Genetics, Oregon Health & Science University, Portland, Oregon, USA

Toshikazu Ushijima (41) Carcinogenesis Division, National Cancer Center Research Institute, Chuo-ku, Tokyo, Japan

Robyn Lynne Ward (79) Lowy Cancer Research Centre, and Prince of Wales Clinical School, University of New South Wales, Kensington, New South Wales, Australia

Xinran Xu (237) Department of Preventive Medicine, Mount Sinai School of Medicine, New York, USA

Yasuhito Yuasa (211) Department of Molecular Oncology, Graduate School of Medical and Dental Sciences, Tokyo Medical and Dental University, Tokyo, Japan

Preface

Epigenetics is a fascinating and rapidly expanding field of modern biology. Over the past decade, the field has witnessed a remarkable improvement in our knowledge of the importance of epigenetic events in the control of both normal cellular processes and abnormal events associated with disease. Both the scientific and medical communities now recognize that epigenetic changes lie at the heart of many complex diseases, most notably cancer.

Epigenetic events have been shown to be associated with virtually every step of tumor development and progression. They are also likely to occur very early in tumor development. The advent and rapid development of new technologies for epigenomics has started to unravel molecular features of cancer cells responsible for cancer development and progression, and to identify novel targets for diagnostics and therapeutics. Accurate measurement of various epigenetic modifications allowed us to evaluate the contribution of environmental, dietary, and lifestyle factors to human cancers. These advances have turned academic, medical, and public attention to the application of epigenetics to cancer prevention, diagnosis, and treatment.

We felt that it is important to deliver many conceptual breakthroughs and technological advances that are likely to revolutionize the traditional concept of cancer and cancer research to a broad scientific and medical community. For this book, we have invited many leading scientists, who have made important contributions to epigenetics and epigenomics and shaped the current trends in the field. We have attempted to "cover" the state-of-the-art in cancer epigenetics and cutting-edge technologies in epigenomics. Our aim was to discuss the state of science and future research needs covering the most recent advances, both conceptual and technological, and to provide novel opportunities for cancer prevention, diagnosis, and treatment. Although this book is intended primarily for academic and professional readers (from basic science to clinical researchers and epidemiologists), we believe that it will appeal to and be used by a wider audience among healthcare workers.

We thank all the authors for their valuable contribution and for making this book what it is. We are much obliged to the reviewers, who spend their precious time reviewing the manuscripts; their constructive criticism and candid opinions significantly improved both the scope and quality of the chapters. Special thanks are due to Drs. Andrea Baccareli, Amir Eden, Robert Dante, Aleksandra Fučić, Koraljka Gall-Trošelj, Anastas Gospodinov, Hector Hernandez-Vargas, Barry Iacopetta, Atsushi Kaneda, Yutaka Kondo, Vladimir Krutovskikh, Saadi Khochbin,

Heinz Linhart, Joel Mason, John Mattick, Kent Nephew, Magali Olivier, Anupam Paliwal, Gerd Pfeifer, Christoph Plass, Hidenobu Soejima, Hiromu Suzuki, Minoru Toyota, Thomas Vaissière, Paolo Vineis, André Verdel, Joseph Wiemels, Nick Wong, and Daniel Worthley. We are also grateful to Sandrine Montigny (from IARC, Lyon) for her excellent secretarial help and final formatting of the chapters. Thanks are also due to Zoe Kruze and Narmada Thangavelu (from Elsevier) for their understanding and patience during (often lengthy) process of preparation and review of the manuscripts. We thank all our colleagues in our respective laboratories (in Lyon and Tokyo) for their understanding, enthusiasm, and support during the preparation of this book.

Zdenko Herceg
Lyon, France
Toshikazu Ushijima
Tokyo, Japan

Epigenetic Changes Induced by Environmental and Dietary/ Lifestyle Factors

1

Induction of Epigenetic Alterations by Dietary and Other Environmental Factors

John C. Mathers,* Gordon Strathdee,† and Caroline L. Relton*,‡

*Human Nutrition Research Centre, Institute for Ageing and Health, Newcastle University, Newcastle upon Tyne, United Kingdom
†Crucible Laboratory, Institute for Ageing and Health, Newcastle University, Newcastle upon Tyne, United Kingdom
‡Sir James Spence Institute, Newcastle University, Royal Victoria Infirmary, Newcastle upon Tyne, United Kingdom

Advances in Genetics, Vol. 71
0065-2660/10 $35.00
DOI: 10.1016/S0065-2660(10)71001-4

ABSTRACT

Dietary and other environmental factors induce epigenetic alterations which may have important consequences for cancer development. This chapter summarizes current knowledge of the impact of dietary, lifestyle, and environmental determinants of cancer risk and proposes that effects of these exposures might be mediated, at least in part, via epigenetic mechanisms. Evidence is presented to support the hypothesis that all recognized epigenetic marks (including DNA methylation, histone modification, and microRNA (miRNA) expression) are influenced by environmental exposures, including diet, tobacco, alcohol, physical activity, stress, environmental carcinogens, genetic factors, and infectious agents which play important roles in the etiology of cancer. Some of these epigenetic modifications change the expression of tumor suppressor genes and oncogenes and, therefore, may be causal for tumorigenesis. Further work is required to understand the mechanisms through which specific environmental factors produce epigenetic changes and to identify those changes which are likely to be causal in the pathogenesis of cancer and those which are secondary, or bystander, effects.

Given the plasticity of epigenetic marks in response to cancer-related exposures, such epigenetic marks are attractive candidates for the development of surrogate endpoints which could be used in dietary or lifestyle intervention studies for cancer prevention. Future research should focus on identifying epigenetic marks which are (i) validated as biomarkers for the cancer under study; (ii) readily measured in easily accessible tissues, for example, blood, buccal cells, or stool; and (iii) altered in response to dietary or lifestyle interventions for which there is convincing evidence for a relationship with cancer risk.   © 2010, Elsevier Inc.

I. INTRODUCTION

Diet and lifestyle are major determinants of chronic disease and premature mortality. A recent quantitative analysis of the impact of 12 modifiable dietary, lifestyle, and metabolic risk factors on mortality in the United States concluded that smoking and high blood pressure are responsible for largest number of deaths with each accounting for about one in five or six deaths in adults (Danaei *et al.*, 2009). Obesity, physical inactivity, alcohol consumption, and poor diet were also major risk factors for premature death, and cardiovascular diseases, cancers, respiratory diseases, and injuries were the most prominent causes of mortality (Danaei *et al.*, 2009). From an etiological perspective, it is difficult to separate the effects of a number of diet and lifestyle factors since they may be interdependent, for example, obesity results from a sustained positive imbalance between dietary energy intake (poor diet) and energy expenditure (physical inactivity;

Mathers, 2010), and dietary factors, including high salt intake, positive energy imbalance (obesity), and heavy alcohol consumption, contribute to risk of hypertension (Stanner, 2005). In addition, it may be helpful to consider clusters of risk factors since some factors contribute to risk of several diseases, for example, obesity is a risk factor for cancer at a number of sites and for cardiovascular diseases. A prospective study of 20,244 middle aged and older people in the United Kingdom found that the combined effects of just four health behaviors (current nonsmoking, not physically inactive, moderate alcohol intake, and plasma vitamin C concentration > 50 μM (a biomarker indicative of intakes of at least five servings of vegetables and fruits daily)) predicted a fourfold difference in total mortality with an estimated impact equivalent to 14 years in chronological age (Khaw et al., 2008). These studies reinforce the importance of understanding the dietary and lifestyle determinants of cancers and other age-related chronic diseases especially those neoplasms for which there is a striking relationship between ageing and cancer incidence (DePinho, 2000). Equally important is the need to use that understanding to develop, and to implement, effective interventions such as those proposed by the World Cancer Research Fund/American Institute for Cancer Research (2007) in the domains of food, nutrition, and physical activity.

A. Dietary, lifestyle, and other environmental determinants of cancer risk

1. Tobacco

For more than half a century, it has been evident that tobacco consumption causes lung cancer. Tobacco also causes tumors of the larynx, pancreas, kidney, and bladder and, in conjunction with alcohol consumption, tobacco use is associated with a high incidence of carcinomas of the oral cavity and esophagus (Stewart and Kleihues, 2003). In most economically developed countries, where tobacco consumption accounts for up to 30% of malignant tumors (Stewart and Kleihues, 2003), there are sustained and progressive disincentives to tobacco smoking. With the corresponding reduction in cigarette-induced lung cancer, lung cancer in individuals who have never smoked tobacco products is an increasing medical and public health issue and research is underway to identify the genetic (and other) causes of this type of lung cancer (Li et al., 2010).

2. Alcohol consumption

In their recent report, the World Cancer Research Fund/American Institute for Cancer Research (2007) recommended that, for those who consume alcoholic drinks, consumption should be limited to no more than two drinks per day for men and one per day for women. This recommendation is based on the convincing evidence that alcoholic drinks are a cause of cancers of the mouth, pharynx, larynx, and esophagus in both sexes and breast (in women) and colorectum (in men). Such drinks are also a probable cause of hepatocellular cancer (both sexes) and of colorectal cancer in women (World Cancer Research Fund/American Institute for Cancer Research, 2007). Even low to moderate alcohol consumption is associated with increased cancer risk in women, especially among current smokers (Allen *et al.*, 2009). In addition, obesity and alcohol consumption combine to increase the risk of morbidity and mortality from liver cirrhosis (Hart *et al.*, 2010; Liu *et al.*, 2010) and these adverse exposures contribute to the development of primary hepatocellular carcinoma (Siegel and Zhu, 2009).

3. Physical activity

Studies in rodents have shown that both voluntary and imposed exercise reduces bowel cancer (Basterfield *et al.*, 2005) and there is now convincing evidence from human epidemiological studies that physical activity protects against colon cancer (World Cancer Research Fund/American Institute for Cancer Research, 2007). In addition, physical activity probably protects against postmenopausal breast cancer and endometrial cancer (World Cancer Research Fund/American Institute for Cancer Research, 2007). As noted below, obesity is also a risk factor for the same cancers (and others) so that it is difficult to separate, etiologically, the adverse effects of sedentary behavior *per se* from the contribution which the latter makes to positive energy balance and, therefore, to obesity-related cancers. However, observations from a population-based, case-control study of colon cancer in Shanghai, China, indicated that commuting physical activity modified significantly the adverse effects of increased adiposity (measured as body mass index (BMI, Hou *et al.*, 2004). The highest colon cancer risk was experienced by those in the highest quintile of BMI and with the lowest physical activity level (Hou *et al.*, 2004).

4. Obesity

Over the past 10–15 years, the evidence that overweight and obesity increase cancer risk at several sites (including several sites within the gastrointestinal tract plus the endometrium, kidney, gallbladder, and breast; postmenopausal) has strengthened considerably (World Cancer Research Fund/American Institute for Cancer Research, 2007). As with cardiovascular disease and type 2 diabetes, abdominal (visceral) adiposity appears to be particularly pathogenic and increases the risk of large bowel cancer (Larsson and Wolk, 2007). Central obesity is also a probable cause of cancers of the pancreas, endometrium, and breast (postmenopausal; World Cancer Research Fund/American Institute for

Cancer Research, 2007). These observations begin to connect, etiologically, several cancers with the metabolic syndrome and may suggest that there are common underlying mechanisms with chronic systemic inflammation as a key potential candidate (Hotamisligil, 2006). Support for the concept that metabolic dysregulation is central to the development of cancer, diabetes, and cardiovascular disease is provided by the recent observation of a common transcriptional signature shared by cancer and inflammatory and metabolic diseases (Hirsch et al., 2010). Of genes showing aberrant expression in cancer, 54 were also implicated in aspects of the metabolic syndrome (Hirsch et al., 2010). Importantly, 11 out of 13 drugs used to treat noncancer diseases also inhibited cellular transformation (Hirsch et al., 2010). These results suggest that interventions which reduce the risk of the metabolic syndrome may also be effective in cancer prevention.

5. Diet

Three decades ago, the pioneering studies of Doll and Peto (1981) (Jenab et al., 2009; Johansson et al., 2010; Spencer et al., 2008) provided strong evidence that dietary factors may be as important as smoking behavior in explaining variation in cancer risk—each explaining about 30% of the variance. However, diet is a very complex exposure and conventional epidemiological approaches are relatively blunt instruments when attempting to dissect out which dietary components, in what doses and over what time periods, enhance risk or protect against cancer development. Such difficulties underlie current attempts to estimate dietary exposure using biomarker approaches (Jenab et al., 2009; Johansson et al., 2010; Spencer et al., 2008). Nevertheless, there is good evidence that diets which contain a high proportion of plant foods (fruits, vegetables, relatively unprocessed cereals, and pulses) are associated with lower risk of several common cancers whereas higher intakes of red, and especially processed, meats appear to be causal for colorectal cancer (World Cancer Research Fund/American Institute for Cancer Research, 2007). In contrast, there is an inverse correlation between higher intakes of fish and colorectal cancer risk (Norat et al., 2005). Very recent data from the large European Prospective Investigation of Cancer and Nutrition (EPIC) study suggest rather weak negative associations between fruit and vegetable intake and overall cancer risk (Boffetta et al., 2010). Recent evidence that above median circulating concentrations of methionine, B_6, and folate several years prior to disease onset combine to reduce dramatically risk of lung cancer (Johansson et al., 2010) point to this area as a high priority for future research into the influence of methyl donors on cancer etiology.

Although there is strong evidence from observational studies that higher intakes of specific micronutrients, for example, vitamin A, folate, selenium, and vitamin D are associated with lower cancer risk, the outcomes of intervention studies testing the antineoplastic effects of such nutrients have often been disappointing (see, e.g., Mathers, 2009 for review of folate and bowel cancer risk). In general, dietary intervention studies for cancer chemoprevention have been carried out in higher risk subjects using relatively large (sometimes unphysiological) doses of individual micronutrients and such designs may have limited relevance in attempts to understand the impact of diet on cancer risk.

6. Stress

In contrast with the many detailed mechanistic studies of the effects of physical stressors such as oxidative stress and inflammation on cancer risk, there has been limited research which attempts to understand how psychosocial stressors modulate cancer biology. Much of the available data relate to associations between psychosocial stress (or ability to cope with such stress) and breast cancer risk. When this topic was reviewed a decade ago, the authors concluded that the evidence for a relationship between psychosocial factors and risk of breast cancer was weak and called for research with a theoretical grounding and greater methodological rigor (Butow *et al.*, 2000). A more recent review (Nielsen and Gronbaek, 2006) concluded that stress does not seem to increase the risk of breast cancer and that there was inadequate evidence on which to make a judgment about the effects of stress on breast cancer progression. Again, these authors drew attention to key methodological issues including the way in which stress was conceptualized and measured (Nielsen and Gronbaek, 2006), issues which will need to be addressed to allow this potentially important field to progress.

7. Other environmental exposures

Since the industrial era, certain occupations (and sites of habitation) have been important source of exposure to cancer-causing chemicals such as polycyclic aromatic hydrocarbons (PAHs; Bosetti *et al.*, 2007). While such exposures have been eliminated from most workplaces, risks remain in some newly industrialized countries where regulations are less restrictive or less rigorously enforced (Stewart and Kleihues, 2003). In addition, some heavy metals notably lead, cadmium, mercury, and arsenic pose threats to human health (Jarup, 2003). The pollution of drinking water by arsenic in the Ganges delta alluvium in Bangladesh has resulted in an estimated 40 million people being at risk of arsenic poisoning-related diseases, including skin cancer (Alam *et al.*, 2002).

    In summary, there is now good knowledge of many of the environmental factors which modulate cancer risk. The challenge is to understand how these exposures affect cancer biology, that is, the accumulation of genetic damage and aberrant gene expression which are fundamental hallmarks of tumor development. Given the impact of a wide variety of environmental exposures on epigenetic marks (see below) and the key role which such marks play in modulating gene expression, an *a priori* case can be made for the importance of understanding the effects of dietary, lifestyle, and other environmental factors on epigenetic marks as a step toward the development of interventions for cancer risk reduction. It has been argued that epigenetic changes are an early event in tumor development (Feinberg et al., 2006) and that changes induced by environmental exposures may only alter risk of cancer initiation but also increase the likelihood that further genetic insults will be acquired (Baylin and Ohm, 2006). Experimental evidence is accumulating that DNA methylation patterns (Belshaw et al., 2008) and gene expression patterns (Polley et al., 2006) are altered even in apparently normal tissues of those at higher cancer risk.

II. INDUCTION OF EPIGENETIC ALTERATIONS BY DIETARY AND OTHER ENVIRONMENTAL FACTORS

To date, there have been few studies which have examined the effects of dietary (and other environmental) factors on epigenetic markings in intervention studies in humans. Much of the available evidence is derived from either observational studies in humans (with their attendant uncertainties about causality and difficulties in characterizing exposure Penn et al., 2009) or animal studies (where, in some cases, experimental conditions and/or exposure doses may be difficult to translate to humans).

A. Impact of dietary factors on epigenetic markings

Perhaps because of the relative ease in making appropriate measurements, much more information is available about the impact of dietary factors on DNA methylation marks than is available for impact on histone modifications or changes in miRNA expression. However, technical advances are facilitating research in both of these areas and enabling a more holistic view of the ways in which dietary factors influence epigenetic marks and processes. Proof of principle that dietary exposures may have lifelong consequences for epigenetic marks comes from recent studies of the adult offspring of women exposed to famine during pregnancy. Methylation of the imprinted gene *IGF2* was lower in adults (approximately 60 years of age) who were periconceptionally exposed to famine during the Dutch Hunger Winter of 1944–1945 (Heijmans et al., 2008). Further analysis showed that changes in DNA methylation after exposure to prenatal famine were common, gene specific, and effects differed according to timing of famine exposure (Tobi et al., 2009). Although such studies have several limitations including lack of knowledge of the specific dietary insults linked with the DNA methylation changes, they demonstrate the potential durability of diet (or stress)-related changes in epigenetic marks and suggest the importance of focusing attention on specific life-course "windows" when the epigenome may be more plastic.

1. Diet, methyl donors, and DNA methylation

Given that S-adenosylmethionine (SAM) is the donor for the methyl groups used in methylation of cytosine residues in DNA, dietary sources of methyl groups, including folate, methionine, betaine, serine, and choline, are primary candidates as potential modulators of DNA methylation. More broadly, availability of dietary factors which influence one-carbon metabolism including B vitamins that are coenzymes in one-carbon metabolism (vitamins B_2, B_6, and B_{12}) are also modulators of DNA methylation (Choi et al., 2009b). When methyl groups are in short supply, there is a competition for the limited resource. For example, when folate supply is low, there is an inverse relationship between concentrations of folate and homocysteine which disappears at high folate concentrations (Reed et al., 2006). Modeling studies predict reduced DNA methylation when folate supply is low and vitamin B_{12} is deficient (Reed et al., 2006).

The plasticity of DNA methylation marks in adult humans has been demonstrated by intervention studies in which folate supply has been manipulated deliberately. For example, in postmenopausal women (aged 49–63 years), hypomethylation of genomic DNA from lymphocytes occurred within a few weeks of feeding a diet providing 56 μg folate/d (compared with an estimated requirement of 200 μg folate/d) and this hypomethylation was reversed when folate intakes were increased (Jacob et al., 1998). In older women (>63 years), folate depletion (118 μg folate/d for 7 weeks) also produced significant ($P < 0.0025$) hypomethylation of lymphocyte DNA but this did not recover significantly when the women were given 415 μg folate/d for 4 weeks (Rampersaud et al., 2000). More recently, Pufulete et al. (2005) supplemented patients with colorectal adenomatous polyps with 400 μg folate/d for 10 weeks and observed a 31% increase ($P < 0.05$) in methylation of genomic DNA in leucocytes. In addition, Pufulete et al. reported that their folic acid supplementation resulted in a 25% increase ($P = 0.09$) in methylation of DNA from the colonic mucosa demonstrating that DNA methylation in tissues other than circulating white cells is modifiable by folic acid supply. However, it should not be assumed that effects seen in one tissue are necessarily representative of all. Pogribny et al. (2008) observed that feeding a diet deficient in several methyl donors (folate, methionine, and choline) resulted in hypomethylation of genomic DNA from the liver but hypermethylation of genomic DNA from the brain in rats.

Such studies establish proof of principle for the plasticity of total DNA methylation marks in response to variation in methyl donor supply and suggest that even relatively short-term alterations in supply, within the physiological range, may produce quite large changes in DNA methylation. However, the outcomes of such studies are difficult to interpret functionally because of the absence of information about where in the genome the labile cytosine residues occur.

To date, there is little evidence from human studies of the effects of methyl donor supply on methylation status of specific genomic sequences. An exception is the recent study by Steegers-Theunissen *et al.* (2009) which found that periconceptional maternal supplementation with 400 μg folic acid/d was associated with increased methylation at some, but not all, CpG sites within the differentially methylated region (DMR) of the *IGF2* gene in offspring aged 17 months. The most widely reported illustration of the effects of methyl donor supply on methylation status of a specific genomic domain which has profound effects on phenotype is the hypermethylation at the A^{vy} allele in offspring of female agouti mice whose diet was supplemented with methyl donors (folate, vitamin B_{12}, betaine, and choline) during pregnancy (Waterland and Jirtle, 2003; Waterland and Michels, 2007). The methylation effects at the A^{vy} allele were similar in all tissues examined (suggesting that the mechanism may have involved altered marking of stem cells early in embryogenesis before tissue differentiation) and persisted into adult life (Waterland and Jirtle, 2003). These changes in epigenetic marks were associated with readily observed changes in offspring coat color and the offspring of the supplemented dams were less likely to become obese (Waterland and Jirtle, 2003). Also in the agouti mouse, maternal supplementation with methyl donors prevented the hypomethylation of nine CpG sites within the A^{vy} allele in the offspring as a consequence of maternal dietary exposure to bisphenol A (BPA) (Dolinoy *et al.*, 2007). Similar effects were observed for CpG residues within the $Cabp^{IAP}$ metastable epiallelle (Dolinoy *et al.*, 2007). Choline deficiency in pregnancy results in hypermethylation of genomic DNA and of the *Igf2* gene which the authors attributed to increased expression of the *Dnmt1* gene (Kovacheva *et al.*, 2007) which encodes the DNA methyltransferase responsible for copying methylation patterns during cell proliferation.

Effects of methyl donors are not restricted to imprinted genes or to metastable epiallelles. Several studies have reported that severe folate deficiency (which increases risk of hepatocellular cancer) induces hypomethylation of the *p53* tumor suppressor gene (see, e.g., Kim *et al.*, 1997). More moderate folate deficiency had no detectable effect on genomic methylation of DNA from liver or colon or on methylation of the *c-myc* gene in the rat colon (Kim *et al.*, 1995).

2. Impact of other dietary factors on DNA methylation

In vitro studies have shown that several bioactive food components, including tea polyphenols (catechin, epicatechin, and (−)-epigallocatechin-3-O-gallate, EGCG) and bioflavonoids (quercetin, fisetin, and myricetin), inhibit DNMT-1-mediated DNA methylation in a dose-dependent manner (Lee *et al.*, 2005). This is effect appears to be due to increased synthesis of S-adenosylhomocysteine (SAH) which is a potent noncompetitive inhibitor of DNMTs (Lee *et al.*, 2005). In the agouti mouse model, maternal dietary supplementation with the polyphenol genistein during pregnancy has similar effects on methylation at the A^{vy} allele and similar effects on phenotype in the offspring (Dolinoy *et al.*, 2006) to those observed with maternal supplementation with methyl donors (Waterland and Jirtle, 2003). Several other dietary compounds (reviewed by Choi *et al.*, 2009b; Mathers and Ford, 2009) have been reported to alter DNA methylation *in vitro*, in animal models or in humans. Because of its potential role in cancer etiology, it is interesting to note that selenium deficiency (which does not affect the activity of enzymes in one-carbon metabolism but alters the transsulphuration pathway; Choi *et al.*, 2009b) causes genomic DNA hypomethylation in rat liver and colon (Davis *et al.*, 2000).

3. Dietary factors and histone modifications

Butyrate was the first food-derived substance shown to affect posttranslational modifications of histones through its action as an inhibitor of class 1 histone deacetylases (HDACs; Davie, 2003; Kruh, 1982; Vidali *et al.*, 1978). Butyrate occurs in millimolar concentrations in the colon as an end-product of bacterial fermentation of carbohydrates and is a potent antineoplastic agent *in vitro* (Dronamraju *et al.*, 2010; Williams *et al.*, 2003) and *in vivo* (Dronamraju *et al.*, 2009). More recently, a number of other dietary components have been identified which modulate the acetylation state of histones or affect the activities of HDACs and/or histone acetyl transferases (reviewed by Delage and Dashwood, 2008) and, therefore, are potential regulators of gene expression by epigenetic mechanisms. Of these, there is strong evidence for effects of organosulphur compounds from garlic (diallyl disulphide, allyl mercaptan, and S-allylmercaptocysteine) and of isothiocyanates (sulphoraphane and 6-methylsulphinylhexyl isothiocyanate) from cruciferous vegetables (Delage and Dashwood, 2008). Interest in the effects of dietary compounds such as resveratrol which activate class III HDACs (sirtuins) is growing rapidly because of their demonstrable role in extending lifespan and in reducing, or delaying, age-related diseases including cancers (Baur, 2010). The polyphenol resveratrol activates PGC1α in a SIRT1-dependent manner (Lagouge *et al.*, 2006) and improves health and survival of mice fed a high-fat diet (Baur *et al.*, 2006). By activating the HDAC SIRT1, resveratrol may protect against damage which results from inflammation associated with high-fat diets (Pfluger *et al.*, 2008).

4. Nutritional modification of miRNA expression

There are reported to be at least 940 miRNAs in the human genome (http://www.mirbase.org/cgi-bin/browse.pl). Each miRNA binds to the 3′-untranslated region of the mRNA of up to 200 gene targets and suppresses expression by affecting mRNA stability and/or targeting the mRNA for degradation (Esquela-Kerscher and Slack, 2006). In this way, miRNAs potentially control the expression of about one-third of human mRNAs and influence almost all genetic pathways by targeting transcription factors, secreted factors, receptors, and transporters (Esquela-Kerscher and Slack, 2006). The potential importance of miRNA in tumorigenesis arises from evidence that mutations or mis-expression of miRNA occurs in various human cancers and that miRNAs repress the expression of cancer-related genes (Esquela-Kerscher and Slack, 2006). Given the major role of diet in the etiology of several cancers (World Cancer Research Fund/American Institute for Cancer Research, 2007), an *a priori* case can be made that changes in miRNA expression in response to dietary exposure may mediate effects on cancer risk. Evidence is now emerging (see Table 1.1) that a wide range of dietary factors including both macronutrients (fat, protein, and alcohol) and micronutrients (folate and vitamin E) alter expression of many miRNA in rodents and in humans. Of particular interest are the findings of Marsit *et al.* (2006) who reported that both folate deficiency and arsenic exposure resulted in a global increase in miRNA expression in the human immortalized lymphoblastoid cell line TK-6. One of their target miRNA (miR-222) was also present at significantly higher levels in peripheral blood cells from people with the lowest 1% versus highest 1% of folate intakes, suggesting that miR-222 may be a novel biomarker of altered folate status.

B. Impact of lifestyle and genetic factors on epigenetic markings

1. Tobacco

One of the most widely studied lifestyle influences on epigenetic patterns is tobacco smoking which has been shown to be associated with specific patterns of gene hypermethylation. A clear association is observed between tobacco smoke and cancer, exemplified by lung cancer, where perturbed epigenetic patterns are a common hallmark. Animal models also suggest that epigenetic changes may be among the first genetic lesions to arise following exposure to cigarette smoke. Mouse models of lung cancer development have shown that even a few weeks of

Table 1.1. Examples of Effects of Dietary Factors on microRNA Expression

Dietary factor	Animal	Tissue	Observation	Reference
n-3 PUFA[a]	Sprague-Dawley rat	Colon	Fish oil supplementation "protected" the colon against azoxymethane(AOM)-induced miRNA dysregulation	Davidson et al. (2009)
Vitamin E	Fisher 344 rats	Liver	Vitamin E deficiency resulted in reduced expression of miR-122a and miR-122b	Gaedicke et al. (2008)
Methyl-deficient diet	Rat	Liver	Methyl-deficient diet led to reduced expression of miR-34a, miR-127, miR-200b, and miR-16a	Tryndyak et al. (2009)
Folate deficiency/ arsenic exposure	Human cell line (TK6)	Lymphoid	Folate deficiency or arsenic exposure both lead to global increases in miRNA expression	Marsit et al. (2006)
Choline-deficient diet	C57BL/6 mice	Liver	Choline-deficient diet resulted in upregulated expression of miR-155, miR-221/222, and miR-21 and downregulation of miR-122	Wang et al. (2009)
Fat	Mice	Adipose tissue	High-fat diet upregulated expression of miR-143 in mesenteric fat	Takanabe et al. (2008)
Fat	C57BL/6 mice	Liver	Maternal high-fat diet during pregnancy and lactation reduced expression of several miRNA including miR-709, miR-194, miR-122, and miR192 in adult offspring	Zhang et al. (2009)
Ethanol	Mouse	Fetal brain	Ethanol treatment of embryos in culture increased expression of several miRNA (including miR-10a, miR-10b, miR-9, and miR-145) and reduced expression of others (including miR200a, miR-496, miR-296, and miR-362)	Wang et al. (2009)
Ethanol	Humans	Colon	Ethanol treatment increased expression of miR-212 in both a cell line (Caco-2) and in colonic mucosal biopsies from patients with alcoholic liver disease	Tang et al. (2008)

Ethanol and methionine– choline-deficient (MCD) diets	C57BL/6 mice	Liver	Both regimes induced fatty livers and resulted in differential expression of the same 5 miRNA. However, expression of miR-705 and miR-1224 was increased by both regimes whereas miR-182, miR-183, and miR-199a-3p were downregulated by alcohol and upregulated by the MCD diet	Dolganiuc et al. (2009)
Protein	FVB/NJ mice	Fetal brain	Feeding a low protein diet during second half of gestation upregulated expression of mmu-mir-27a and 27b and reduced expression of mmu-mir-330 in fetal offspring	Goyal et al. (2010)

[a] 11.5 g fish oil + 3.5 g corn oil/100 g diet.

exposure to tobacco smoke or tobacco smoke condensate can increase the methylation levels of numerous genes (Phillips and Goodman, 2009). In this model system, epigenetic changes occurred prior to any overt histopathological changes (Phillips and Goodman, 2009). In a human context, tobacco smoking has been shown to have a strong influence on DNA methylation levels of specific cancer-associated genes (*RASSF1A* and *MTHFR*), although this relationship was not observed in all genes investigated (Vaissiere *et al.*, 2009).

The effects of tobacco smoke on DNA methylation may not be restricted solely to the hypermethylation of genes. Liu *et al.* (2007) have recently shown that cigarette smoke extract induced expression of the prometastatic oncogene, synuclein-gamma, through promoter demethylation. Analysis of the mechanism involved showed that exposure to cigarette smoke extract resulted in decreased expression of one of the DNMT enzymes, DNMT3b. The effects of cigarette smoke extract on synuclein-gamma expression could be replicated by direct downregulation of DNMT3b and was reversed by enforced expression of DNMT3b (but not DNMT1 or DNMT3a).

Tobacco smoke has also been implicated in the modulation of histone modifications. Hussain *et al.* (2009) have shown that exposure of A549 lung carcinoma cells to cigarette smoke condensate resulted in increased tumorigenesis in nude mice and repression of the DKK-1 gene. This gene repression was not associated with DNA methylation changes but with reduced histone acetylation and increased trimethylation of H3K27. This latter modification was linked to the recruitment of elements of the polycomb repression complex. Altered histone acetylation of proinflammatory genes in lung tissue from mice exposed to cigarette smoke has also been demonstrated, providing evidence for a role of chromatin modifications in abnormal and sustained inflammatory response in smokers (Marwick *et al.*, 2004; Yang *et al.*, 2008).

The influence of maternal cigarette smoking during pregnancy on epigenetic patterns in offspring has also been reported (Breton *et al.*, 2009; Terry *et al.*, 2008). Such methylation-sensitive exposures during early development have been used as evidence for a role of epigenetic mechanisms in the developmental programming of adult disease (Waterland and Michels, 2007). Although the focus here has largely been on cardiometabolic disease and cancer does not feature prominently amongst those diseases believed to be programmed in early development, there is some evidence that breast and prostate cancer have early life antecedents (Eriksson *et al.*, 2007; Xu *et al.*, 2009).

2. Alcohol consumption

Excessive alcohol intake plays an important contributory role in several common cancers. Animal studies have shown that chronic alcohol consumption is associated with reduced genomic DNA methylation in the colon (Sauer *et al.*, 2010), itself an established hallmark of colorectal neoplasia (Feinberg *et al.*, 1988). Other alcohol-related cancers including esophageal cancer (Toh *et al.*, 2010) and hepatocellular cancer (Hernandez-Vargas *et al.*, 2010) also involve epigenetic alterations. Alcohol is also a well-established risk factor for breast cancer in women and a link between genome-wide DNA hypomethylation and the development of breast cancer has been reported (Choi *et al.*, 2009a). The molecular actions of ethanol are believed to include site-specific changes to histone modifications as well as perturbation of one-carbon metabolism and consequent interference with methyl group donation (reviewed by Seitz and Stickel, 2007; Shukla *et al.*, 2008).

Although most attention has focused on the link between alcohol consumption and DNA methylation, this exposure is also believed to influence other epigenetic modifications. Chromatin remodeling in response to alcohol consumption which influences abnormal plasticity in reward-related learning processes may contribute to persistent behavior change and long-term addiction problems (Guerri and Pascual, 2010). A recent study demonstrated that ethanol exposure during adolescence induces changes in the acetylation of histones H3 and H4 in the brain (Pascual *et al.*, 2009). Although histone modifications have been postulated to explain the observed link between alcohol consumption and elevated cancer risk, there is little empirical evidence to support this at present.

miRNA expression has recently been linked to alcohol consumption in a case-control study of head and neck cancer (HNSCC). Expression of miRNA-375 was shown to increase with alcohol consumption (Avissar *et al.*, 2009), although the etiological significance of this observation requires further investigation. Several studies have reported altered expression of miRNAs in hepatocellular cancer, although it is recognized that this altered expression may be secondary to malignant transformation rather than causal (Tryndyak *et al.*, 2009). miRNA profiling of liver tumor tissue is associated with risk factors, including alcohol consumption (Ladeiro *et al.*, 2008) but this association may be confounded and requires prospective analysis of tissues taken prediagnosis or use of *in vivo* models to assist in establishing causality.

3. Physical activity

Physical inactivity is strongly associated with enhanced risk of colorectal cancer (Basterfield *et al.*, 2005). Emerging evidence suggests that several forms of epigenetic alteration may play a role in mediating exercise-induced changes in gene expression. Altered gene expression in colonic tissue is observed in animals engaging in differing levels of physical activity and it has been postulated that this is mediated in part by changes to epigenetic patterns. Differentially expressed genes included those involved in signal transduction, immune response, cytoskeleton, vascularization, and in methyl group donation (Buehlmeyer *et al.*, 2008).

Increased histone acetylation at lysine 36 (H3K36) was observed in human muscle biopsy tissue following exercise (McGee *et al*., 2009), providing evidence that chromatin remodeling may be important in mediating skeletal muscle adaptations to exercise. A recent study by Radom-Aizik *et al*. (2010) provided evidence for miRNA involvement in exercise-associated changes in gene expression. Exercise was shown to alter the expression levels of 38 miRNAs in circulating neutrophils in humans.

4. Obesity

Several strands of evidence point to a role for epigenetic mechanisms in the development of obesity (reviewed by Mathers, 2010). The phenotypic characteristics of imprinting disorders are associated with increased adiposity; Prader-Willi syndrome (PWS), caused by the absence of the paternally derived region, commonly by inappropriate methylation, of chromosome 15q11–q13, is characterized by appetite dysregulation and morbid obesity in childhood (Cassidy and Driscoll, 2009). Several genes have recently been described as being epigenetically regulated (reviewed by Campion *et al*., 2009), including genes involved in appetite regulation, adipogenesis, inflammation, insulin signaling, and energy metabolism. As noted earlier, the potentially pleiotropic effects of these pathways in the pathogenesis of both metabolic disease and cancer have been mooted and these may involve common epigenetic mechanisms. In a rodent model study of overfeeding, Plagemann *et al*. (2009) demonstrated that neonatal overfeeding led to an obese phenotype with hallmarks of the metabolic syndrome and in these animals the appetite regulatory gene proopiomelanocortin (*Pomc*) was hypermethylated. These observations demonstrate dietary induction of epigenetic changes with functional consequences for appetite regulation and body composition.

Adiposity is a risk factor for several forms of cancer, one example being renal cell carcinoma. A link has been reported, albeit in a small study with limited power, between methylation of the *RASSF1A* tumor suppressor gene and adiposity in normal kidney tissue, providing evidence that adiposity enhances *RASSF1A* promoter methylation and thereby elevates cancer risk (Peters *et al*., 2007).

With respect to other possible epigenetic mechanisms, miRNAs are potential regulators of adipogenesis. Seventy miRNAs were shown to be up- or downregulated in mature adipocytes when compared with preadipocytes (Ortega *et al*., 2010). A recent study has also demonstrated that human adipose tissue miRNA expression correlates with adipocyte phenotype, parameters of obesity, and glucose metabolism (Kloting *et al*., 2009).

In summary, there is emerging evidence that several epigenetic mechanisms might contribute to the pathogenesis of obesity. In addition, obesity or the metabolic comorbidities associated with this condition may themselves alter epigenetic patterns, possibly through inflammatory pathways and have downstream influences on the pathogenesis of cancer. Delineating whether the relationship between epigenetic perturbation and obesity is causal will be central to understanding whether this mechanism can explain the association between obesity and cancer risk.

5. Stress

Perhaps one of the most remarkable examples of environmental influences on the epigenome comes from studies of the effects of maternal grooming on offspring in rats. Natural variation in both licking and grooming (LG) and arched back nursing (ABN) occurs in rats. Offspring of mothers who exhibit high levels of LG and ABN are less fearful as adults and show reduced hypothalamic-pituitary-adrenal (HPA) responses to stress. Cross-fostering studies showed that this was a programmed, and not an inherited, response (Francis et al., 1999). These differences appear to be due to altered expression of the glucocorticoid receptor (GR) in the hippocampus, as reversal of the differential GR expression in rats nursed by high and low LG and ABN mothers resulted in abrogation of the differences in stress responses in adulthood (Meaney, 2001). Searching for a mechanistic explanation of how maternal behavior during the first week of life could result in behavioral differences that persist into adulthood, Weaver et al. (2004) demonstrated dramatic increases in methylation at specific CpG sites in the GR gene in pups nursed by low LG and ABN mothers. This increased DNA methylation was also associated with a corresponding reduction of histone acetylation. Subsequent treatment with an HDAC inhibitor (trichostatin A, TSA) reversed the reduced histone acetylation and also the increased DNA methylation seen at the GR gene in pups nursed by low LG and ABN mothers and led to increased GR expression. TSA treatment also eliminated the differences in HPA responses, implying a direct correlation between the induced epigenetic changes and control of adult rat behavior (Weaver et al., 2004).

This observed association between maternal care and epigenetic regulation of offspring stress response prompted a study in humans to assess the relationship between prenatal exposure to maternal mood and methylation status of the human GR gene (NR3C1; Oberlander et al., 2008). Methylation status of the human NR3C1 gene in newborns was sensitive to prenatal maternal mood (Oberlander et al., 2008), suggesting that epigenetic mechanisms may be important in mediating the influence of stress not only upon neurocognitive outcomes but potentially more broadly.

6. Genetic factors

Twin and family-based studies have demonstrated that variation in epigenetic patterns is not only heritable but also shows Mendelian patterns of inheritance (Sandovici *et al.*, 2003; Silva and White, 1988). A study of pedigrees from the Centre d'Etude du Polymorphisme Humain (CEPH) collection evaluated variation of chromatin states between families, demonstrating familial clustering and supporting the idea that genetic inheritance can determine the epigenetic state of the chromatin (Kadota *et al.*, 2007). In a separate study of CEPH pedigrees, familial correlation in the stability of DNA methylation patterns overtime was estimated (Bjornsson *et al.*, 2008). Many families showed clustering of DNA methylation changes and both increased and decreased methylation was observed (Bjornsson *et al.*, 2008). A heritability estimate of $h^2 = 0.743$ ($P = 0.003$) was deduced. A study of DNA methylation levels at the imprinted *IGF2* locus produced a very similar heritability estimate of $h^2 = 0.75$ (Heijmans *et al.*, 2007).

It is likely that a proportion of interindividual variation in epigenetic patterns can be attributed to common genetic variants (polymorphisms). Indeed, Heijmans *et al.* (2007) estimated that 6.5% of the variance in DNA methylation at the *IGF2* locus could be explained by five single nucleotide polymorphisms (SNPs). This could occur through a specific SNP directly abolishing a CpG site, or through other local, *cis* effects. In an elegant and novel assessment of parent of origin effects of sequence variants, Kong *et al.* (2009) identified a SNP which conferred risk when maternally transmitted and protection when paternally transmitted. The SNP was tightly correlated with the methylation status of a nearby CTCF-binding site, introducing the possibility that methylation status can modulate the penetrance of a specific SNP through altering the dosage of the two parental chromosomes.

The proteins that establish and maintain epigenetic patterns across the genome are encoded in the DNA sequence and harbor genetic variations in the form of SNPs, insertions, deletions, and other functionally relevant alterations. Several polymorphisms in epigenetic genes (e.g., DNA methyltransferases, methyl-CpG-binding domain proteins, histone acetyltransferases, histone deacteylases) have been associated with human cancers (reviewed by Miremadi *et al.*, 2007). Although various associations have been reported between polymorphisms in this gene family and cancer risk, little has been done to date to correlate polymorphisms with epigenetic patterns of histone modification or DNA methylation. The most comprehensive study to date by Cebrian *et al.* (2006) examined common variants in 12 genes coding for DNA methyltransferases, histone acetyltransferases, histone deacetyltransferases, histone methyltrasferases, and methyl-CpG-binding domain proteins, for association with breast cancer. This large case-control study reported some evidence of a link between histone methylation and polymorphisms in histone methyltransferase genes (Cebrian *et al.*, 2006). There is scope for considerably more work in this area in the future.

In addition to the role of germ-line genetic variants, the acquisition of somatic mutations in epigenetic genes is also of relevance to the pathogenesis of cancer. It is plausible that environmental mutagens might act through the induction somatic mutations in epigenetic genes, although to date there is little evidence to support this postulated mechanistic pathway.

C. Impact of other environmental factors on epigenetic markings

In this section, we will examine some of the other environmental stimuli which are known to elicit epigenetic changes. One of the most intriguing potential roles of epigenetic regulation of genes is the possibility that these could function as sensors for environmental exposures, altering gene expression patterns in temporary or even permanent ways in response to various environmental exposures. It is becoming increasingly obvious that many different environmental exposures lead to altered epigenetic marks on the genome (Table 1.2). These environmental exposures fall into many categories, including well-established carcinogens (e.g., cigarette smoke, radiation), exposure to infectious agents (e.g., *Helicobacter pylori*), chemicals which alter hormonal signaling (e.g., endocrine disruptors), and, as noted above, different forms of behavior (e.g., maternal LG behavior in rats). As awareness of the potential interplay between environment and epigenetics increases, it is likely many more environmental agents capable of modulating our genome through epigenetic factors will be identified. However, the extent to which these epigenetic changes represent an adaptive response to specific environmental stimuli, or are a new class of genotoxic agents, is still an open question.

1. Ionizing radiation

Ionizing radiation (IR) is a well-established mutagen which induces cancers, especially leukemia and lymphomas. It is also a mainstay of cancer treatment, where its use is strongly linked with the development of secondary tumors in cancer survivors. Increasing evidence suggests that the genotoxic effects of IR are due in part to epigenetic effects and, interestingly, much of this appears to occur not in the directly irradiated cells, but as a bystander effect in nonirradiated cells (Ilnytskyy *et al.*, 2009; Koturbash *et al.*, 2008; Tamminga *et al.*, 2008). For example in their 2008 report, Taminga and colleagues examined bystander effects in shielded testes tissue in mice following cranial irradiation and showed dramatic reductions in global DNA methylation in the nonirradiated tissue (Tamminga *et al.*, 2008). This may have been a direct effect on the DNA methylation machinery, or potential secondary insult to unrepaired DNA lesions, which were also detected. Remarkably though, the pronounced DNA

Table 1.2. Nondietary Environmental Exposures Which Modulate Epigenetic Regulation

Category	Exposure	Main epigenetic targets identified	References
Carcinogens	Cigarette smoke	DNA methylation, histone acetylation, histone methylation	Liu et al. (2007), Hussain et al. (2009), Phillips and Goodman (2009)
	Ionizing radiation	DNA methylation	Giotopoulos et al. (2006), Ilnytskyy et al. (2009), Koturbash et al. (2008), Tamminga et al. (2008), Belinsky et al. (2004)
Toxic compounds	Nickel	DNA methylation, histone acetylation, histone methylation, histone phosphorylation	Lee et al. (1995), Broday et al. (2000), Ke et al. (2008), Chen et al. (2010)
	Arsenic	DNA methylation, histone methylation	Zhong and Mass (2001), Reichard et al. (2007), Zhou et al. (2008), Jensen et al. (2009)
	Cadmium	DNA methylation	Takiguchi et al. (2003), Jiang et al. (2008)
	Mercury	DNA methylation	Onishchenko et al. (2008)
	Polycyclic aromatic hydrocarbons	DNA methylation	Pavanello et al. (2009), Perera et al. (2009)
	Particulate matter/Benzene	DNA methylation, miRNA expression	Bollati et al. (2007, 2010), Baccarelli et al. (2009), Tarantini et al. (2009), Jardim et al. (2009)
Endocrine disruptors	Vinclozolin	DNA methylation	Anway et al. (2005), Skinner and Anway (2007)
	Bisphenol A	DNA methylation	Prins et al. (2008)
	Diethylstilbestrol	DNA methylation	Li et al. (1997, 2003)
Infectious agents	Helicobacter pylori	DNA methylation	Suzuki et al. (2009), Niwa et al. (2010), Shin et al. (2010)
	Viruses (Epstein-Barr, polyoma, adenovirus)	DNA methylation	Tsai et al. (2002, 2006), McCabe et al. (2006)
Behavior	Maternal grooming	DNA methylation, histone acetylation	Weaver et al. (2004), Oberlander et al. (2008)
	Physical activity	Histone acetylation, miRNA expression	McGee et al. (2009), Radom-Aizik et al. (2010)

hypomethylation persisted into the next generation, where widespread loss of global DNA methylation was seen in multiple tissues. In agreement with these studies, Giotopoulos *et al.* (2006) also found that IR induced DNA hypomethylation. Interestingly this study reported that induction of hypomethylation may be genotype dependent since the effect was found in the CBA/H strain, which is susceptible to IR-induced AML, but not in C57BL/6 mice which are resistant to IR-induced AML.

IR may also have more acute effects on DNA methylation. Leng *et al.* (2008) demonstrated that increased methylation in lung cancer patients was associated with reduced capacity to repair double strand breaks and that this may in part be mediated by polymorphisms in genes involved in double strand break repair. Furthermore, the same group subsequently showed that transformation induced by double strand break inducing agents was associated with increased gene methylation and upregulation of DNMT1 in a lung epithelial cell line model (Damiani *et al.*, 2008). In addition, knockdown of DNMT1 prior to carcinogen treatment prevented both gene hypermethylation and cell transformation (Damiani *et al.*, 2008).

IR may also induce gene-specific hypermethylation effects. For example, secondary myeloid leukemias in patients treated with IR and other cytotoxic agents exhibit significantly higher levels of CpG island methylation than primary myeloid leukemias (Figueroa *et al.*, 2009). Similarly, Bennett *et al.* (2010) found that hypermethylation of a subset of CpG islands in HNSCC patients was very strongly associated with prior radiotherapy. Some of these effects may also be gene specific, for example, Belinsky *et al.* (2004) found increased gene promoter methylation of the p16 and GATA5 genes in lung cancer patients exposed to plutonium versus nonexposed lung cancer patients (Lyon *et al.*, 2007).

2. Heavy metals

A number of metal compounds present in the environment have been implicated in the modulation of various epigenetic alterations (reviewed in Salnikow and Zhitkovich, 2008). Nickel compounds, which are carcinogenic without any apparent induction of DNA damage, appear to modulate the epigenome at many levels. Initially, Nickel ions were reported to induce hypermethylation of a transgene in CHO cells (Lee *et al.*, 1995). However, this proved to be one of many epigenetic targets for nickel ions, which have also been shown to suppress histone H4 acetylation (Broday *et al.*, 2000). *In vitro* evidence suggests that this is achieved through direct inhibition of histone acetyltransferase enzymes (Broday *et al.*, 2000). Similarly, a recent study has shown that nickel ions can inhibit the activity of the histone demethylase enzyme JMJD1A by replacing iron ions within the enzyme's catalytic domain (Chen *et al.*, 2010). Nickel has also been shown to induce phosphorylation of histone H3S10 (Ke *et al.*, 2008) and the ability of nickel ions to induce heterochromatinization has recently been proposed as one of its primary mechanisms of carcinogenesis (Ellen *et al.*, 2009).

Similar to nickel, arsenic-containing compounds were originally identified as potential modulators of DNA methylation and were found to produce region-specific DNA hypermethylation and hypomethylation in cells in culture (Zhong and Mass, 2001). Similarly a recent report has shown that long-term exposure to low levels of arsenic containing compounds can lead to global DNA hypomethylation (Reichard *et al.*, 2007). However, subsequent studies have suggested effects on other epigenetic targets. Exposure of A549 lung carcinoma cells to arsenic had pleiotropic effects on histone methylation, including increased and decreased methylation at specific histone H3 residues. This included increased levels of H3K9me2, which appeared to be primarily related to increased levels of expression of the histone methyltransferase G9a following arsenic exposure (Zhou *et al.*, 2008). Similarly arsenic exposure was shown to induce the noncanonical wnt pathway, through upregulation of WNT5a, due to induction of a number of different histone modifications in the WNT5a gene promoter (Jensen *et al.*, 2009). The targeting of epigenetic modifications leads to a permanent change in gene expression, since even following subsequent removal of arsenic, activation of WNT5a was stably maintained (Jensen *et al.*, 2009). Arsenic has been well studied in both *in vitro* and human systems. Hypermethylation of tumor suppressor genes in subjects exposed to high levels of arsenic have been reported (Chanda *et al.*, 2006; Zhang *et al.*, 2007). As noted earlier, Marsit *et al.* (2006) observed global increases in miRNA expression when TK-6 cells (an immortalized human lymphoblastoid cell line) were exposed to arsenic.

Ecologic epidemiological studies have demonstrated a link between methylmercury exposure and elevated risk of leukemia mortality (Yorifuji *et al.*, 2008). The vast majority of studies on methylmercury exposure have focused upon neurological outcomes and included among these is a study linking prenatal methylmercury exposure to epigenetic suppression of the *BDNF* gene (Onishchenko *et al.*, 2008). A recent study of brain tissue from polar bears reported decreased genome-wide methylation levels in animals exposed to environmental mercury. It is therefore possible that epigenetic alterations in cancer-related genes may arise through exposure to this environmental pollutant.

Other metallic compounds found in the environment may also modify epigenetic marks and indeed exposure to both cadmium and lead has been associated with alterations to DNA methylation (Jiang *et al.*, 2008; Takiguchi *et al.*, 2003; Wu *et al.*, 2008). A recent study by Wright *et al.* (2010) reported a correlation between global levels of DNA methylation, measured using the *LINE-1* assay, in peripheral blood DNA and cumulative lead exposure.

3. Air pollution

PAHs are among the most carcinogenic molecules yet identified and are present in a variety of environmental sources including car exhaust fumes, as biproducts of industrial activities, as contaminants in certain foodstuffs and as a component of cigarette smoke. Recently Pavanello *et al.* (2009) have shown that coke-oven workers exposed to high levels of PAHs exhibited multiple abnormalities in DNA methylation including increased genome-wide methylation at repetitive sequences as well as gene-specific areas of both hypermethylation and hypomethylation. The extent of these abnormalities correlated with exposure levels to PAHs, implying that PAHs were indeed likely to be the causative agent. Perera *et al.* (2009) studied the potential epigenetic effects of transplacental exposure to PAHs from car exhaust fumes and identified multiple gene-specific hypermethylation events in exposed individuals. The strongest correlation was with methylation of the acyl-CoA synthetase long-chain family member 3 (*ACSL3*) gene promoter. An expanded study confirmed the strong correlation between increased PAH exposure and *ACSL3* hypermethylation, which was also found to be associated with the development of asthma in children exposed to high PAH levels *in utero* (Perera *et al.*, 2009). The ability of PAHs to modulate the epigenome may be partly through their interaction with NF-kB signaling (Tian, 2009). PAHs activate signal transduction pathways through binding to the aryl hydrocarbon receptor transcription factor. This signaling pathway interacts directly with NF-kB signaling, a key controller of inflammatory responses, which has also been associated with modulation of the epigenome (Tian, 2009). PAHs may not be the only air pollutant derived from car exhaust fumes which can modulate the epigenome. Two recent studies have associated increased levels of exposure to particulate matter to global reductions in DNA methylation at repetitive elements and gene-specific sequences (Baccarelli *et al.*, 2009; Tarantini *et al.*, 2009) and similarly increased exposure to benzene has also been associated with reduced global methylation (Bollati *et al.*, 2007). Jardim *et al.* (2009) treated human primary bronchial epithelial cells in culture with a high dose of diesel exhaust particles (DEP; 10 $\mu g/cm^2$) and observed altered expression of a large panel of miRNA. The 12 most altered miRNA indicated that DEP exposure is associated with inflammatory response pathways and with a strong tumorigenic disease signature (Jardim *et al.*, 2009). Also, from studies of workers in an electric-furnace steel plant, Bollati *et al.* (2010) observed that expression of two miRNAs (miR-222 and miR-21) was increased significantly after 3 d exposure to particulate matter. Increased expression of miR-222 correlated significantly with lead exposure. However, since all these air pollutants can be derived from traffic-related pollution, and thus exposure levels to the different pollutants may be interrelated, future studies will be required to determine the specific abilities of the individual pollutants to induce epigenetic changes.

4. Endocrine disruptors

Endocrine disruptors are a group of chemicals that antagonize or activate steroid hormone signaling and have become of particular interest recently due to their impact on human fertility, function as non-DNA damaging carcinogens, and also because they induce transgenerational epigenetic changes (Anway and Skinner, 2006). For example, exposure of pregnant rats to vinclozolin (a commonly used antifungal agent) is associated with various reproductive abnormalities, particularly male infertility, as well as increased levels of tumor development (Skinner and Anway, 2007). Reduced sperm count in male offspring is associated with multiple abnormalities in the pattern of DNA methylation in male offspring. Remarkably, after breeding from these mice, both the reduced fertility and the DNA methylation abnormalities persisted into the F4 generation (Anway *et al.*, 2005).

Another endocrine active agent BPA has also been implicated as a modulator of the epigenome. *In utero* exposure of rats to BPA leads to increased frequency of cancer, particularly prostate cancer (Prins *et al.*, 2008) and prostatic tissue from exposed animals contained multiple abnormalities in DNA methylation. These included hypomethylation of the phosphodiesterase type 4 variant 4, which was associated with overexpression of the enzyme. Another endocrine disruptor shown to induce epigenetic changes is diethylstilbestrol (DES), which was given to millions of pregnant women to reduce the risk of miscarriage. However, female offspring of treated mothers showed increased frequency of tumors of the reproductive tract (Herbst *et al.*, 1971). Mice neonatally exposed to DES also showed an increased level of tumor susceptibility and analysis of uterine tissue in exposed mice identified consistent changes in methylation status of specific genes (Li *et al.*, 1997, 2003). Among the changes observed was hypomethylation of the oncogene *c-fos*, which was associated with stable upregulation of gene expression (Li *et al.*, 2003).

5. Infectious agents

Several infectious agents have been linked with the development of specific types of cancer and in many cases these are associated with specific epigenetic changes. In some instances these epigenetic alterations may be induced directly by the infectious agent, while in others the epigenetic changes may be secondary to the induction of inflammation. *H. pylori* infection is a key step in the development of gastric cancer (Suzuki *et al.*, 2009) and is associated with multiple DNA methylation abnormalities. For example, Shin *et al.* (2010) have recently shown that hypermethylation of a group of three gene promoters was associated with *H. pylori* infection of gastric mucosa. This hypermethylation was seen even in patients without any pathological evidence of gastric cancer. Furthermore Niwa *et al.* (2010) have provided evidence that increased CpG island hypermethylation is related to the induction of inflammation by *H. pylori* infection. The authors used a genome-wide screen to identify a panel of genes which were hypermethylated in infected, but not in uninfected, gastric mucosa. Using a gerbil model system the authors then showed that aberrant methylation of these genes was evident 5–10 weeks after *H. pylori* infection. Clearance of the infection resulted in some reduction in the extent of methylation of these genes but it remained elevated compared with that in noninfected animals. Importantly, the authors found that treatment with inflammation suppressing drugs at the same time as *H. pylori* infection prevented the appearance of the abnormal CpG island methylation, even though colonization of the gastric mucosa by *H. pylori* was unaffected which suggests that inflammation may contribute to the development of differential epigenetic marks.

However, not all epigenetic alterations induced by infectious agents appear to be modulated by inflammation. Epstein-Barr virus (EBV) is another infectious agent associated with the development of malignancy. One of the viral proteins expressed during EBV latency has been shown to downregulate expression of the metastasis suppressing E-cadherin gene (Fahraeus *et al.*, 1992). Tsai *et al.* (2002) demonstrated that downregulation of E-cadherin was due to hypermethylation of the E-cadherin promoter. Furthermore they also demonstrated that LMP1 acted to upregulate expression of DNMT enzymes in a c-jun kinase-dependent pathway (Tsai *et al.*, 2006). This mechanism may well not be unique to EBV as it has subsequently been demonstrated that both human polyoma virus and adenovirus are able to induce expression of DNMT1 (McCabe *et al.*, 2006).

III. CONCLUSIONS

A. Potential mechanisms linking environmental exposures to epigenetic change

It is now clear that many different environmental exposures can alter epigenetic marks and result in changes in gene expression. There is considerable potential for these effects to have significant impact in the development of cancer and the evidence that some epigenetic changes can be passed through the germ-line raises the possibility that these effects may persist in multiple generations. However, for most of these environmental exposures, there is limited understanding of the mechanisms through which these epigenetic changes occur although several potential mechanisms can be proposed:

1. *Induction of DNA synthesis (synthesis as a component of DNA repair or increased cellular proliferation).* The fidelity of replication of epigenetic marks is much lower than that of the DNA sequence itself; therefore any agent that leads to increased DNA synthesis may inadvertently lead to increased diversity of epigenetic marks.

2. *Altered gene expression.* In some cases, changes in epigenetic marks may not be the cause of altered gene expression, but rather a consequence (e.g., as in response to chemicals such as endocrine disruptors which alter signal transduction pathways or as a consequence of the obese state precipitating chronic inflammation). However, because epigenetic marks are replicated across cell generations, such epigenetic marks and associated changes in gene expression could remain even when the originating signal is no longer present.

3. *Direct targeting of epigenetic machinery.* Some chemical exposures may lead to epigenetic changes through directly altering the activity of important elements of the epigenetic machinery. For example, the polyphenol ECGC is a potent inhibitor of DNA methyltransferase enzymes whereas some viral infections directly upregulate expression of DNMT enzymes

4. *Adaptive responses to changes in the environment.* In the main, the above mechanisms suggest that the environmental stimuli act largely as genotoxic agents, altering the genome not by changing the DNA sequence, but through altering the epigenetic marks which control the genome. However, the discovery that even normal aspects of behavior (e.g., maternal behavior in rats) can also induce long-term epigenetic changes suggests that there may be adaptive mechanism in the genome which allow remodeling of the genome in response to the environment in which an organism is placed. This concept that the genome integrates intrinsic and extrinsic signals to regulate gene expression by epigenetic mechanisms was proposed by Jaenisch and Bird (2003) and the idea that the genome may "learn from experience" was amplified by Mathers (2010).

Clearly much work remains to be done to determine which, if any, of these mechanisms play significant roles in the modulation of the epigenome by environmental exposures.

B. Importance for cancer etiology and prevention

Despite the strong evidence from observational epidemiology that diet, lifestyle, and other environmental exposures are major modulators of cancer risk, there has been limited understanding of the mechanisms through which such exposures have their effects on the molecular steps in tumorigenesis. It is now becoming apparent that altered epigenetic marks may play a fundamental role in determining not only susceptibility to cancer (Feinberg *et al.*, 2006) but also the rate of progression of tumor development (Baylin and Ohm, 2006). As illustrated in Fig. 1.1, it is possible that many environmental factors interact with the genome to produce altered epigenomics marks which change the expression of tumor suppressor genes and oncogenes. For example, genes which

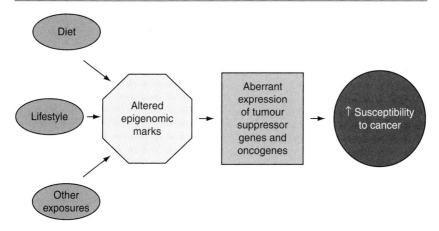

Figure 1.1. Overview of potential role of epigenetic modulation by dietary and other environmental factors in cancer development.

are critical for DNA repair such as *MGMT* and *MLH1* are readily silenced by promoter methylation (Menigatti *et al.*, 2009; Shen *et al.*, 2005). Cells in which DNA repair genes are silent accumulate DNA damage much faster than those with functional repair systems (Jones and Baylin, 2002). Environmental factors which cause silencing of such genes though epigenetic mechanisms would be expected to increase cancer risk.

Although observational studies provide evidence of associations between diet, lifestyle, and other environmental exposures and cancer risk, intervention studies are needed to provide evidence of causality. However, progress in developing such evidence is limited severely by the lack of robust biomarkers of cancer risk which could be used as surrogate endpoints. Given the apparent plasticity of epigenetic marks in response to cancer-related exposures, epigenetic marks are attractive candidates for the development of surrogate endpoints (Belshaw *et al.*, 2008) which could be used in (short term) dietary or lifestyle intervention studies. Future research should focus on identifying epigenetic marks which are (i) validated as biomarkers for the cancer under study, (ii) readily measured in easily accessible tissues, for example, blood, buccal cells, or stool, and (iii) altered in response to dietary or lifestyle interventions for which there is convincing evidence for a relationship with the cancer of interest.

Acknowledgments

Research on the topic of this review in our laboratories is funded by the BBSRC and EPSRC through the Centre for Integrated Systems Biology of Ageing and Nutrition (CISBAN) (BB/C008200/1), by the BBSRC (grant nos. BH071163, BH081097, and BH090948) by EU FP6 funding from NuGO

"The European Nutrigenomics Organisation; linking genomics, nutrition, and health research" (NuGO; CT-2004-505944), and through the Centre for Brain Ageing & Vitality which is funded under the Lifelong Health and Wellbeing cross-council initiative by the MRC, BBSRC, EPSRC, and ESRC.

References

Alam, M. G., Allinson, G., Stagnitti, F., Tanaka, A., and Westbrooke, M. (2002). Arsenic contamination in Bangladesh groundwater: A major environmental and social disaster. *Int. J. Environ. Health Res.* **12**, 235–253.

Allen, N. E., Beral, V., Casabonne, D., Kan, S. W., Reeves, G. K., Brown, A., and Green, J. (2009). Moderate alcohol intake and cancer incidence in women. *J. Natl. Cancer Inst.* **101**, 296–305.

Anway, M. D., and Skinner, M. K. (2006). Epigenetic transgenerational actions of endocrine disruptors. *Endocrinology* **147**, S43–S49.

Anway, M. D., Cupp, A. S., Uzumcu, M., and Skinner, M. K. (2005). Epigenetic transgenerational actions of endocrine disruptors and male fertility. *Science* **308**, 1466–1469.

Avissar, M., McClean, M. D., Kelsey, K. T., and Marsit, C. J. (2009). MicroRNA expression in head and neck cancer associates with alcohol consumption and survival. *Carcinogenesis* **30**, 2059–2063.

Baccarelli, A., Wright, R. O., Bollati, V., Tarantini, L., Litonjua, A. A., Suh, H. H., Zanobetti, A., Sparrow, D., Vokonas, P. S., and Schwartz, J. (2009). Rapid DNA methylation changes after exposure to traffic particles. *Am. J. Respir. Crit. Care Med.* **179**, 572–578.

Basterfield, L., Reul, J. M., and Mathers, J. C. (2005). Impact of physical activity on intestinal cancer development in mice. *J. Nutr.* **135**, 3002S–3008S.

Baur, J. A. (2010). Resveratrol, sirtuins, and the promise of a DR mimetic. *Mech. Ageing Dev.* **131**, 261–269.

Baur, J. A., Pearson, K. J., Price, N. L., Jamieson, H. A., Lerin, C., Kalra, A., Prabhu, V. V., Allard, J. S., Lopez-Lluch, G., Lewis, K., *et al.* (2006). Resveratrol improves health and survival of mice on a high-calorie diet. *Nature* **444**, 337–342.

Baylin, S. B., and Ohm, J. E. (2006). Epigenetic gene silencing in cancer—A mechanism for early oncogenic pathway addiction? *Nat. Rev. Cancer* **6**, 107–116.

Belinsky, S. A., Klinge, D. M., Liechty, K. C., March, T. H., Kang, T., Gilliland, F. D., Sotnic, N., Adamova, G., Rusinova, G., and Telnov, V. (2004). Plutonium targets the p16 gene for inactivation by promoter hypermethylation in human lung adenocarcinoma. *Carcinogenesis* **25**, 1063–1067.

Belshaw, N. J., Elliott, G. O., Foxall, R. J., Dainty, J. R., Pal, N., Coupe, A., Garg, D., Bradburn, D. M., Mathers, J. C., and Johnson, I. T. (2008). Profiling CpG island field methylation in both morphologically normal and neoplastic human colonic mucosa. *Br. J. Cancer* **99**, 136–142.

Bennett, K. L., Lee, W., Lamarre, E., Zhang, X., Seth, R., Scharpf, J., Hunt, J., and Eng, C. (2010). HPV status-independent association of alcohol and tobacco exposure or prior radiation therapy with promoter methylation of FUSSEL18, EBF3, IRX1, and SEPT9, but not SLC5A8, in head and neck squamous cell carcinomas. *Genes Chromosom. Cancer* **49**, 319–326.

Bjornsson, H. T., Sigurdsson, M. I., Fallin, M. D., Irizarry, R. A., Aspelund, T., Cui, H., Yu, W., Rongione, M. A., Ekstrom, T. J., Harris, T. B., *et al.* (2008). Intra-individual change over time in DNA methylation with familial clustering. *JAMA* **299**, 2877–2883.

Boffetta, P., Couto, E., Wichmann, J., Ferrari, P., Trichopoulos, D., Bueno-de-Mesquita, H. B., van Duijnhoven, F. J., Buchner, F. L., Key, T., Boeing, H., *et al.* (2010). Fruit and vegetable intake and overall cancer risk in the European Prospective Investigation into Cancer and Nutrition (EPIC). *J. Natl. Cancer Inst.* **102**, 529–537.

Bollati, V., Baccarelli, A., Hou, L., Bonzini, M., Fustinoni, S., Cavallo, D., Byun, H. M., Jiang, J., Marinelli, B., Pesatori, A. C., *et al.* (2007). Changes in DNA methylation patterns in subjects exposed to low-dose benzene. *Cancer Res.* **67**, 876–880.

Bollati, V., Marinelli, B., Apostoli, P., Bonzini, M., Nordio, F., Hoxha, M., Pegoraro, V., Motta, V., Tarantini, L., Cantone, L., *et al.* (2010). Exposure to metal-rich particulate matter modifies the expression of candidate microRNAs in peripheral blood leukocytes. *Environ. Health Perspect.* **118**, 763–768.

Bosetti, C., Boffetta, P., and La Vecchia, C. (2007). Occupational exposures to polycyclic aromatic hydrocarbons, and respiratory and urinary tract cancers: A quantitative review to 2005. *Ann. Oncol.* **18**, 431–446.

Breton, C. V., Byun, H. M., Wenten, M., Pan, F., Yang, A., and Gilliland, F. D. (2009). Prenatal tobacco smoke exposure affects global and gene-specific DNA methylation. *Am. J. Respir. Crit. Care Med.* **180**, 462–467.

Broday, L., Peng, W., Kuo, M. H., Salnikow, K., Zoroddu, M., and Costa, M. (2000). Nickel compounds are novel inhibitors of histone H4 acetylation. *Cancer Res.* **60**, 238–241.

Buehlmeyer, K., Doering, F., Daniel, H., Kindermann, B., Schulz, T., and Michna, H. (2008). Alteration of gene expression in rat colon mucosa after exercise. *Ann. Anat.* **190**, 71–80.

Butow, P. N., Hiller, J. E., Price, M. A., Thackway, S. V., Kricker, A., and Tennant, C. C. (2000). Epidemiological evidence for a relationship between life events, coping style, and personality factors in the development of breast cancer. *J. Psychosom. Res.* **49**, 169–181.

Campion, J., Milagro, F. I., and Martinez, J. A. (2009). Individuality and epigenetics in obesity. *Obes. Rev.* **10**, 383–392.

Cassidy, S. B., and Driscoll, D. J. (2009). Prader-Willi syndrome. *Eur. J. Hum. Genet.* **17**, 3–13.

Cebrian, A., Pharoah, P. D., Ahmed, S., Ropero, S., Fraga, M. F., Smith, P. L., Conroy, D., Luben, R., Perkins, B., Easton, D. F., *et al.* (2006). Genetic variants in epigenetic genes and breast cancer risk. *Carcinogenesis* **27**, 1661–1669.

Chanda, S., Dasgupta, U. B., Guhamazumder, D., Gupta, M., Chaudhuri, U., Lahiri, S., Das, S., Ghosh, N., and Chatterjee, D. (2006). DNA hypermethylation of promoter of gene p53 and p16 in arsenic-exposed people with and without malignancy. *Toxicol. Sci.* **89**, 431–437.

Chen, H., Giri, N. C., Zhang, R., Yamane, K., Zhang, Y., Maroney, M., and Costa, M. (2010). Nickel ions inhibit histone demethylase JMJD1A and DNA repair enzyme ABH2 by replacing the ferrous iron in the catalytic centers. *J. Biol. Chem.* **285**, 7374–7383.

Choi, J. Y., James, S. R., Link, P. A., McCann, S. E., Hong, C. C., Davis, W., Nesline, M. K., Ambrosone, C. B., and Karpf, A. R. (2009a). Association between global DNA hypomethylation in leukocytes and risk of breast cancer. *Carcinogenesis* **30**, 1889–1897.

Choi, S.-W., Corrocher, R., and Friso, S. (2009b). Nutrients and DNA methylation. *In* "Nutrients and Epigenetic" (S.-W. Choi and S. Friso, eds.), pp. 105–126. CRC Press Taylor & Francis Group, Boca Raton, FL.

Damiani, L. A., Yingling, C. M., Leng, S., Romo, P. E., Nakamura, J., and Belinsky, S. A. (2008). Carcinogen-induced gene promoter hypermethylation is mediated by DNMT1 and causal for transformation of immortalized bronchial epithelial cells. *Cancer Res.* **68**, 9005–9014.

Danaei, G., Ding, E. L., Mozaffarian, D., Taylor, B., Rehm, J., Murray, C. J., and Ezzati, M. (2009). The preventable causes of death in the United States: Comparative risk assessment of dietary, lifestyle, and metabolic risk factors. *PLoS Med.* **6**, e1000058.

Davidson, L. A., Wang, N., Shah, M. S., Lupton, J. R., Ivanov, I., and Chapkin, R. S. (2009). n-3 Polyunsaturated fatty acids modulate carcinogen-directed non-coding microRNA signatures in rat colon. *Carcinogenesis* **30**, 2077–2084.

Davie, J. R. (2003). Inhibition of histone deacetylase activity by butyrate. *J. Nutr.* **133**, 2485S–2493S.

Davis, C. D., Uthus, E. O., and Finley, J. W. (2000). Dietary selenium and arsenic affect DNA Methylation in vitro in Cac-2 cells and in vivo in rat liver and colon. *J. Nutr.* **130,** 2903–2909.

Delage, B., and Dashwood, R. H. (2008). Dietary manipulation of histone structure and function. *Annu. Rev. Nutr.* **28,** 347–366.

DePinho, R. A. (2000). The age of cancer. *Nature* **408,** 248–254.

Dolganiuc, A., Petrasek, J., Kodys, K., Catalano, D., Mandrekar, P., Velayudham, A., and Szabo, G. (2009). MicroRNA expression profile in Lieber-DeCarli diet-induced alcoholic and methionine choline deficient diet-induced nonalcoholic steatohepatitis models in mice. *Alcohol. Clin. Exp. Res.* **33,** 1704–1710.

Dolinoy, D. C., Weidman, J. R., Waterland, R. A., and Jirtle, R. L. (2006). Maternal genistein alters coat color and protects Avy mouse offspring from obesity by modifying the fetal epigenome. *Environ. Health Perspect.* **114,** 567–572.

Dolinoy, D. C., Huang, D., and Jirtle, R. L. (2007). Maternal nutrient supplementation counteracts bisphenol A-induced DNA hypomethylation in early development. *Proc. Natl. Acad. Sci. USA* **104,** 13056–13061.

Doll, R., and Peto, R. (1981). The causes of cancer: Quantitative estimates of avoidable risks of cancer in the United States today. *J. Natl. Cancer Inst.* **66,** 1191–1308.

Dronamraju, S. S., Coxhead, J. M., Kelly, S. B., Burn, J., and Mathers, J. C. (2009). Cell kinetics and gene expression changes in colorectal cancer patients given resistant starch: A randomised controlled trial. *Gut* **58,** 413–420.

Dronamraju, S. S., Coxhead, J. M., Kelly, S. B., and Mathers, J. C. (2010). Differential antineoplastic effects of butyrate in cells with and without a functioning DNA mismatch repair. *Nutr. Cancer* **62,** 105–115.

Ellen, T. P., Kluz, T., Harder, M. E., Xiong, J., and Costa, M. (2009). Heterochromatinization as a potential mechanism of nickel-induced carcinogenesis. *Biochemistry* **48,** 4626–4632.

Eriksson, M., Wedel, H., Wallander, M. A., Krakau, I., Hugosson, J., Carlsson, S., and Svardsudd, K. (2007). The impact of birth weight on prostate cancer incidence and mortality in a population-based study of men born in 1913 and followed up from 50 to 85 years of age. *Prostate* **67,** 1247–1254.

Esquela-Kerscher, A., and Slack, F. J. (2006). Oncomirs—MicroRNAs with a role in cancer. *Nat. Rev. Cancer* **6,** 259–269.

Fahraeus, R., Chen, W., Trivedi, P., Klein, G., and Obrink, B. (1992). Decreased expression of E-cadherin and increased invasive capacity in EBV-LMP-transfected human epithelial and murine adenocarcinoma cells. *Int. J. Cancer* **52,** 834–838.

Feinberg, A. P., Gehrke, C. W., Kuo, K. C., and Ehrlich, M. (1988). Reduced genomic 5-methylcytosine content in human colonic neoplasia. *Cancer Res.* **48,** 1159–1161.

Feinberg, A. P., Ohlsson, R., and Henikoff, S. (2006). The epigenetic progenitor origin of human cancer. *Nat. Rev. Genet.* **7,** 21–33.

Figueroa, M. E., Skrabanek, L., Li, Y., Jiemjit, A., Fandy, T. E., Paietta, E., Fernandez, H., Tallman, M. S., Greally, J. M., Carraway, H., *et al.* (2009). MDS and secondary AML display unique patterns and abundance of aberrant DNA methylation. *Blood* **114,** 3448–3458.

Francis, D., Diorio, J., Liu, D., and Meaney, M. J. (1999). Nongenomic transmission across generations of maternal behavior and stress responses in the rat. *Science* **286,** 1155–1158.

Gaedicke, S., Zhang, X., Schmelzer, C., Lou, Y., Doering, F., Frank, J., and Rimbach, G. (2008). Vitamin E dependent microRNA regulation in rat liver. *FEBS Lett.* **582,** 3542–3546.

Giotopoulos, G., McCormick, C., Cole, C., Zanker, A., Jawad, M., Brown, R., and Plumb, M. (2006). DNA methylation during mouse hemopoietic differentiation and radiation-induced leukemia. *Exp. Hematol.* **34,** 1462–1470.

Goyal, R., Goyal, D., Leitzke, A., Gheorghe, C. P., and Longo, L. D. (2010). Brain renin-angiotensin system: fetal epigenetic programming by maternal protein restriction during pregnancy. *Reprod. Sci.* **17,** 227–238.

Guerri, C., and Pascual, M. (2010). Mechanisms involved in the neurotoxic, cognitive, and neurobehavioral effects of alcohol consumption during adolescence. *Alcohol* **44,** 15–26.

Hart, C. L., Morrison, D. S., Batty, G. D., Mitchell, R. J., and Davey Smith, G. (2010). Effect of body mass index and alcohol consumption on liver disease: Analysis of data from two prospective cohort studies. *BMJ* **340,** c1240.

Heijmans, B. T., Kremer, D., Tobi, E. W., Boomsma, D. I., and Slagboom, P. E. (2007). Heritable rather than age-related environmental and stochastic factors dominate variation in DNA methylation of the human IGF2/H19 locus. *Hum. Mol. Genet.* **16,** 547–554.

Heijmans, B. T., Tobi, E. W., Stein, A. D., Putter, H., Blauw, G. J., Susser, E. S., Slagboom, P. E., and Lumey, L. H. (2008). Persistent epigenetic differences associated with prenatal exposure to famine in humans. *Proc. Natl. Acad. Sci. USA* **105,** 17046–17049.

Herbst, A. L., Ulfelder, H., and Poskanzer, D. C. (1971). Adenocarcinoma of the vagina. Association of maternal stilbestrol therapy with tumor appearance in young women. *N. Engl. J. Med.* **284,** 878–881.

Hernandez-Vargas, H., Lambert, M. P., Le Calvez-Kelm, F., Gouysse, G., McKay-Chopin, S., Tavtigian, S. V., Scoazec, J. Y., and Herceg, Z. (2010). Hepatocellular carcinoma displays distinct DNA methylation signatures with potential as clinical predictors. *PLoS ONE* **5,** e9749.

Hirsch, H. A., Iliopoulos, D., Joshi, A., Zhang, Y., Jaeger, S. A., Bulyk, M., Tsichlis, P. N., Shirley Liu, X., and Struhl, K. (2010). A transcriptional signature and common gene networks link cancer with lipid metabolism and diverse human diseases. *Cancer Cell* **17,** 348–361.

Hotamisligil, G. S. (2006). Inflammation and metabolic disorders. *Nature* **444,** 860–867.

Hou, L., Ji, B. T., Blair, A., Dai, Q., Gao, Y. T., and Chow, W. H. (2004). Commuting physical activity and risk of colon cancer in Shanghai, China. *Am. J. Epidemiol.* **160,** 860–867.

Hussain, M., Rao, M., Humphries, A. E., Hong, J. A., Liu, F., Yang, M., Caragacianu, D., and Schrump, D. S. (2009). Tobacco smoke induces polycomb-mediated repression of Dickkopf-1 in lung cancer cells. *Cancer Res.* **69,** 3570–3578.

Ilnytskyy, Y., Koturbash, I., and Kovalchuk, O. (2009). Radiation-induced bystander effects in vivo are epigenetically regulated in a tissue-specific manner. *Environ. Mol. Mutagen.* **50,** 105–113.

Jacob, R. A., Gretz, D. M., Taylor, P. C., James, S. J., Pogribny, I. P., Miller, B. J., Henning, S. M., and Swendseid, M. E. (1998). Moderate folate depletion increases plasma homocysteine and decreases lymphocyte DNA methylation in postmenopausal women. *J. Nutr.* **128,** 1204–1212.

Jaenisch, R., and Bird, A. (2003). Epigenetic regulation of gene expression: How the genome integrates intrinsic and environmental signals. *Nat. Genet.* **33,** 25–254.

Jardim, M. J., Fry, R. C., Jaspers, I., Dailey, L., and Diaz-Sanchez, D. (2009). Disruption of microRNA expression in human airway cells by diesel exhaust particles is linked to tumorigenesis-associated pathways. *Environ. Health Perspect.* **117,** 1745–1751.

Jarup, L. (2003). Hazards of heavy metal contamination. *Br. Med. Bull.* **68,** 167–182.

Jenab, M., Slimani, N., Bictash, M., Ferrari, P., and Bingham, S. A. (2009). Biomarkers in nutritional epidemiology: Applications, needs and new horizons. *Hum. Genet.* **125,** 507–525.

Jensen, T. J., Wozniak, R. J., Eblin, K. E., Wnek, S. M., Gandolfi, A. J., and Futscher, B. W. (2009). Epigenetic mediated transcriptional activation of WNT5A participates in arsenical-associated malignant transformation. *Toxicol. Appl. Pharmacol.* **235,** 39–46.

Jiang, G., Xu, L., Song, S., Zhu, C., Wu, Q., Zhang, L., and Wu, L. (2008). Effects of long-term low-dose cadmium exposure on genomic DNA methylation in human embryo lung fibroblast cells. *Toxicology* **244,** 49–55.

Johansson, M., Relton, C., Ueland, P., Vollset, S., Midttun, Q., Nygård, O., Slimani, N., Boffetta, P., Jenab, M., Clavel-Chapelon, F., *et al.* (2010). Serum B vitamin levels and risk of lung cancer. *JAMA* **303,** 2377–2385.

Jones, P. A., and Baylin, S. B. (2002). The fundamental role of epigenetic events in cancer. *Nat. Rev. Genet.* **3,** 415–428.

Kadota, M., Yang, H. H., Hu, N., Wang, C., Hu, Y., Taylor, P. R., Buetow, K. H., and Lee, M. P. (2007). Allele-specific chromatin immunoprecipitation studies show genetic influence on chromatin state in human genome. *PLoS Genet.* **3,** e81.

Ke, Q., Li, Q., Ellen, T. P., Sun, H., and Costa, M. (2008). Nickel compounds induce phosphorylation of histone H3 at serine 10 by activating JNK-MAPK pathway. *Carcinogenesis* **29,** 1276–1281.

Khaw, K. T., Wareham, N., Bingham, S., Welch, A., Luben, R., and Day, N. (2008). Combined impact of health behaviours and mortality in men and women: The EPIC-Norfolk prospective population study. *PLoS Med.* **5,** e12.

Kim, Y. I., Christman, J. K., Fleet, J. C., Cravo, M. L., Salomon, R. N., Smith, D., Ordovas, J., Selhub, J., and Mason, J. B. (1995). Moderate folate deficiency does not cause global hypomethylation of hepatic and colonic DNA or c-myc-specific hypomethylation of colonic DNA in rats. *Am. J. Clin. Nutr.* **61,** 1083–1090.

Kim, Y., Pogribn, I., Basnakian, A., Miller, J., Selhub, J., James, S., and Mason, J. (1997). Folate deficiency in rats induces DNA strand breaks and hypomethylation within the p53 tumor suppressor gene[1-5]. *Am. J. Clin. Nutr.* **65,** 46–52.

Kloting, N., Berthold, S., Kovacs, P., Schon, M. R., Fasshauer, M., Ruschke, K., Stumvoll, M., and Bluher, M. (2009). MicroRNA expression in human omental and subcutaneous adipose tissue. *PLoS ONE* **4,** e4699.

Kong, A., Steinthorsdottir, V., Masson, G., Thorleifsson, G., Sulem, P., Besenbacher, S., Jonasdottir, A., Sigurdsson, A., Kristinsson, K. T., Jonasdottir, A., *et al.* (2009). Parental origin of sequence variants associated with complex diseases. *Nature* **462,** 868–874.

Koturbash, I., Kutanzi, K., Hendrickson, K., Rodriguez-Juarez, R., Kogosov, D., and Kovalchuk, O. (2008). Radiation-induced bystander effects in vivo are sex specific. *Mutat. Res.* **642,** 28–36.

Kovacheva, V. P., Mellott, T. J., Davison, J. M., Wagner, N., Lopez-Coviella, I., Schnitzler, A. C., and Blusztajn, J. K. (2007). Gestational choline deficiency causes global and Igf2 gene DNA hypermethylation by up-regulation of Dnmt1 expression. *J. Biol. Chem.* **282,** 31777–31788.

Kruh, J. (1982). Effects of sodium butyrate, a new pharmacological agent, on cells in culture. *Mol. Cell. Biochem.* **42,** 65–82.

Ladeiro, Y., Couchy, G., Balabaud, C., Bioulac-Sage, P., Pelletier, L., Rebouissou, S., and Zucman-Rossi, J. (2008). MicroRNA profiling in hepatocellular tumors is associated with clinical features and oncogene/tumor suppressor gene mutations. *Hepatology* **47,** 1955–1963.

Lagouge, M., Argmann, C., Gerhart-Hines, Z., Meziane, H., Lerin, C., Daussin, F., Messadeq, N., Milne, J., Lambert, P., Elliott, P., *et al.* (2006). Resveratrol improves mitochondrial function and protects against metabolic disease by activating SIRT1 and PGC-1alpha. *Cell* **127,** 1109–1122.

Larsson, S. C., and Wolk, A. (2007). Obesity and colon and rectal cancer risk: A meta-analysis of prospective studies. *Am. J. Clin. Nutr.* **86,** 556–565.

Lee, Y. W., Klein, C. B., Kargacin, B., Salnikow, K., Kitahara, J., Dowjat, K., Zhitkovich, A., Christie, N. T., and Costa, M. (1995). Carcinogenic nickel silences gene expression by chromatin condensation and DNA methylation: A new model for epigenetic carcinogens. *Mol. Cell. Biol.* **15,** 2547–2557.

Lee, W. J., Shim, J. Y., and Zhu, B. T. (2005). Mechanisms for the inhibition of DNA methyltransferases by tea catechins and bioflavonoids. *Mol. Pharmacol.* **68,** 1018–1030.

Leng, S., Stidley, C. A., Willink, R., Bernauer, A., Do, K., Picchi, M. A., Sheng, X., Frasco, M. A., Van Den Berg, D., Gilliland, F. D., *et al.* (2008). Double-strand break damage and associated DNA repair genes predispose smokers to gene methylation. *Cancer Res.* **68,** 3049–3056.

Li, S., Washburn, K. A., Moore, R., Uno, T., Teng, C., Newbold, R. R., McLachlan, J. A., and Negishi, M. (1997). Developmental exposure to diethylstilbestrol elicits demethylation of estrogen-responsive lactoferrin gene in mouse uterus. *Cancer Res.* **57,** 4356–4359.

Li, S., Hansman, R., Newbold, R., Davis, B., McLachlan, J. A., and Barrett, J. C. (2003). Neonatal diethylstilbestrol exposure induces persistent elevation of *c-fos* expression and hypomethylation in its exon-4 in mouse uterus. *Mol. Carcinog.* **38,** 78–84.

Li, Y., Sheu, C. C., Ye, Y., de Andrade, M., Wang, L., Chang, S. C., Aubry, M. C., Aakre, J. A., Allen, M. S., Chen, F., *et al.* (2010). Genetic variants and risk of lung cancer in never smokers: A genome-wide association study. *Lancet Oncol.* **11,** 321–330.

Liu, H., Zhou, Y., Boggs, S. E., Belinsky, S. A., and Liu, J. (2007). Cigarette smoke induces demethylation of prometastatic oncogene synuclein-gamma in lung cancer cells by downregulation of DNMT3B. *Oncogene* **26,** 5900–5910.

Liu, B., Balkwill, A., Reeves, G., and Beral, V. (2010). Body mass index and risk of liver cirrhosis in middle aged UK women: Prospective study. *BMJ* **340,** c912.

Lyon, C. M., Klinge, D. M., Liechty, K. C., Gentry, F. D., March, T. H., Kang, T., Gilliland, F. D., Adamova, G., Rusinova, G., Telnov, V., *et al.* (2007). Radiation-induced lung adenocarcinoma is associated with increased frequency of genes inactivated by promoter hypermethylation. *Radiat. Res.* **168,** 409–414.

Marsit, C. J., Eddy, K., and Kelsey, K. T. (2006). MicroRNA responses to cellular stress. *Cancer Res.* **66,** 10843–10848.

Marwick, J. A., Kirkham, P. A., Stevenson, C. S., Danahay, H., Giddings, J., Butler, K., Donaldson, K., Macnee, W., and Rahman, I. (2004). Cigarette smoke alters chromatin remodeling and induces proinflammatory genes in rat lungs. *Am. J. Respir. Cell Mol. Biol.* **31,** 633–642.

Mathers, J. C. (2009). Folate intake and bowel cancer risk. *Genes Nutr.* **4,** 173–178.

Mathers, J. C. (2010). Nutrition, epigenomics and the development of obesity: How the genome learns from experience. *In* "Obesity Prevention: The Role of Brain and Society on Individual Behavior" (L. Dube, ed.), pp 191–201. Academic Press, Oxford.

Mathers, J. C., and Ford, D. (2009). Nutrition, epigenetics and aging. *In* "Nutrients and Epigenetics" (S.-W. Chio and S. Friso, eds.), pp 175–206. CRC Press Taylor & Francis Group, Boca Raton, FL.

McCabe, M. T., Low, J. A., Imperiale, M. J., and Day, M. L. (2006). Human polyomavirus BKV transcriptionally activates DNA methyltransferase 1 through the pRb/E2F pathway. *Oncogene* **25,** 2727–2735.

McGee, S. L., Fairlie, E., Garnham, A. P., and Hargreaves, M. (2009). Exercise-induced histone modifications in human skeletal muscle. *J. Physiol.* **587,** 5951–5958.

Meaney, M. J. (2001). Maternal care, gene expression, and the transmission of individual differences in stress reactivity across generations. *Annu. Rev. Neurosci.* **24,** 1161–1192.

Menigatti, M., Truninger, K., Gebbers, J. O., Marbet, U., Marra, G., and Schar, P. (2009). Normal colorectal mucosa exhibits sex- and segment-specific susceptibility to DNA methylation at the hMLH1 and MGMT promoters. *Oncogene* **28,** 899–909.

Miremadi, A., Oestergaard, M. Z., Pharoah, P. D., and Caldas, C. (2007). Cancer genetics of epigenetic genes. *Hum. Mol. Genet.* **16**(Spec No 1), R28–R49.

Nielsen, N. R., and Gronbaek, M. (2006). Stress and breast cancer: A systematic update on the current knowledge. *Nat. Clin. Pract. Oncol.* **3,** 612–620.

Niwa, T., Tsukamoto, T., Toyoda, T., Mori, A., Tanaka, H., Maekita, T., Ichinose, M., Tatematsu, M., and Ushijima, T. (2010). Inflammatory processes triggered by *Helicobacter pylori* infection cause aberrant DNA methylation in gastric epithelial cells. *Cancer Res.* **70,** 1430–1440.

Norat, T., Bingham, S., Ferrari, P., Slimani, N., Jenab, M., Mazuir, M., Overvad, K., Olsen, A., Tjonneland, A., Clavel, F., *et al.* (2005). Meat, fish, and colorectal cancer risk: The European Prospective Investigation into cancer and nutrition. *J. Natl. Cancer Inst.* **97,** 906–916.

Oberlander, T. F., Weinberg, J., Papsdorf, M., Grunau, R., Misri, S., and Devlin, A. M. (2008). Prenatal exposure to maternal depression, neonatal methylation of human glucocorticoid receptor gene (NR3C1) and infant cortisol stress responses. *Epigenetics* **3,** 97–106.

Onishchenko, N., Karpova, N., Sabri, F., Castren, E., and Ceccatelli, S. (2008). Long-lasting depression-like behavior and epigenetic changes of BDNF gene expression induced by perinatal exposure to methylmercury. *J. Neurochem.* **106,** 1378–1387.

Ortega, F. J., Moreno-Navarrete, J. M., Pardo, G., Sabater, M., Hummel, M., Ferrer, A., Rodriguez-Hermosa, J. I., Ruiz, B., Ricart, W., Peral, B., et al. (2010). MiRNA expression profile of human subcutaneous adipose and during adipocyte differentiation. *PLoS ONE* **5,** e9022.

Pascual, M., Boix, J., Felipo, V., and Guerri, C. (2009). Repeated alcohol administration during adolescence causes changes in the mesolimbic dopaminergic and glutamatergic systems and promotes alcohol intake in the adult rat. *J. Neurochem.* **108,** 920–931.

Pavanello, S., Bollati, V., Pesatori, A. C., Kapka, L., Bolognesi, C., Bertazzi, P. A., and Baccarelli, A. (2009). Global and gene-specific promoter methylation changes are related to anti-B[a]PDE-DNA adduct levels and influence micronuclei levels in polycyclic aromatic hydrocarbon-exposed individuals. *Int. J. Cancer* **125,** 1692–1697.

Penn, L., White, M., Oldroyd, J., Walker, M., Alberti, K. G., and Mathers, J. C. (2009). Prevention of type 2 diabetes in adults with impaired glucose tolerance: the European Diabetes Prevention RCT in Newcastle upon Tyne, UK. *BMC Public Health* **9,** 342.

Perera, F., Tang, W. Y., Herbstman, J., Tang, D., Levin, L., Miller, R., and Ho, S. M. (2009). Relation of DNA methylation of 5'-CpG island of ACSL3 to transplacental exposure to airborne polycyclic aromatic hydrocarbons and childhood asthma. *PLoS ONE* **4,** e4488.

Peters, I., Vaske, B., Albrecht, K., Kuczyk, M. A., Jonas, U., and Serth, J. (2007). Adiposity and age are statistically related to enhanced RASSF1A tumor suppressor gene promoter methylation in normal autopsy kidney tissue. *Cancer Epidemiol. Biomarkers Prev.* **16,** 2526–2532.

Pfluger, P. T., Herranz, D., Velasco-Miguel, S., Serrano, M., and Tschop, M. H. (2008). Sirt1 protects against high-fat diet-induced metabolic damage. *Proc. Natl. Acad. Sci. USA* **105,** 9793–9798.

Phillips, J. M., and Goodman, J. I. (2009). Inhalation of cigarette smoke induces regions of altered DNA methylation (RAMs) in SENCAR mouse lung. *Toxicology* **260,** 7–15.

Plagemann, A., Harder, T., Brunn, M., Harder, A., Roepke, K., Wittrock-Staar, M., Ziska, T., Schellong, K., Rodekamp, E., Melchior, K., et al. (2009). Hypothalamic proopiomelanocortin promoter methylation becomes altered by early overfeeding: An epigenetic model of obesity and the metabolic syndrome. *J. Physiol.* **587,** 4963–4976.

Pogribny, I. P., Karpf, A. R., James, S. R., Melnyk, S., Han, T., and Tryndyak, V. P. (2008). Epigenetic alterations in the brains of Fisher 344 rats induced by long-term administration of folate/methyl-deficient diet. *Brain Res.* **1237,** 25–34.

Polley, A. C., Mulholland, F., Pin, C., Williams, E. A., Bradburn, D. M., Mills, S. J., Mathers, J. C., and Johnson, I. T. (2006). Proteomic analysis reveals field-wide changes in protein expression in the morphologically normal mucosa of patients with colorectal neoplasia. *Cancer Res.* **66,** 6553–6562.

Prins, G. S., Tang, W. Y., Belmonte, J., and Ho, S. M. (2008). Perinatal exposure to oestradiol and bisphenol A alters the prostate epigenome and increases susceptibility to carcinogenesis. *Basic Clin. Pharmacol. Toxicol.* **102,** 134–138.

Pufulete, M., Al-Ghnaniem, R., Khushal, A., Appleby, P., Harris, N., Gout, S., Emery, P. W., and Sanders, T. A. (2005). Effect of folic acid supplementation on genomic DNA methylation in patients with colorectal adenoma. *Gut* **54,** 648–653.

Radom-Aizik, S., Zaldivar, F. P., Jr., Oliver, S. R., Galassetti, P. R., and Cooper, D. M. (2010). Evidence for microRNA involvement in exercise-associated neutrophil gene expression changes. *J. Appl. Physiol.* **109,** 252–261.

Rampersaud, G. C., Kauwell, G. P. A., Hutson, A. D., Cerda, J. J., and Bailey, L. B. (2000). Genomic DNA methylation decreases in response to moderate folate depletion in elderly women. *Am. J. Clin. Nutr.* **72,** 998–1003.

Reed, M. C., Nijhout, H. F., Neuhouser, M. L., Gregory, J. F., 3rd., Shane, B., James, S. J., Boynton, A., and Ulrich, C. M. (2006). A mathematical model gives insights into nutritional and genetic aspects of folate-mediated one-carbon metabolism. *J. Nutr.* **136**, 2653–2661.

Reichard, J. F., Schnekenburger, M., and Puga, A. (2007). Long term low-dose arsenic exposure induces loss of DNA methylation. *Biochem. Biophys. Res. Commun.* **352**, 188–192.

Salnikow, K., and Zhitkovich, A. (2008). Genetic and epigenetic mechanisms in metal carcinogenesis and cocarcinogenesis: Nickel, arsenic, and chromium. *Chem. Res. Toxicol.* **21**, 28–44.

Sandovici, I., Leppert, M., Hawk, P. R., Suarez, A., Linares, Y., and Sapienza, C. (2003). Familial aggregation of abnormal methylation of parental alleles at the IGF2/H19 and IGF2R differentially methylated regions. *Hum. Mol. Genet.* **12**, 1569–1578.

Sauer, J., Jang, H., Zimmerly, E. M., Kim, K. C., Liu, Z., Chanson, A., Smith, D. E., Mason, J. B., Friso, S., and Choi, S. W. (2010). Ageing, chronic alcohol consumption and folate are determinants of genomic DNA methylation, p16 promoter methylation and the expression of p16 in the mouse colon. *Br. J. Nutr.* **104**, 24–30.

Seitz, H. K., and Stickel, F. (2007). Molecular mechanisms of alcohol-mediated carcinogenesis. *Nat. Rev. Cancer* **7**, 599–612.

Shen, L., Kondo, Y., Rosner, G. L., Xiao, L., Hernandez, N. S., Vilaythong, J., Houlihan, P. S., Krouse, R. S., Prasad, A. R., Einspahr, J. G., et al. (2005). MGMT promoter methylation and field defect in sporadic colorectal cancer. *J. Natl Cancer Inst.* **97**, 1330–1338.

Shin, C. M., Kim, N., Jung, Y., Park, J. H., Kang, G. H., Kim, J. S., Jung, H. C., and Song, I. S. (2010). Role of *Helicobacter pylori* infection in aberrant DNA methylation along multistep gastric carcinogenesis. *Cancer Sci.* Feb 18. [Epub ahead of print].

Shukla, S. D., Velazquez, J., French, S. W., Lu, S. C., Ticku, M. K., and Zakhari, S. (2008). Emerging role of epigenetics in the actions of alcohol. *Alcohol. Clin. Exp. Res.* **32**, 1525–1534.

Siegel, A. B., and Zhu, A. X. (2009). Metabolic syndrome and hepatocellular carcinoma: Two growing epidemics with a potential link. *Cancer* **115**, 5651–5661.

Silva, A. J., and White, R. (1988). Inheritance of allelic blueprints for methylation patterns. *Cell* **54**, 145–152.

Skinner, M. K., and Anway, M. D. (2007). Epigenetic transgenerational actions of vinclozolin on the development of disease and cancer. *Crit. Rev. Oncog.* **13**, 75–82.

Spencer, J. P., Abd El Mohsen, M. M., Minihane, A. M., and Mathers, J. C. (2008). Biomarkers of the intake of dietary polyphenols: Strengths, limitations and application in nutrition research. *Br. J. Nutr.* **99**, 12–22.

Stanner, S. (ed.) (2005). *In* Cardiovascular Disease: Diet, Nutrition and Emerging Risk Factors. The Report of a British Nutrition Foundation Task Force. Blackwell Publishing, Oxford.

Steegers-Theunissen, R. P., Obermann-Borst, S. A., Kremer, D., Lindemans, J., Siebel, C., Steegers, E. A., Slagboom, P. E., and Heijmans, B. T. (2009). Periconceptional maternal folic acid use of 400 microg per day is related to increased methylation of the IGF2 gene in the very young child. *PLoS ONE* **4**, e7845.

Stewart, B. W., and Kleihues, P. (2003). World Cancer Report. IARC, Lyon.

Suzuki, H., Iwasaki, E., and Hibi, T. (2009). *Helicobacter pylori* and gastric cancer. *Gastric Cancer* **12**, 79–87.

Takanabe, R., Ono, K., Abe, Y., Takaya, T., Horie, T., Wada, H., Kita, T., Satoh, N., Shimatsu, A., and Hasegawa, K. (2008). Up-regulated expression of microRNA-143 in association with obesity in adipose tissue of mice fed high-fat diet. *Biochem. Biophys. Res. Commun.* **376**, 728–732.

Takiguchi, M., Achanzar, W. E., Qu, W., Li, G., and Waalkes, M. P. (2003). Effects of cadmium on DNA-(Cytosine-5) methyltransferase activity and DNA methylation status during cadmium-induced cellular transformation. *Exp. Cell Res.* **286**, 355–365.

Tamminga, J., Koturbash, I., Baker, M., Kutanzi, K., Kathiria, P., Pogribny, I. P., Sutherland, R. J., and Kovalchuk, O. (2008). Paternal cranial irradiation induces distant bystander DNA damage in the germline and leads to epigenetic alterations in the offspring. *Cell Cycle* **7**, 1238–1245.

Tang, Y., Banan, A., Forsyth, C. B., Fields, J. Z., Lau, C. K., Zhang, L. J., and Keshavarzian, A. (2008). Effect of alcohol on miR-212 expression in intestinal epithelial cells and its potential role in alcoholic liver disease. *Alcohol. Clin. Exp. Res.* **32**, 355–364.

Tarantini, L., Bonzini, M., Apostoli, P., Pegoraro, V., Bollati, V., Marinelli, B., Cantone, L., Rizzo, G., Hou, L., Schwartz, J., et al. (2009). Effects of particulate matter on genomic DNA methylation content and iNOS promoter methylation. *Environ. Health Perspect.* **117**, 217–222.

Terry, M. B., Ferris, J. S., Pilsner, R., Flom, J. D., Tehranifar, P., Santella, R. M., Gamble, M. V., and Susser, E. (2008). Genomic DNA methylation among women in a multiethnic New York City birth cohort. *Cancer Epidemiol. Biomarkers Prev.* **17**, 2306–2310.

Tian, Y. (2009). Ah receptor and NF-kappaB interplay on the stage of epigenome. *Biochem. Pharmacol.* **77**, 670–680.

Tobi, E. W., Lumey, L. H., Talens, R. P., Kremer, D., Putter, H., Stein, A. D., Slagboom, P. E., and Heijmans, B. T. (2009). DNA methylation differences after exposure to prenatal famine are common and timing- and sex-specific. *Hum. Mol. Genet.* **18**, 4046–4053.

Toh, Y., Oki, E., Ohgaki, K., Sakamoto, Y., Ito, S., Egashira, A., Saeki, H., Kakeji, Y., Morita, M., Sakaguchi, Y., et al. (2010). Alcohol drinking, cigarette smoking, and the development of squamous cell carcinoma of the esophagus: Molecular mechanisms of carcinogenesis. *Int. J. Clin. Oncol.* **15**, 135–144.

Tryndyak, V. P., Ross, S. A., Beland, F. A., and Pogribny, I. P. (2009). Down-regulation of the microRNAs miR-34a, miR-127, and miR-200b in rat liver during hepatocarcinogenesis induced by a methyl-deficient diet. *Mol. Carcinog.* **48**, 479–487.

Tsai, C. N., Tsai, C. L., Tse, K. P., Chang, H. Y., and Chang, Y. S. (2002). The Epstein-Barr virus oncogene product, latent membrane protein 1, induces the downregulation of E-cadherin gene expression via activation of DNA methyltransferases. *Proc. Natl. Acad. Sci. USA* **99**, 10084–10089.

Tsai, C. L., Li, H. P., Lu, Y. J., Hsueh, C., Liang, Y., Chen, C. L., Tsao, S. W., Tse, K. P., Yu, J. S., and Chang, Y. S. (2006). Activation of DNA methyltransferase 1 by EBV LMP1 Involves c-Jun NH (2)-terminal kinase signaling. *Cancer Res.* **66**, 11668–11676.

Vaissiere, T., Hung, R. J., Zaridze, D., Moukeria, A., Cuenin, C., Fasolo, V., Ferro, G., Paliwal, A., Hainaut, P., Brennan, P., et al. (2009). Quantitative analysis of DNA methylation profiles in lung cancer identifies aberrant DNA methylation of specific genes and its association with gender and cancer risk factors. *Cancer Res.* **69**, 243–252.

Vidali, G., Boffa, L. C., Bradbury, E. M., and Allfrey, V. G. (1978). Butyrate suppression of histone deacetylation leads to accumulation of multiacetylated forms of histones H3 and H4 and increased DNase I sensitivity of the associated DNA sequences. *Proc. Natl. Acad. Sci. USA* **75**, 2239–2243.

Wang, B., Majumder, S., Nuovo, G., Kutay, H., Volinia, S., Patel, T., Schmittgen, T. D., Croce, C., Ghoshal, K., and Jacob, S. T. (2009). Role of microRNA-155 at early stages of hepatocarcinogenesis induced by choline-deficient and amino acid-defined diet in C57BL/6 mice. *Hepatology* **50**, 1152–1161.

Waterland, R. A., and Jirtle, R. L. (2003). Transposable elements: Targets for early nutritional effects on epigenetic gene regulation. *Mol. Cell. Biol.* **23**(15), 5293–5300.

Waterland, R. A., and Michels, K. B. (2007). Epigenetic epidemiology of the developmental origins hypothesis. *Annu. Rev. Nutr.* **27**, 363–388.

Weaver, I. C., Cervoni, N., Champagne, F. A., D'Alessio, A. C., Sharma, S., Seckl, J. R., Dymov, S., Szyf, M., and Meaney, M. J. (2004). Epigenetic programming by maternal behavior. *Nat. Neurosci.* **7**, 847–854.

Williams, E. A., Coxhead, J. M., and Mathers, J. C. (2003). Anti-cancer effects of butyrate: Use of micro-array technology to investigate mechanisms. *Proc. Nutr. Soc.* **62**, 107–115.

World Cancer Research Fund / American Institute for Cancer Research. (2007). "Food, Nutrition, Physical Activity, and the Prevention of Cancer: a Global Perspective." Washington DC: AICR.

Wright, R. O., Schwartz, J., Wright, R. J., Bollati, V., Tarantini, L., Park, S. K., Hu, H., Sparrow, D., Vokonas, P., and Baccarelli, A. (2010). Biomarkers of lead exposure and DNA methylation within retrotransposons. *Environ. Health Perspect.* **118**, 790–795.

Wu, J., Basha, M. R., Brock, B., Cox, D. P., Cardozo-Pelaez, F., McPherson, C. A., Harry, J., Rice, D. C., Maloney, B., Chen, D., *et al.* (2008). Alzheimer's disease (AD)-like pathology in aged monkeys after infantile exposure to environmental metal lead (Pb): Evidence for a developmental origin and environmental link for AD. *J. Neurosci.* **28**, 3–9.

Xu, X., Dailey, A. B., Peoples-Sheps, M., Talbott, E. O., Li, N., and Roth, J. (2009). Birth weight as a risk factor for breast cancer: A meta-analysis of 18 epidemiological studies. *J. Womens Health (Larchmt)* **18**, 1169–1178.

Yang, S. R., Valvo, S., Yao, H., Kode, A., Rajendrasozhan, S., Edirisinghe, I., Caito, S., Adenuga, D., Henry, R., Fromm, G., *et al.* (2008). IKK alpha causes chromatin modification on pro-inflammatory genes by cigarette smoke in mouse lung. *Am. J. Respir. Cell Mol. Biol.* **38**, 689–698.

Yorifuji, T., Tsuda, T., Takao, S., and Harada, M. (2008). Long-term exposure to methylmercury and neurologic signs in Minamata and neighboring communities. *Epidemiology* **19**, 3–9.

Zhang, A. H., Bin, H. H., Pan, X. L., and Xi, X. G. (2007). Analysis of p16 gene mutation, deletion and methylation in patients with arseniasis produced by indoor unventilated-stove coal usage in Guizhou, China. *J. Toxicol. Environ. Health A* **70**, 970–975.

Zhang, J., Zhang, F., Didelot, X., Bruce, K. D., Cagampang, F. R., Vatish, M., Hanson, M., Lehnert, H., Ceriello, A., and Byrne, C. D. (2009). Maternal high fat diet during pregnancy and lactation alters hepatic expression of insulin like growth factor-2 and key microRNAs in the adult offspring. *BMC Genomics* **10**, 478.

Zhong, C. X., and Mass, M. J. (2001). Both hypomethylation and hypermethylation of DNA associated with arsenite exposure in cultures of human cells identified by methylation-sensitive arbitrarily-primed PCR. *Toxicol. Lett.* **122**, 223–234.

Zhou, X., Sun, H., Ellen, T. P., Chen, H., and Costa, M. (2008). Arsenite alters global histone H3 methylation. *Carcinogenesis* **29**, 1831–1836.

2 Induction of Epigenetic Alterations by Chronic Inflammation and Its Significance on Carcinogenesis

Tohru Niwa and Toshikazu Ushijima

Carcinogenesis Division, National Cancer Center Research Institute, Chuo-ku, Tokyo, Japan

Advances in Genetics, Vol. 71
Copyright 2010, Elsevier Inc. All rights reserved.

0065-2660/10 $35.00
DOI: 10.1016/S0065-2660(10)71002-6

ABSTRACT

Chronic inflammation is deeply involved in development of human cancers, such as gastric and liver cancers. Induction of cell proliferation, production of reactive oxygen species, and direct stimulation of epithelial cells by inflammation-inducing factors have been considered as mechanisms involved. Inflammation-related cancers are known for their multiple occurrences, and aberrant DNA methylation is known to be present even in noncancerous tissues. Importantly, for some cancers, the degree of accumulation has been demonstrated to be correlated with risk of developing cancers. This indicates that inflammation induces aberrant epigenetic alterations in a tissue early in the process of carcinogenesis, and accumulation of such alterations forms "an epigenetic field for cancerization." This also suggests that inhibition of induction of epigenetic alterations and removal of the accumulated alterations are novel approaches to cancer prevention. Disturbances in cytokine and chemokine signals and induction of cell proliferations are important mechanisms of how inflammation induces aberrant DNA methylation. Aberrant DNA methylation is induced in specific genes, and gene expression levels, the presence of RNA polymerase II (active or stalled), and trimethylation of H3K4 are involved in the specificity. Expression of DNA methyltransferases (DNMTs) is not necessarily induced by inflammation, and local imbalance between DNMTs and factors that protect genes from DNA methylation seems to be important.   © 2010, Elsevier Inc.

I. INTRODUCTION

Chronic inflammation is deeply involved in development and progression of human cancers, contributing up to 25% of them (Hussain and Harris, 2007). As mechanisms of how chronic inflammation induces irreversible genetic/epigenetic alterations, acceleration of cell proliferation and production of reactive oxygen species (ROS) have been mainly considered. At the same time, as involvement of epigenetic alterations in development and progression of cancers became apparent, induction of epigenetic alterations has joined the mechanisms of how chronic inflammation induces cancers.

In this chapter, we will describe the relationship between inflammation and cancers before the epigenetic era, epigenetic alterations induced by chronic inflammations, and how epigenetic alterations are induced.

II. TRADITIONAL VIEWS ON HOW INFLAMMATION LEADS TO CANCERS

Specific types of inflammation are closely associated with cancer development and progression, and the association had been attributed mainly to induction of cell proliferation and mutations.

A. Types of inflammation associated with cancer development and progression

Epidemiological data demonstrate a close connection between specific types of chronic inflammation and cancer (Hussain and Harris, 2007). Hepatitis due to infection of hepatitis B and C viruses is responsible for the majority of hepato-cellular carcinomas (Gomaa et al., 2008). Inflammation mainly involving the intrahepatic biliary tract, induced by the liver fluke, a parasite, elevates risk of cholangiocarcinoma (Shin et al., 2010). Chronic gastritis induced by *Helicobacter pylori* infection is the major risk factor of human gastric cancers with hazard ratios of 2.2–21 (Ekstrom et al., 2001).

Exposure to chemicals can also induce chronic inflammation and cancers. Reflux of gastric acids to the esophagus can lead to reflux esophagitis associated with metaplasia (Barrett's esophagus), and the esophagitis is associated with increased risk of esophageal cancers with hazard ratios of 2.2–10.6 (Solaymani-Dodaran et al., 2004). Inhalation of asbestos fibers causes chronic lung and pleural inflammation, and is a definite inducer of mesotheliomas (Bianchi and Bianchi, 2007). Chronic inflammation due to some immunological defects, such as ulcerative colitis (UC) and Crohn's disease, is associated with increased risk of colon cancers (UC, 5.7-fold; Crohn's disease, 2.5-fold; Ekbom et al., 1990a,b). It is also noteworthy that these inflammation-related cancers are known for their multiple occurrences (Cotran et al., 1989; Choi and Zelig, 1994; Nakajima et al., 2006b), suggesting irreversible genetic/epigenetic alterations are accumulated in normal-appearing tissues exposed to these kinds of inflammation.

On the other hand, there are other types of inflammation that are not associated with cancers, such as asthmatic bronchitis, rheumatoid arthritis, and atopic dermatitis.

B. Traditionally known molecular mechanisms of how chronic inflammation leads to cancers

Acute inflammation is induced upon infection of a tissue with a microorganism, in which neutrophils infiltrate to eliminate the microorganism and damaged cells. If the elimination fails, the acute inflammation will make a transition into a chronic phase, in which lymphocytes and macrophages will dominate (Cotran *et al.*, 1989). During the acute and chronic phases of inflammation, strong cell proliferation is induced. Not only compensatory cell proliferation in response to severe tissue damage but also inflammatory cytokines (e.g., IL1B and TNF) and prostanoids (e.g., prostaglandin E2) induce cell proliferation and inhibit apoptosis (Castellone *et al.*, 2005; Kim *et al.*, 2005). Accelerated cell proliferation leads to increased spontaneous mutations even if the mutation rate is not affected, and some investigators suggest that abnormally increased cell proliferation may be accompanied by increased mutation rates (Tomatis, 1993). Besides the accelerated cell proliferation, production of ROS by infiltrating inflammatory cells leads to DNA strand breaks and production of 8-hydroxyguanine that eventually leads to G to T transversions (Federico *et al.*, 2007). Peroxydation of proteins and lipids by ROS also leads to accelerated cell proliferation due to cellular damage (Federico *et al.*, 2007).

Direct stimulation of epithelial cells by inflammation-related factors is also considered to be a mechanism of how specific types of inflammation promote carcinogenesis. For instance, CagA protein directly injected by *H. pylori* affects polarity and junctions of epithelial cells via perturbation of PAK1/MAPK kinase signals (Saadat *et al.*, 2007). Physical stimulus by asbestos leads to upregulation of specific transcription factors in epithelial cells (Fig. 2.1) (Heintz *et al.*, 1993).

III. EPIGENETIC ALTERATION INDUCED BY INFLAMMATION AND ITS SIGNIFICANCE ON CARCINOGENESIS

In addition to mutations, epigenetic alterations are now recognized to be induced by chronic inflammation, and the induction of epigenetic alterations is an important mechanism of how inflammation leads to carcinogenesis.

A. Aberrant DNA methylation induced by inflammation

1. Association between inflammation and aberrant DNA methylation

In normal cells, most CpG islands (CGIs) are kept unmethylated (Yamashita *et al.*, 2009). In contrast, in noncancerous tissues exposed to inflammation, aberrant DNA methylation can be detected (Table 2.1). Such an association was first identified in colonic mucosae of patients with UC (Hsieh *et al.*, 1998). Methylation of promoter CGI of *CDKN2A* (*p16*) was detected in 12.7% of UC samples without dysplasia, and the incidence increased to 70% in UC samples with dysplasia and to 100% in those with carcinomas. Methylation involved other CGIs, such as a promoter CGI (*MYOD1*) and gene body CGIs [*CDKN2A*

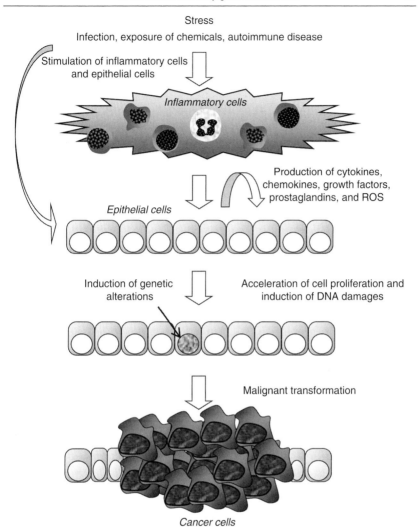

Figure 2.1. Schematic representation of a traditional carcinogenic scenario during inflammation. Stresses stimulate inflammatory cells and epithelial cells to produce inflammation-related factors, such as cytokines, chemokines, growth factors, prostaglandins, and ROS. These factors induce DNA damage in the epithelial cells directly or accelerate cell proliferation, both of which lead to induction of genetic alterations. If such genetic alterations are induced in critical genes, for example, tumor-suppressor genes and oncogenes, multistep carcinogenesis is promoted.

Table 2.1. Inflammation Associated with Aberrant DNA Methylation and Genes Methylated

Disease	Type of cancer induced	Methylated genes	Reference
HBV-asociated hepatitis	Liver cancer	CDKN2A (p16; P) CYP7B1 (P), HOXA11 (P), OCIAD2 (P), RASGRF2 (P), RRAD (P), RUNX3 (P), SMOC2 (P), TBX5 (P), TUBB6 (P), ZNF141 (P), ZNF382 (P)	Narimatsu et al. (2004) Deng et al. (in press)
HCV-associated hepatitis	Liver cancer	CDKN2A (p16; P) RUNX3 (P) CD38 (P), CYP7B1 (P), CYP24A1 (P), JAKMIP1 (P), NPR1 (P), RRAD (P), RUNX3 (P), TBX5 (P), TUBB6 (P), ZNF382 (P)	Narimatsu et al. (2004) Nishida et al. (2008) Deng et al. (in press)
Hemochromatosis	Liver cancer	APC (P), CDKN2A (p16; P), CCND2 (P), GSTP1 (P), RASSF1A (P), SOCS1 (P)	Lehmann et al. (2007)
H.pylori-associated gastritis	Stomach cancer	ARPC1B (B), CDKN2A (p16; P; B), FLNC (P) HAND1 (P), HRASLS (P) ,LOX (P), THBD (P)	Maekita et al. (2005)
Pancreatitis	Pancreatic cancer	APC (P), CDH1 (P), CDKN2A (p16; P), COX2 (P), MLH1 (P)	Perri et al. (2007)
		APC (P), BRCA1 (P), CDKN2A (p16; P), CDKN2B (P), GSTP1(P)	Peng et al. (2006)
Reflux esophagitis (Barret's esophagus)	Esophageal cancer	CDKN2A (p16; P)	Klump et al. (1998)
		APC (P), CDKN2A (p16; P), ESR1 (P)	Eads et al. (2000)
		APC (P), CDKN2A (p16; P), CRBP1 (P), HPP1 (P), MGMT (P), 3-0ST-2 (P), RIZ1 (P), RUNX3 (P), TIMP3 (P),	Schulmann et al. (2005)
		GPX3 (P), GPX7 (P), GSTM2 (P), GSTM3 (P)	Peng et al. (2008)
		APC (P), CDKN2A (p16; P)	Wang et al. (2009)
Ulcerative colitis	Colon cancer	CDKN2A (p16; P)	Hsieh et al. (1998)
		CDKN2A (p16; B), ESR1 (B), MYOD1 (P)	Issa et al. (2001)
		CDKN2A (p14; B)	Sato et al. (2002)

P, promoter region of the gene; B, gene body (exon and intron).

(p14 and p16) and *ESR1* (*ERα*)] (Issa *et al.*, 2001; Sato *et al.*, 2002). In addition, Barrett's esophagus frequently contains methylation of promoter CGIs of *CDKN2A* (p16), *HPP1*, and *RUNX3* (Eads *et al.*, 2000; Klump *et al.*, 1998; Schulmann *et al.*, 2005; Wang *et al.*, 2009). Livers with hepatitis due to infection of hepatitis viruses or hemochromatosis contain frequent methylation of promoter CGIs of *CDKN2A* (p16), *RASSF1A*, and *RUNX3* (Deng *et al.*, in press; Lehmann *et al.*, 2007; Narimatsu *et al.*, 2004; Nishida *et al.*, 2008).

In human gastric mucosae with inflammation due to *H. pylori* infection, aberrant methylation of multiple genes is present (Maekita *et al.*, 2006). Among the multiple CGIs methylated, the methylation level of a promoter CGI of a tumor-suppressor gene, *CDKN2A* (p16), was relatively low ($<0.3\%$) compared to those of other nontumor suppressor genes such as *HAND1* and *THBD* (0.8–11.2%). The resistance of tumor-suppressor genes to DNA methylation induction is likely to be a general rule (Takeshima and Ushijima, 2010). The frequency of cells with aberrant methylation in gastric mucosae, which approximately equals the methylation level, is much higher than that of mutations induced by *H. pylori* infection, which is estimated to be one per 10^4–10^5 cells (Touati *et al.*, 2003). This suggests that induction of aberrant DNA methylation in gastric mucosae is an important carcinogenic mechanism by *H. pylori* infection.

2. Causal role of inflammation in induction of aberrant DNA methylation

The presence of aberrant DNA methylation in tissues with inflammation strongly indicates a causal role of inflammation in induction of aberrant DNA methylation. To experimentally demonstrate the causal role, we used an animal model of Mongolian gerbils in which *H. pylori* infection induces gastritis, aberrant DNA methylation of CGIs in gastric mucosal epithelial cells, and finally gastric cancers (Niwa *et al.*, 2010). Suppression of inflammation by an immunosuppressive reagent, cyclosporin A, did not affect colonization of *H. pylori*, but markedly repressed methylation induction by *H. pylori* infection. Therefore, although *H. pylori* infection is important to trigger inflammation capable of inducing aberrant DNA methylation, some inflammation processes appear to be critical in induction of aberrant DNA methylation.

3. Use of DNA methylation as a marker of past exposure to inflammation

Aberrant DNA methylation in tissue stem cells is expected to remain even after its inducing stimulus has disappeared. Indeed, *H. pylori* infection induces aberrant DNA methylation of specific CGIs, which will remain even after eradication of *H. pylori* (Nakajima *et al.*, 2009; Niwa *et al.*, 2010). Smoking duration was shown to be correlated with methylation levels of specific genes in esophageal mucosae, where smoking is a risk factor of esophageal cancers (Oka *et al.*, 2009). These data suggest that some of the epigenetic alterations induced by inflammation remain and accumulate in noncancerous tissues, and that the accumulated alterations may serve as a marker of past exposure to inflammation.

B. DNA hypomethylation induced by inflammation

Global DNA hypomethylation, defined as a decrease in the content of 5-methylcytosine in the genome, is a hallmark of cancers (Feinberg *et al.*, 2006), and is often associated with hypomethylation of normally methylated repetitive sequences, such as LINE1, *Alu*, and Satα (Rollins *et al.*, 2006). Global hypomethylation has been shown to be involved in chromosomal instability and cancer development (Gaudet *et al.*, 2003; Holm *et al.*, 2005). In connection with inflammation, chronic inflammation is known to induce global DNA hypomethylation. In colonic mucosae of patients with UC, a marked decrease in the 5-methylcytosine content in the genome is present (Gloria *et al.*, 1996). In gastric mucosae of individuals infected with *H. pylori*, significant decreases in the methylation levels of *Alu* and Satα, but not LINE1, are present, whereas global hypomethylation is present only in some individuals (Yoshida *et al.*, in press).

C. Histone modification alterations induced by inflammation

Since aberrant DNA methylation is frequently observed in tissues with inflammation, histone modification alterations are expected to be present. However, so far, there is a very limited number of studies that assessed histone modification alterations in epithelial cells exposed to inflammation. Hahn *et al.* (2008) showed alteration of trimethylation of histone H3 at lysine 27 (H3K27me3) in mice exposed to ileocolitis due to genetic deficiency of *Gpx1* and *Gpx2*. One of the reasons why few reports are available in this field is the technical difficulty in analyzing histone modifications by chromatin immunoprecipitation. It requires complete separation of individual cells, which itself is difficult in tissue samples and can require time-consuming, and thus sample-degrading, steps.

D. An epigenetic field for cancerization

Aberrant DNA methylation and hypomethylation are present in noncancerous tissues that are undergoing or underwent inflammation, and methylation levels of specific genes correlate with risk of cancer development (Nakajima *et al.*, 2006a; Kaise *et al.*, 2008; Ushijima, 2007), as above discussed. Genes methylated in noncancerous tissues involve both tumor-suppressor genes, such as *CDKN2A*, and other passenger genes, such as *FLNc* and *THBD* (Maekita *et al.*, 2006). The condition that a significant number of cells in a tissue has already accumulated aberrant methylation of tumor-suppressor genes (and passenger genes) has a high risk of developing cancers is referred to as "an epigenetic field for cancerization" or "an epigenetic field defect."

For cancer prevention, the presence of an epigenetic field defect indicates that inhibition of induction of epigenetic alterations is likely to be effective, and that the effectiveness of a method can be assessed by measuring the degree of epigenetic field defects. It also suggests that removal of accumulated epigenetic alterations might be effective for prevention of cancer development, although we need to be cautious not to remove physiologically necessary DNA methylation or any other epigenetic modifications.

IV. MECHANISMS FOR INDUCTION OF EPIGENETIC ALTERATIONS BY INFLAMMATION

Mechanisms for induction of epigenetic alterations by inflammation are still unclear. Inflammation is characterized by disturbance of cytokine signals and induction of cell proliferation, but it is still unclear how these abnormalities lead to induction of aberrant DNA methylation. Although it had been unclear whether there is gene specificity in methylation induction by inflammation, the presence of specificity and its mechanisms are becoming clear.

A. Disturbance of cytokine signals

In tissues with inflammation, many inflammatory cytokines and chemokines are produced by inflammatory cells and epithelial cells, depending on the types of inflammation. These cytokines and chemokines are essential for inflammation, and they are likely to be involved in induction of epigenetic alterations. In gerbil gastric mucosae infected with *H. pylori*, expression of *Cxcl2* (a functional homolog of human *IL8*), *Il1b*, *Nos2*, and *Tnf* is associated with induction of DNA methylation (Niwa *et al.*, 2010). Three of these four factors are also upregulated in human hepatitis and UC (Cappello *et al.*, 1992; Llorente *et al.*, 1996; McLaughlan *et al.*, 1997). Especially for *IL1B*, its promoter polymorphism responsible for overproduction of IL1B protein is associated with risk of human gastric cancers and with methylation of multiple genes in gastric cancers (Chan *et al.*, 2007; El-Omar *et al.*, 2000). However, so far, it is unclear how individual cytokines and chemokines change the molecular machinery of DNA methylation in epithelial cells exposed to them.

B. Induction of cell proliferation

Inflammation induces proliferation of epithelial cells to compensate loss of damaged cells and repair tissue organization. It is well known that aberrant DNA methylation of some genes is induced by aging (Issa *et al.*, 1994). Therefore, one of the important mechanisms of how inflammation induces aberrant DNA methylation is acceleration of cell proliferation (Issa *et al.*, 2001). By analysis of *Gpx1/Gpx2* KO mice, some genes that undergo age-dependent DNA methylation were shown to have increased levels of DNA methylation, and induction of cell proliferation was indicated as an accelerator of age-dependent DNA methylation. However, notably, the majority ($\sim$ 70%) of genes that undergo age-dependent methylation were not methylated by inflammation (Hahn *et al.*, 2008), and methylation induced by inflammation was suggested to have different target genes from aging.

C. Mechanisms for gene specificity in induction of aberrant DNA methylation

There is clear gene specificity in induction of aberrant DNA methylation according to tissue types and inducers (Costello *et al.*, 2000; Nakajima *et al.*, 2009; Oka *et al.*, 2009). As determinants of the susceptibility, gene expression, the presence of RNA polymerase II (Pol II) (active or stalled), and the presence of specific histone modifications, such as trimethylation of lysine 27 of histone H3 (H3K27me3), are known (Takeshima and Ushijima, 2010; Takeshima *et al.*, 2009).

Genes with low gene expression are susceptible to induction of DNA methylation in cell lines (De Smet *et al.*, 2004; Song *et al.*, 2002) and in human tissues (Nakajima *et al.*, 2009). Susceptibility was also observed in a genome-wide analysis (Takeshima *et al.*, 2009). Genes with H3K27me3 in embryonic stem cells and/or normal epithelial cells were shown to tend to be methylated in cancers for specific genes (Ohm *et al.*, 2007; Schlesinger *et al.*, 2007; Widschwendter *et al.*, 2007), and the association was also confirmed in a genome-wide study (Gal-Yam *et al.*, 2008). On the other hand, the presence of Pol II, active or stalled, is associated with resistance to induction of DNA methylation (Genes A and B in Fig. 2.2; Takeshima *et al.*, 2009).

It can be speculated that inflammation induces decreased expression of some genes (Gene C) and eliminates stalled Pol II (Genes C and D), and recruits a polycomb complex to induce H3K27me3. This could lead to DNA methylation of genes that are not susceptible without inflammation. Genes that physiologically have H3K27me3 can also be methylated by inflammation (Gene E). For example, in ileocolitis of *Gpx1/Gpx2* mice, aberrant DNA methylation was

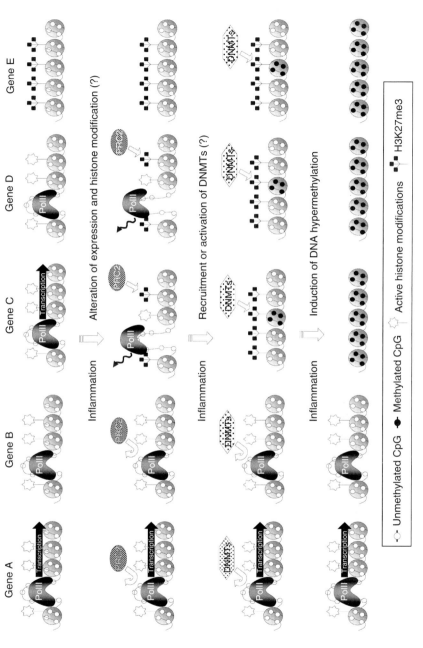

Figure 2.2. (Continued)

induced in epithelial cells, and 59% of the methylated genes had an H3K27me3 mark in normal epithelial cells (Hahn *et al.*, 2008). However, it is still unclear why genes without the H3K27me3 mark were also methylated.

D. Mechanisms for methylation induction at specific genes

To methylate DNA molecules, DNA methyltransferases (DNMTs) are essential as final effectors, and their overexpression has been suggested to be involved in induction of aberrant DNA methylation. Indeed, in the liver and pancreases with inflammation, overexpression of DNMT1 mRNA or protein was observed (Peng *et al.*, 2005; Sun *et al.*, 1997). In contrast, in gastric mucosae with *H. pylori* infection, expression of *DNMT1*, *DNMT3A*, and *DNMT3B* is not increased in humans and gerbils (Nakajima *et al.*, 2009; Niwa *et al.*, 2010). Therefore, aberrant methylation is likely to be induced by a local imbalance between DNMTs and factors that protect genes from DNA methylation, in addition to global overexpression of DNMTs.

EZH2, which is a component of polycomb repressive complex 2 (PRC2) and has a function of a methylase of H3K27, is reported to interact with DNMTs (DNMT3A and DNMT3B) directly (Vire *et al.*, 2006). In addition, BMI1 is a component of PRC1 that recognizes H3K27me3 marks, and was shown to have indirect interaction with DNMT1 through DNMT1-associated protein 1 (DMAP1) (Negishi *et al.*, 2007). These factors (EZH2, BMI1, and DMAP1) appear to be necessary for the maintenance of some CGI methylation in both normal and cancer cells (Negishi *et al.*, 2007; Vire *et al.*, 2006). Thus, changes in expression or localization of these factors by inflammation may link inflammation and aberrant methylation induction (Fig2.2, gene C and D).

V. CONCLUSIONS

Chronic inflammation appears to induce aberrant epigenetic alterations, and the induction is likely to be one of the major mechanisms of how chronic inflammation induces cancers. However, the molecular mechanisms underlying induction

Figure 2.2. Scheme of aberrant DNA methylation induction by inflammation. Genes with H3K27me3, which have little expression, have high risk of being methylated by inflammation (Genes E), and recruitment of DNMTs to the H3K27me3 mark is likely to be one of its mechanisms. In contrast, genes with active histone modification and Pol II are resistant to methylation induction, regardless of their expression levels (Genes A and B). Active histone marks and Pol II localization might prevent recruitment of DNMTs. If localization of active histone modification and Pol II is disturbed by inflammation, it leads to recruitment of PRC2 and DNMTs, which can result in de novo methylation (Genes C and D).

of aberrant DNA methylation are largely unknown. Research focusing on individual cytokines and chemokines and local balance between DNMTs and protecting factors, such as the presence of Pol II, along with histone modification alterations induced by inflammation, are expected to offer clues to uncover the mechanisms.

Acknowledgment

This study was supported by Grants-in-Aid for Cancer Research from the Ministry of Health, Labour and Welfare, Japan.

References

Bianchi, C., and Bianchi, T. (2007). Malignant mesothelioma: Global incidence and relationship with asbestos. *Ind. Health* **45,** 379–387.

Cappello, M., Keshav, S., Prince, C., Jewell, D. P., and Gordon, S. (1992). Detection of mRNAs for macrophage products in inflammatory bowel disease by in situ hybridisation. *Gut* **33,** 1214–1219.

Castellone, M. D., Teramoto, H., Williams, B. O., Druey, K. M., and Gutkind, J. S. (2005). Prostaglandin E2 promotes colon cancer cell growth through a Gs-axin-beta-catenin signaling axis. *Science* **310,** 1504–1510.

Chan, A. O., Chu, K. M., Huang, C., Lam, K. F., Leung, S. Y., Sun, Y. W., Ko, S., Xia, H. H., Cho, C. H., Hui, W. M., *et al.* (2007). Association between Helicobacter pylori infection and interleukin 1beta polymorphism predispose to CpG island methylation in gastric cancer. *Gut* **56,** 595–597.

Choi, P. M., and Zelig, M. P. (1994). Similarity of colorectal cancer in Crohn's disease and ulcerative colitis: Implications for carcinogenesis and prevention. *Gut* **35,** 950–954.

Costello, J. F., Fruhwald, M. C., Smiraglia, D. J., Rush, L. J., Robertson, G. P., Gao, X., Wright, F. A., Feramisco, J. D., Peltomaki, P., Lang, J. C., *et al.* (2000). Aberrant CpG-island methylation has non-random and tumour-type-specific patterns. *Nat. Genet.* **24,** 132–138.

Cotran, R. S., Kumar, V., and Robbins, S. L. (1989). Robbins pathologic basis of disease W.B. Saunders company, Philadelphia.

De Smet, C., Loriot, A., and Boon, T. (2004). Promoter-dependent mechanism leading to selective hypomethylation within the 5′ region of gene MAGE-A1 in tumor cells. *Mol. Cell. Biol.* **24,** 4781–4790.

Deng, Y. B., Nagae, G., Midorikawa, Y., Yagi, K., Tsutsumi, S., Yamamoto, S., Hasegawa, K., Kokudo, N., Aburatani, H., and Kaneda, A. (2010). Identification of genes preferentially methylated in hepatitis C virus-related hepatocellular carcinoma. *Cancer Sci.* (in press).

Eads, C. A., Lord, R. V., Kurumboor, S. K., Wickramasinghe, K., Skinner, M. L., Long, T. I., Peters, J. H., DeMeester, T. R., Danenberg, K. D., Danenberg, P. V., *et al.* (2000). Fields of aberrant CpG island hypermethylation in Barrett's esophagus and associated adenocarcinoma. *Cancer Res.* **60,** 5021–5026.

Ekbom, A., Helmick, C., Zack, M., and Adami, H. O. (1990a). Increased risk of large-bowel cancer in Crohn's disease with colonic involvement. *Lancet* **336,** 357–359.

Ekbom, A., Helmick, C., Zack, M., and Adami, H. O. (1990b). Ulcerative colitis and colorectal cancer: A population-based study. *N. Engl. J. Med.* **323,** 1228–1233.

Ekstrom, A. M., Held, M., Hansson, L. E., Engstrand, L., and Nyren, O. (2001). Helicobacter pylori in gastric cancer established by CagA immunoblot as a marker of past infection. *Gastroenterology* **121,** 784–791.

El-Omar, E. M., Carrington, M., Chow, W. H., McColl, K. E., Bream, J. H., Young, H. A., Herrera, J., Lissowska, J., Yuan, C. C., Rothman, N., *et al.* (2000). Interleukin-1 polymorphisms associated with increased risk of gastric cancer. *Nature* **404**, 398–402.

Federico, A., Morgillo, F., Tuccillo, C., Ciardiello, F., and Loguercio, C. (2007). Chronic inflammation and oxidative stress in human carcinogenesis. *Int. J. Cancer* **121**, 2381–2386.

Feinberg, A. P., Ohlsson, R., and Henikoff, S. (2006). The epigenetic progenitor origin of human cancer. *Nat. Rev. Genet.* **7**, 21–33.

Gal-Yam, E. N., Egger, G., Iniguez, L., Holster, H., Einarsson, S., Zhang, X., Lin, J. C., Liang, G., Jones, P. A., and Tanay, A. (2008). Frequent switching of Polycomb repressive marks and DNA hypermethylation in the PC3 prostate cancer cell line. *Proc. Natl. Acad. Sci. USA* **105**, 12979–12984.

Gaudet, F., Hodgson, J. G., Eden, A., Jackson-Grusby, L., Dausman, J., Gray, J. W., Leonhardt, H., and Jaenisch, R. (2003). Induction of tumors in mice by genomic hypomethylation. *Science* **300**, 489–492.

Gloria, L., Cravo, M., Pinto, A., de Sousa, L. S., Chaves, P., Leitao, C. N., Quina, M., Mira, F. C., and Soares, J. (1996). DNA hypomethylation and proliferative activity are increased in the rectal mucosa of patients with long-standing ulcerative colitis. *Cancer* **78**, 2300–2306.

Gomaa, A. I., Khan, S. A., Toledano, M. B., Waked, I., and Taylor-Robinson, S. D. (2008). Hepatocellular carcinoma: epidemiology, risk factors and pathogenesis. *World J. Gastroenterol.* **14**, 4300–4308.

Hahn, M. A., Hahn, T., Lee, D. H., Esworthy, R. S., Kim, B. W., Riggs, A. D., Chu, F. F., and Pfeifer, G. P. (2008). Methylation of polycomb target genes in intestinal cancer is mediated by inflammation. *Cancer Res.* **68**, 10280–10289.

Heintz, N. H., Janssen, Y. M., and Mossman, B. T. (1993). Persistent induction of c-fos and c-jun expression by asbestos. *Proc. Natl. Acad. Sci. USA* **90**, 3299–3303.

Holm, T. M., Jackson-Grusby, L., Brambrink, T., Yamada, Y., Rideout, W. M., 3rd., and Jaenisch, R. (2005). Global loss of imprinting leads to widespread tumorigenesis in adult mice. *Cancer Cell* **8**, 275–285.

Hsieh, C. J., Klump, B., Holzmann, K., Borchard, F., Gregor, M., and Porschen, R. (1998). Hypermethylation of the p16INK4a promoter in colectomy specimens of patients with long-standing and extensive ulcerative colitis. *Cancer Res.* **58**, 3942–3945.

Hussain, S. P., and Harris, C. C. (2007). Inflammation and cancer: An ancient link with novel potentials. *Int. J. Cancer* **121**, 2373–2380.

Issa, J. P., Ahuja, N., Toyota, M., Bronner, M. P., and Brentnall, T. A. (2001). Accelerated age-related CpG island methylation in ulcerative colitis. *Cancer Res.* **61**, 3573–3577.

Issa, J. P., Ottaviano, Y. L., Celano, P., Hamilton, S. R., Davidson, N. E., and Baylin, S. B. (1994). Methylation of the oestrogen receptor CpG island links ageing and neoplasia in human colon. *Nat. Genet.* **7**, 536–540.

Kaise, M., Yamasaki, T., Yonezawa, J., Miwa, J., Ohta, Y., and Tajiri, H. (2008). CpG island hypermethylation of tumor-suppressor genes in H. pylori-infected non-neoplastic gastric mucosa is linked with gastric cancer risk. *Helicobacter* **13**, 35–41.

Kim, S. F., Huri, D. A., and Snyder, S. H. (2005). Inducible nitric oxide synthase binds, S-nitrosylates, and activates cyclooxygenase-2. *Science* **310**, 1966–1970.

Klump, B., Hsieh, C. J., Holzmann, K., Gregor, M., and Porschen, R. (1998). Hypermethylation of the CDKN2/p16 promoter during neoplastic progression in Barrett's esophagus. *Gastroenterology* **115**, 1381–1386.

Lehmann, U., Wingen, L. U., Brakensiek, K., Wedemeyer, H., Becker, T., Heim, A., Metzig, K., Hasemeier, B., Kreipe, H., and Flemming, P. (2007). Epigenetic defects of hepatocellular carcinoma are already found in non-neoplastic liver cells from patients with hereditary haemochromatosis. *Hum. Mol. Genet.* **16**, 1335–1342.

Llorente, L., Richaud-Patin, Y., Alcocer-Castillejos, N., Ruiz-Soto, R., Mercado, M. A., Orozco, H., Gamboa-Dominguez, A., and Alcocer-Varela, J. (1996). Cytokine gene expression in cirrhotic and non-cirrhotic human liver. *J. Hepatol.* **24,** 555–563.

Maekita, T., Nakazawa, K., Mihara, M., Nakajima, T., Yanaoka, K., Iguchi, M., Arii, K., Kaneda, A., Tsukamoto, T., Tatematsu, M., *et al.* (2006). High levels of aberrant DNA methylation in Helicobacter pylori-infected gastric mucosae and its possible association with gastric cancer risk. *Clin. Cancer Res.* **12,** 989–995.

McLaughlan, J. M., Seth, R., Vautier, G., Robins, R. A., Scott, B. B., Hawkey, C. J., and Jenkins, D. (1997). Interleukin-8 and inducible nitric oxide synthase mRNA levels in inflammatory bowel disease at first presentation. *J. Pathol.* **181,** 87–92.

Nakajima, T., Maekita, T., Oda, I., Gotoda, T., Yamamoto, S., Umemura, S., Ichinose, M., Sugimura, T., Ushijima, T., and Saito, D. (2006a). Higher methylation levels in gastric mucosae significantly correlate with higher risk of gastric cancers. *Cancer Epidemiol. Biomarkers Prev.* **15,** 2317–2321.

Nakajima, T., Oda, I., Gotoda, T., Hamanaka, H., Eguchi, T., Yokoi, C., and Saito, D. (2006b). Metachronous gastric cancers after endoscopic resection: How effective is annual endoscopic surveillance? *Gastric Cancer* **9,** 93–98.

Nakajima, T., Yamashita, S., Maekita, T., Niwa, T., Nakazawa, K., and Ushijima, T. (2009). The presence of a methylation fingerprint of Helicobacter pylori infection in human gastric mucosae. *Int. J. Cancer* **124,** 905–910.

Narimatsu, T., Tamori, A., Koh, N., Kubo, S., Hirohashi, K., Yano, Y., Arakawa, T., Otani, S., and Nishiguchi, S. (2004). p16 promoter hypermethylation in human hepatocellular carcinoma with or without hepatitis virus infection. *Intervirology* **47,** 26–31.

Negishi, M., Saraya, A., Miyagi, S., Nagao, K., Inagaki, Y., Nishikawa, M., Tajima, S., Koseki, H., Tsuda, H., Takasaki, Y., *et al.* (2007). Bmi1 cooperates with Dnmt1-associated protein 1 in gene silencing. *Biochem. Biophys. Res. Commun.* **353,** 992–998.

Nishida, N., Nagasaka, T., Nishimura, T., Ikai, I., Boland, C. R., and Goel, A. (2008). Aberrant methylation of multiple tumor suppressor genes in aging liver, chronic hepatitis, and hepatocellular carcinoma. *Hepatology* **47,** 908–918.

Niwa, T., Tsukamoto, T., Toyoda, T., Mori, A., Tanaka, H., Maekita, T., Ichinose, M., Tatematsu, M., and Ushijima, T. (2010). Inflammatory Processes Triggered by Helicobacter pylori Infection Cause Aberrant DNA Methylation in Gastric Epithelial Cells. *Cancer Res.* **70,** 1430–1440.

Ohm, J. E., McGarvey, K. M., Yu, X., Cheng, L., Schuebel, K. E., Cope, L., Mohammad, H. P., Chen, W., Daniel, V. C., Yu, W., *et al.* (2007). A stem cell-like chromatin pattern may predispose tumor suppressor genes to DNA hypermethylation and heritable silencing. *Nat. Genet.* **39,** 237–242.

Oka, D., Yamashita, S., Tomioka, T., Nakanishi, Y., Kato, H., Kaminishi, M., and Ushijima, T. (2009). The presence of aberrant DNA methylation in noncancerous esophageal mucosae in association with smoking history: A target for risk diagnosis and prevention of esophageal cancers. *Cancer* **115,** 3412–3426.

Peng, D. F., Kanai, Y., Sawada, M., Ushijima, S., Hiraoka, N., Kosuge, T., and Hirohashi, S. (2005). Increased DNA methyltransferase 1 (DNMT1) protein expression in precancerous conditions and ductal carcinomas of the pancreas. *Cancer Sci.* **96,** 403–408.

Rollins, R. A., Haghighi, F., Edwards, J. R., Das, R., Zhang, M. Q., Ju, J., and Bestor, T. H. (2006). Large-scale structure of genomic methylation patterns. *Genome Res.* **16,** 157–163.

Saadat, I., Higashi, H., Obuse, C., Umeda, M., Murata-Kamiya, N., Saito, Y., Lu, H., Ohnishi, N., Azuma, T., Suzuki, A., *et al.* (2007). Helicobacter pylori CagA targets PAR1/MARK kinase to disrupt epithelial cell polarity. *Nature* **447,** 330–333.

Sato, F., Harpaz, N., Shibata, D., Xu, Y., Yin, J., Mori, Y., Zou, T. T., Wang, S., Desai, K., Leytin, A., et al. (2002). Hypermethylation of the p14(ARF) gene in ulcerative colitis-associated colorectal carcinogenesis. *Cancer Res.* **62**, 1148–1151.

Schlesinger, Y., Straussman, R., Keshet, I., Farkash, S., Hecht, M., Zimmerman, J., Eden, E., Yakhini, Z., Ben-Shushan, E., Reubinoff, B. E., et al. (2007). Polycomb-mediated methylation on Lys27 of histone H3 pre-marks genes for de novo methylation in cancer. *Nat. Genet.* **39**, 232–236.

Schulmann, K., Sterian, A., Berki, A., Yin, J., Sato, F., Xu, Y., Olaru, A., Wang, S., Mori, Y., Deacu, E., et al. (2005). Inactivation of p16, RUNX3, and HPP1 occurs early in Barrett's-associated neoplastic progression and predicts progression risk. *Oncogene* **24**, 4138–4148.

Shin, H. R., Oh, J. K., Masuyer, E., Curado, M. P., Bouvard, V., Fang, Y. Y., Wiangnon, S., Sripa, B., and Hong, S. T. (2010). Epidemiology of cholangiocarcinoma: An update focusing on risk factors. *Cancer Sci.* **101**, 579–585.

Solaymani-Dodaran, M., Logan, R. F., West, J., Card, T., and Coupland, C. (2004). Risk of oesophageal cancer in Barrett's oesophagus and gastro-oesophageal reflux. *Gut* **53**, 1070–1074.

Song, J. Z., Stirzaker, C., Harrison, J., Melki, J. R., and Clark, S. J. (2002). Hypermethylation trigger of the glutathione-S-transferase gene (GSTP1) in prostate cancer cells. *Oncogene* **21**, 1048–1061.

Sun, L., Hui, A. M., Kanai, Y., Sakamoto, M., and Hirohashi, S. (1997). Increased DNA methyl-transferase expression is associated with an early stage of human hepatocarcinogenesis. *Jpn. J. Cancer Res.* **88**, 1165–1170.

Takeshima, H., and Ushijima, T. (2010). Methylation destiny: Moira takes account of histones and RNA polymerase II. *Epigenetics* **5**, 89–95.

Takeshima, H., Yamashita, S., Shimazu, T., Niwa, T., and Ushijima, T. (2009). The presence of RNA polymerase II, active or stalled, predicts epigenetic fate of promoter CpG islands. *Genome Res.* **19**, 1974–1982.

Tomatis, L. (1993). Cell proliferation and carcinogenesis: A brief history and current view based on an IARC workshop report. International Agency for Research on Cancer. *Environ. Health Perspect.* **101**(Suppl 5), 149–151.

Touati, E., Michel, V., Thiberge, J. M., Wuscher, N., Huerre, M., and Labigne, A. (2003). Chronic Helicobacter pylori infections induce gastric mutations in mice. *Gastroenterology* **124**, 1408–1419.

Ushijima, T. (2007). Epigenetic field for cancerization. *J. Biochem. Mol. Biol.* **40**, 142–150.

Vire, E., Brenner, C., Deplus, R., Blanchon, L., Fraga, M., Didelot, C., Morey, L., Van Eynde, A., Bernard, D., Vanderwinden, J. M., et al. (2006). The Polycomb group protein EZH2 directly controls DNA methylation. *Nature* **439**, 871–874.

Wang, J. S., Guo, M., Montgomery, E. A., Thompson, R. E., Cosby, H., Hicks, L., Wang, S., Herman, J. G., and Canto, M. I. (2009). DNA promoter hypermethylation of p16 and APC predicts neoplastic progression in Barrett's esophagus. *Am. J. Gastroenterol.* **104**, 2153–2160.

Widschwendter, M., Fiegl, H., Egle, D., Mueller-Holzner, E., Spizzo, G., Marth, C., Weisenberger, D. J., Campan, M., Young, J., Jacobs, I., et al. (2007). Epigenetic stem cell signature in cancer. *Nat. Genet.* **39**, 157–158.

Yamashita, S., Hosoya, K., Gyobu, K., Takeshima, H., and Ushijima, T. (2009). Development of a novel output value for quantitative assessment in methylated DNA immunoprecipitation-CpG island microarray analysis. *DNA Res.* **16**, 275–286.

Yoshida, T., Yamashita, S., Takamura-Enya, T., Niwa, T., Ando, T., Enomoto, S., Maekita, T., Nakazawa, K., Tatematsu, M., Ichinose, M., et al. Alu and Satalpha hypomethylation in Helicobacter pylori-infected gastric mucosae. Int J Cancer. (in press)

3

In Utero Life and Epigenetic Predisposition for Disease

Kent L. Thornburg,*,† Jackilen Shannon,‡,§
Philippe Thuillier,‡,§,¶ and Mitchell S. Turker§,‖

*Department of Medicine, Division of Cardiovascular Medicine, Oregon Health & Science University, Portland, Oregon, USA
†Heart Research Center, Oregon Health & Science University, Portland, Oregon, USA
‡Department of Public Health and Preventative Medicine, Oregon Health & Science University, Portland, Oregon, USA
§Center for Research on Occupational and Environmental Toxicology, Oregon Health & Science University, Portland, Oregon, USA
¶OHSU Knight Cancer Institute, Oregon Health & Science University, Portland, Oregon, USA
‖Department of Molecular & Medical Genetics, Oregon Health & Science University, Portland, Oregon, USA

Advances in Genetics, Vol. 71   0065-2660/10 $35.00
Copyright 2010, Elsevier Inc. All rights reserved.   DOI: 10.1016/S0065-2660(10)71003-8

ABSTRACT

Regulatory regions of the human genome can be modified through epigenetic processes during prenatal life to make an individual more likely to suffer chronic diseases when they reach adulthood. The modification of chromatin and DNA contributes to a larger well-documented process known as "programming" whereby stressors in the womb give rise to adult onset diseases, including cancer. It is now well known that death from ischemic heart disease is related to birth weight; the lower the birth weight, the higher the risk of death from cardiovascular disease as well as type 2 diabetes and osteoporosis. Recent epidemiological data link rapid growth in the womb to metabolic disease and obesity and also to breast and lung cancers. There is increasing evidence that "marked" regions of DNA can become "unmarked" under the influence of dietary nutrients. This gives hope for reversing propensities for cancers and other diseases that were acquired in the womb. For several cancers, the size and shape of the placenta are associated with a person's cardiovascular and cancer risks as are maternal body mass index and height. The features of placental growth and nutrient transport properties that lead to adult disease have been little studied. In conclusion, several cancers have their origins in the womb, including lung and breast cancer. More research is needed to determine the epigenetic processes that underlie the programming of these diseases.   © 2010, Elsevier Inc.

I. INTRODUCTION TO PROGRAMMING

Over the past 20 years, evidence for a link between environmental conditions in the womb and disease risk in offspring has been growing (Gluckman *et al.*, 2008). The hypothesis that the intrauterine environment affects disease risk in adulthood now enjoys widespread support. While the associations between fetal growth and later adult disease may partly represent the pleiotropic effects of genes transmitted from mother to child, maternally mediated modulation of gene expression in offspring, through the environment that a mother's body provides, appears to be more important than a purely heritable genetic risk. The association between an environmental stress in the womb and disease outcome in later life is called "programming" (Barker, 1998; Thornburg and Louey, 2005). Many stressors that lead to fetal programming have been identified including nutritional factors like over- and undernutrition, high corticosteroid exposure, and fetal hypoxia. In animal models and human studies alike, malnutrition in its various forms affects a host of development processes that manifest as disease in adulthood.

       The programming story began when David Barker's team in the United Kingdom showed that the risk for mortality from ischemic heart disease was inversely related to the birth weight of residents of Hertfordshire, United Kingdom

(Barker *et al.*, 1989). This relationship showed a graded effect across the entire birth weight spectrum. The low birth weight–high adult disease risk relationship is now known to be very strong for many chronic conditions, including hypertension, coronary artery disease, type 2 diabetes, and osteoporosis (Gluckman *et al.*, 2008). In animal models, any insult that reduces the flow of nutrients from mother to fetus leads to programmed offspring that suffer cardiovascular and metabolic disturbances for life (McMillen and Robinson, 2005).

While the link between nutrient flow and later disease is now documented in dozens of human and animal studies (Gluckman *et al.*, 2008), the mechanisms by which environmental stressors in the womb alter the developing embryo and fetus remain a mystery (Fig. 3.1). Figure 3.1 shows, in diagrammatic form, the major steps and modifiers to which the genome is sensitive and that lead to programming. On the one hand, developmental plasticity allows an organism to adapt to environmental conditions to improve its odds of survival, but on the other hand, it allows for the modification of the epigenome in ways

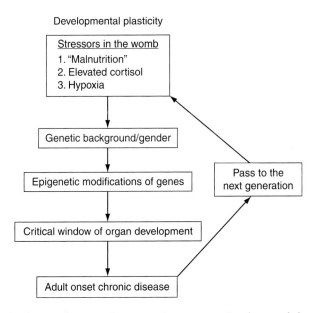

Figure 3.1. Flow diagram illustrating the process of programming. Developmental plasticity allows a number of gene expression options during development. Stressors like malnutrition, excess cortisol, or hypoxia may lead to changes in gene expression in the embryo that predispose the offspring to disease in later life. The effect on the fetus will depend on its gender, its stage of gestation, the nutrient environment, and its genetic background. Female offspring may give birth to offspring that are programmed, repeating the cycle in the next generation.

that increase the risk for chronic adult onset diseases. While it is now certain that epigenetic mechanisms are important in mediating the enduring effects of fetal malnutrition (Burdge *et al.*, 2007), the extent to which epigenetic processes underlie even common chronic diseases in humans is unknown. Also, the extent to which epigenetic modifications can be reversed after birth is not known.

In addition to epigenetic mechanisms, another accommodation to malnutrition in the fetus is the "trading off" of anatomic structures that then predispose to later disease. Perhaps the best example is the reduction in nephron number that accompanies a nutritional stress. Because nephron number in each kidney is set before birth in many animals and in people, an inadequate nephron endowment at birth cannot be reversed in later life. The variation in nephron number in the apparently normal human kidney is large, ranging from some 300,000 to greater than 1.8 million (Zandi-Nejad *et al.*, 2006). Brenner hypothesized that a reduction in nephron number predicts the risk for adult onset systemic hypertension; animal studies generally support this hypothesis (Zandi-Nejad *et al.*, 2006).

Additional compromised structural changes in the fetus are associated with future disease states and include reductions in liver size (Barker 2002; Gentili *et al.*, 2009), skeletal muscle endowment (Baker *et al.*, 2010), elastin in blood vessels (Martyn and Greenwald, 1997), and numbers of working cardio-myocytes in the heart (Jonker *et al.*, 2010). However, many aspects of program-ming do not appear to be related to changes in organ structure. For example, changes in appetite, brain function, and tissue metabolism that accompany over- and undernutrition during fetal life appear to involve intricate but permanent changes in hormonal or cellular processes. For these, the mechanisms by which genes and environment interact must yet be determined.

It is fortunate that the relationship between fetal weight and disease outcome in adults was robust enough to be detected in Barker's early studies because it was profoundly important in the discovery of the programming process. It is now known, however, that many stressors lead to programming without affecting fetal weight at birth. Thus, in any given birth weight category in a population, there would be a range of growth trajectories which different fetuses travel to gain that final weight, which raises the possibility that even cohorts with similar birth weights could have distinct disease predispositions. As the field becomes more sophisticated, it will undoubtedly become possible to unravel the varied pathways of fetal growth that leads to different disease out-comes. That process of discovery will necessarily include determining the effects of variations in the genome that predispose to programming effects. There is increasing evidence that genetic background modifies the effects of birth weight on later disease. Three examples will illustrate different degrees to which the genome and birth weight interact:

1. Eriksson *et al.* (2002) found that Pro12Pro and the Pro12Ala polymorphisms of the PPAR-γ2 gene affected insulin resistance in 152 elderly people depending upon their body size at birth. The Pro12Pro polymorphism of the PPAR-γ2 gene was associated with increased insulin resistance ($P < 0.002$) and elevated insulin concentrations ($P < 0.003$) only in individuals who had low birth weight. Individuals with the Pro12Ala polymorphism did not show the relationship.

    Thus, there is an urgent need for studies of polymorphisms and their relationship to prenatal nutritional factors that are known to be associated with enduring risks for disease.

2. In another study, (Dennison *et al.*, 2001) bone mineral density in the spine was found to be higher among people of genotype BB who were in the lowest third of the birth weight distribution compared in those with the Bb or bb genotype ($P < 0.01$) after adjustment for age, sex, and current weight. However, people with the genotype BB and who were also in the highest birth weight category had the lowest bone mineral density compared with people with the Bb or bb genotype ($P = 0.04$).

3. In a cohort study of female twins (4000 subjects), significant intrapair correlation was found between birth weight and bone mass, even between monozygous twins (Antoniades *et al.*, 2003). This suggests that the intrauterine environment dominates the relationship between birth weight and bone mass compared to genomic inheritance.

II. PROGRAMMING AND CANCER

There is increasing evidence that the high birth weight baby is also programmed for vulnerability to adult onset diseases including cancer. For studies reported thus far, cancers have not been linked to low birth weight conditions. Macrosomic babies born to diabetic mothers have increased risks for obesity and type 2 diabetes and the metabolic syndrome in adulthood. Only in the past few years has evidence arisen linking adult onset cancers with intrauterine environment. The possibility that the origins of metastatic disease and metabolic disease are linked through common mechanisms that regulate prenatal growth has not been investigated thus far but would appear to be a fertile area of study.

In a recent Finnish study, adult onset lung cancer was linked to newborns with a large ponderal index (weight/length^3) if the mother's height was below the median. These data suggest that risks carried by birth size are modified by maternal phenotype (Eriksson *et al.*, 2010). These studies also showed that the surface area of the chorionic plate of the delivered placenta was also related to the disease (Barker *et al.*, 2010).

The relationship between birth weight and breast cancer has been studied more extensively than for other cancers. In several studies, birth weight has been shown to be positively associated with rates of breast cancer. There has been growing interest in the potential association between maternal dietary habits and adult breast cancer risk in offspring. In a study from the Finnish Birth Cohort, Barker *et al.* (2008a) showed that the width and roundness of a woman's hips predicted breast cancer in her daughters. A similar relationship was found for ovarian cancer (Barker *et al.*, 2008b). The hazard ratio for breast cancer in daughters was 3.7 (95% CI: 2.1–6.6) if the distance between the iliac crests was greater than 30 cm and if they were born at or after 40 weeks gestation. The shape of the female body pelvis is determined by the growth factors and hormones to which the pelvis is exposed during puberty. High pubertal levels of estrogen cause the pelvis to widen and become rounded.

If peak plasma estrogen levels are related to a girl's nutrition during and preceding puberty, a high calorie diet during that period development could lead to pelvic changes and play an important role in her cancer risk and the risk of her offspring. Women with wide, round hips may impart a higher estrogen exposure to their embryos in the womb. One potential explanation for the relationship between pelvic size/shape and breast cancer in daughters is the exposure of early fetal breast stem cells to maternal estrogen. This hypothesis has been suggested (Barker *et al.*, 2008a). While estrogen is known to be carcinogenic, a direct toxic effect of the hormone on the breast cell is not the only potential explanation for its actions. It is also possible that the high estrogen levels of the mother lead to epigenetic changes in affected breast tissues and that these changes lead to vulnerability for the cancer in later life. The role of epigenetic regulation of estrogen receptors is being investigated (Leader *et al.*, 2006).

Several studies have been based on the hypothesis that a highly estrogenic environment will result in epigenetic modifications and increased breast cancer risk (Hilakivi-Clarke and de Assis 2006). Human studies have mostly targeted indirect measures of the fetal estrogen environment, including birth size, gestational age, and birth weight (as reviewed by Ruder *et al.*, 2008). Thus, the field has not yet provided mechanistic answers to question regarding the role played by estrogens in the developing breast. A few studies have attempted to characterize maternal dietary intake, as it relates directly to adult breast cancer risk, they provide is evidence for increased risk in offspring with maternal consumption of fat and phytoestrogens such as genistein (Ruder *et al.*, 2008).

There is evidence that other steroid hormones are influential in programming offspring through epigenetic mechanisms. For example, Meaney's laboratory (McGowan *et al.*, 2008) has shown that the programmed changes in brain function in newborn rat pups that occur in response to maternal licking are mediated by glucocorticoid actions. Increased levels of pup licking/grooming during the first week of life leads to increased expression of the glucocorticoid receptor (GR) in the hippocampus, augmented glucocorticoid feedback sensitivity, and suppressed hypothalamic pituitary stress responses compared to offspring reared by mothers who did not groom their offspring (Francis *et al.*, 1999).

The epigenetic pathways in this sequence of events are known. 5-Hydroxytryptamine receptors in the rat pup hippocampus are activated by maternal licking and grooming. This leads to increased cAMP and phosphorylation of the transcription factor, NGFI-A, and the recruitment of histone acetyltranferase/creb-binding protein to the glucocorticoid exon 17 promoter. Acetylation of histone tails at the binding site facilitates its demethylation. In pups whose mothers do not groom, this gene regulation process is reduced, leading to differential epigenetic programming of the promoter region of the GR gene between pups. The HDAC inhibitor, TSA, increases histone acetylation and facilitates demethylation and epigenetic activation of the gene in the offspring of the low grooming mothers. On the other hand, the administration of methionine to adult offspring of attentive mothers leads to increased S-adenosylmethionine (SAM) inhibition of demethylation, increased DNA methylation, and reduced activity of the GR exon 17 promoter. Thus, the surprising part of the story is that the epigenetic state is reversible in the adult under specific dietary conditions.

The influence of the GR is highly significant for many aspects of programming because the GR is important to a number of developmental processes and it is so often subject to epigenetic regulation. For example, modification of GR expression has been documented in lung, liver, adrenal gland, and kidney in the offspring of animals that were malnourished during pregnancy (Brennan *et al.*, 2005; Gnanalingham *et al.*, 2005; Whorwood *et al.*, 2001).

III. DYNAMIC CHANGES IN THE EPIGENOME DURING MAMMALIAN DEVELOPMENT

A longstanding conceptual problem in developmental biology is that only a small fraction of the genome is transcribed into RNA in a given cell type, and the specific fraction that is transcribed is different for different cell types. The epigenome explains how cell-specific gene expression occurs. The main components of the epigenome are DNA methylation and histone tail modifications, most notably lysine acetylation and methylation (Vaissiere *et al.*, 2008). Lysine acetylation marks actively expressed regions of the genome, whereas histone methylation can mark transcribed or repressed regions depending upon the specific lysine residue that is modified.

While the genome is stable throughout life, the epigenome is dynamic (Turker, 1999). Establishment of an individual's epigenome begins after fertilization when gamete-specific DNA methylation patterns are removed at different rates (Fulka *et al.*, 2008). Paternal genome DNA methylation in humans and rodents is actively removed shortly after fertilization, before the first cell division, via an active process that directly removes methyl groups from cytosine bases or the methylated cytosine bases *in toto*. The specific mechanism has not yet been elucidated. In contrast, DNA methylation of the maternal genome is removed via a passive process that is believed to occur over several cell divisions because the maintenance DNA methyltransferase contributed by the oocyte (DNMT1o) cannot move into the nucleus of the developing embryo. The exception to this rule is at the eight-cell stage when DNMT1o moves into the nucleus, apparently to protect maternally transmitted imprinted loci (Howell *et al.*, 2001).

The first cellular differentiation step occurs at about the time of blastocyst implantation when the trophoectoderm and inner cell mass (ICM) cells emerge (Howell *et al.*, 2001). At that time widespread genomic DNA methylation returns, with higher levels in the ICM cells. A simplistic view of this process is that it is required to convert gamete-specific methylation patterns to those that are initially compatible with totipotency and eventually consistent with differentiated cell types (Gopalakrishnan *et al.*, 2008; Turker, 1999). Importantly, formation of somatic cell DNA methylation patterns continues during embryonic development, infancy, and childhood, as cellular growth and differentiation proceed (Gopalakrishnan *et al.*, 2008). DNA methylation changes continue, albeit at a much slower pace, until death (Issa, 1999, 2000), though adult DNA methylation patterns are quite stable relative to the dynamic changes that occur during development.

The parental gametic genomes also exhibit different levels of histone modifications, in large part because paternal gametes use protamines instead of histones to package DNA. Shortly after fertilization, the protamines are replaced by maternally contributed histones, which are then available for modification. Early modifications during the first few cell divisions include methylation of histone 3 lysine 9 and histone 3 lysine 27; both are repressive marks. An additional modification that occurs is histone 3 lysine 4 methylation, which is an activating mark (Albert and Peters, 2009; Corry *et al.*, 2009). Current thought is that the bivalent modification of both lysine 27 and lysine 4 methylation at a variety of loci allows for rapid activation or repression, respectively, during differentiation by removing a specific modification (Bernstein *et al.*, 2006). The tight relation between DNA methylation and histone modifications (Vaissiere *et al.*, 2008), and the presence of embryonic specific patterns of histone modifications (Balch *et al.*, 2007; Ohm *et al.*, 2007), likely leads to additional locus-specific changes in histone modifications, as fetal development continues into childhood development.

IV. ENVIRONMENTAL PERTURBATION OF THE EPIGENOME AND THE DEVELOPMENTAL ORIGINS OF DISEASE

As a general rule, it's easier to move an object in motion than at rest. This rule could help provide a basis for the nascent field of environmental epigenetics, which simply means that environmental exposures can alter the epigenome, particularly during the dynamic phases of embryonic development (Baccarelli and Bollati, 2009, 2010). In other words, dynamic changes programmed to occur in the epigenome during development and then stabilize as an optimal epigenome in the adult could lead to a suboptimal epigenome if environmental exposures "moved" the developing epigenome. Alternatively, though not exclusively, the altered epigenome could at times represent an adaptive response to environmental exposure that conditions the fetus to give rise to an adult better adjusted to the conditions that it faced during development (Tang and Ho, 2007). According to one theory the fetus "assumes" that any unusual environmental condition it faces *in utero* will continue after birth, and therefore adapts itself for this condition (Gluckman and Hanson, 2004).

The adaptive response concept is consistent with the programming hypothesis (Barker, 1990), as demonstrated by the established relation between nutritional deprivation during fetal development and increased incidence of obesity, diabetes, and heart disease in the adult (Roseboom *et al.*, 2001). One explanation for this phenomenon, termed the thrifty phenotype, is that the epigenome of a nutritionally deprived fetus is reset to more efficiently utilize food resources to improve its chances for survival in a challenged environment (Ross and Milner, 2007). Experimental evidence consistent with an epigenetic basis for this hypothesis is beginning to emerge. For example, blood DNA in persons conceived in the Netherlands during a war-induced famine in winter 1944–1945 exhibited decreased levels of DNA methylation at the imprinted *IFG2* locus (Heijmans *et al.*, 2009). This locus is normally expressed only from the maternally inherited allele and it plays an important role in growth and development. According to the adaptive hypothesis, increased expression throughout life would lead to a normal weight in the adult under conditions in which food resources were scarce, but would instead lead to obesity if food was readily available. Interestingly, famine exposure late in pregnancy had no effect on *IGF2* methylation, suggesting that the relevant epigenome had stabilized by that time (Heijmans *et al.*, 2009). Other genes with increased or decreased DNA methylation associated with nutritional deprivation *in utero* have also been reported, in some cases with gender-specific changes (Dobosy *et al.*, 2008; Tobi *et al.*, 2009), suggesting that nutritional deprivation *in utero* leads to widespread modification of the epigenome.

The programming hypothesis, as put forward by Barker, assumes that an adaptive epigenetic response could subsequently have maladaptive consequences if environmental conditions change. However, the possibility also exists that pathological epigenetic changes could occur from toxic environmental exposures early in life (Baccarelli and Bollati, 2009, 2010). For example, the estrogenic drug, diethylstilbestrol (DES), was widely prescribed until 1971 based on the erroneous belief that it could prevent miscarriage (Langston, 2008); instead it turned into a "biological time bomb" (Herbst *et al.*, 1971) causing a variety of reproductive problems in the daughters of exposed women (Rubin, 2007). The observed pathologies in daughters of exposed women may have been caused by epigenetic mechanisms, as borne out by results from rodent models (Bromer *et al.*, 2009; Li *et al.*, 1997; McLachlan, 2006; Miller *et al.*, 1998; Ruden *et al.*, 2005).

A story similar to that for DES may be playing out in the modern environment with the now ubiquitous endocrine disrupting chemical bisphenol-A (BPA), which has been shown to decrease DNA methylation of *Hoxa10* gene in the uteri of female mice exposed *in utero* (Bromer *et al.*, 2010). This effect is opposite to increased methylation induced by DES on this gene (Bromer *et al.*, 2009). BPA exposure *in utero* also alters coat color in the agouti mouse model (Dolinoy *et al.*, 2007b), which is a sensitive detector of changes in genomic DNA methylation (Dolinoy *et al.*, 2007a). A variety of other environmental chemicals are believed to have epigenetic effects, and in some cases may have aberrant effects that last not only over the lifetime of an individual but also over subsequent generations (Baccarelli and Bollati, 2009, 2010). Information describing the pathways by which these chemicals perturb the developing epigenome is clearly needed.

An additional, and again not exclusive, explanation for altered epigenomes in the environmentally exposed, developing fetus is based on the common observation that DNA methylation patterns are variable, even, for example, in identical twins (Fraga *et al.*, 2005; Petronis *et al.*, 2003). Thus, the epigenome of each cell is somewhat unique. A concept termed "epigenetic gambling" suggests that this variation allows preferential amplification of cells whose epigenomes can best survive in particular environments (Martin, 2009). The epigenetic gambling concept could help explain both adaptive changes in the epigenome, as envisioned for the programming hypothesis, and pathological changes that could occur from chemical exposures because in both cases selective processes (e.g., a toxic environmental exposure) coupled with variable epigenomes could lead to individuals with markedly different epigenomes as adults.

V. MECHANISMS INVOLVED IN DIETARY MODULATION OF EPIGENETIC MODIFICATION

While animal and human studies have linked dietary factors to epigenetic regulation, it has been challenging to determine the exact mechanisms that nutrients play in gene regulation. A number of biologically active constituents of

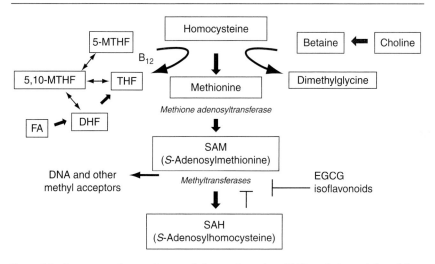

Figure 3.2. A summary of one-carbon metabolism and its role in DNA methylation (adapted from Johnson and Belshaw, 2008).

food have been discovered including vitamins B6 and B12, methionine, and folate (Chmurzynska, 2010). These compounds are potential methyl donors and deficiencies in these nutrients can affect DNA methylation status in mammals.

One carbon metabolism (Fig. 3.2) represents the best understood mechanism of diet-regulated DNA methylation. In this scheme, methionine is generated by methylation of homocysteine via the folate- and B12-dependent methionine synthase reaction and the betaine-homocysteine methyltransferase reaction. Transfer of adenosine to methionine by methionine adenosyltransferase gives the product, S-adenosylmethionine (SAM), which is the principal donor for a host of intracellular methyl transfer reactions. As a consequence of methyl group transfer, SAM is converted to S-adenosylhomocysteine (SAH), which binds with high affinity to methyltransferases and induces potent product inhibition (De Cabo *et al.*, 1995). The ratio of SAM:SAH is, therefore, a crucial determinant of the methylation capacity. Perturbations in this system may be caused by dietary imbalances affecting the supply of methyl donors such as folic acid (FA) or by genetic polymorphisms affecting regulatory enzymes. Folate is not synthesized and is thus provided only by the diet. It has the potential to affect the SAM:SAH ratio and thereby an individual's methylation status.

Epidemiological data suggest that low folate status is associated with a higher risk of colorectal cancer (Giovannucci, 2002) while data for other cancers remain unconclusive. As illustrated in Fig. 3.2, low folate status is likely to result in elevated levels of homocysteine and SAH and lead to DNA hypomethylation.

Colonic adenocarcinoma cell lines and human adenocarcinoma cells (Stempak *et al.*, 2005) demonstrate demethylation and remethylation of p53 when folate levels in the medium are altered (Wasson *et al.*, 2006).

The application of animal studies to human disease is not always straightforward. Rats maintained on diets deficient in folate showed no methylation changes in the colon either genome wide or at specific genes (Choi *et al.*, 2003; Duthie *et al.*, 2000; Kim, 2005). In contrast, similar feeding studies deficient in methyl donors including folate displayed high variation in gene-specific DNA hypo- and hypermethylation in the liver (Pogribny *et al.*, 1995, 2006). These differences in effect are probably attributed to modulation of expression of the DNA methylation machinery including DNMT enzymes and methyl CpG-binding proteins (Ghoshal *et al.*, 2006). The SAM:SAH ratio may become different in colon versus liver in response to a methyl donor deficiency (Kim, 2005). A different effect was observed in mice in response to a folate deficient diet. Folate and choline deficiency in the Apc$^{Min/+}$ mouse resulted in DNA hypomethylation in the small intestine and was correlated with tumor multiplicity (Sibani *et al.*, 2002), though dietary factors may also play a role because a genetic deficiency in DNA methylation leads to decreased tumor formation in the Apc$^{Min/+}$ mouse (Yamada *et al.*, 2005). These variations in the results between animal models have added to the complexity and challenge in translating animal data to the human population.

Several folate deficiency studies have shown a correlation between DNA methylation status and cancers in colon (Pufulete *et al.*, 2003, 2005), cervix (Fowler *et al.*, 1998), and lung (Piyathilake *et al.*, 2000). The variability in response to folate deficiency prompted the investigation of polymorphisms in the one carbon metabolism and association of the methylenetetrahydrofolate reducatse gene (MTHFR) responsible for the reduction of 5,10-methylenetetrahydrofolate (5,10-MTHF) to 5-methyltetrahydrofolate (5-MTHF) (Fig. 3.2). Two main polymorphisms in MTHFR have been identified; the carriers respond differently to folate deficiency with regard to polyp risk in the colon but not to other cancers (Giovannucci, 2002; Martinez *et al.*, 2006).

Other substances including selenium (Davis and Uthus, 2002) and phytochemicals, such as epigallocatechin-3 gallate (Fang *et al.*, 2003), alter DNA methylation status; both act through the inhibition of DNA methyltransferase (DNMT). Phytoestrogens, including the isoflavonoid genestein, may be beneficial prostate cancer prevention through its epigenetic actions (reviewed in Molinié and Georgel, 2009). The list of dietary compounds that have purported epigenetic regulatory actions is growing rapidly; most phytochemicals seem to regulate DNA methylation but others are known to affect histone acetylation.

While the mechanisms underlying global dietary regulation of DNA methylation remain unclear, even less is known about the regulation of histone acetylation by nutrients. However, this latter topic has been studied more intensely over the last few years, perhaps due to the now recognized interacting roles that methylation and acetylation play in regulating gene expression. Histone acetylation leads to chromatin remodeling and a derepression of transcription. Histone deacetylase (HDAC) inhibitors have become a recent focus for cancer prevention and therapy by virtue of their ability to "reactivate" the expression of epigenetically silenced genes, including those involved in differentiation, cell cycle regulation, apoptosis, angiogenesis, invasion, and metastasis (Dashwood *et al.*, 2006). Several natural HDAC inhibitors have been shown to affect the growth and survival of tumor cells *in vitro* and *in vivo*. Among them three dietary chemopreventive agents, butyrate, diallyl disulfide, and sulforaphane, have HDAC inhibitory activity (Myzak and Dashwood, 2006). These compounds have been shown to protect against prostate cancer. Other dietary agents such as biotin, lipoic acid, garlic organosulfur compounds, and metabolites of vitamin E have structural features compatible with HDAC inhibition.

It is becoming clear that many dietary compounds and regiments acting as methyl donors or as inhibitors of enzymatic activity can be used to regulate epigenetic modifications. It is also clear that these same compounds have chemopreventive activities. The correlation between the two activities are becoming stronger although more studies will be needed to include the genetic effect of genetic polymorphisms on these events and how they may influence human disease, starting at conception.

VI. MATERNAL DIET AND RISK OF CANCER IN OFFSPRING: HUMAN STUDIES

Determining the association between maternal diet and the risk of cancer in offspring in humans presents a number of challenges, not least of which is logistic; connecting dietary habits during pregnancy to disease some years if not decades later is very difficult. However, once it was known that nutrient intake by a mother could alter fetal development and augment disease risk for life, the need for such research became apparent. Early studies focused on gestational dietary exposure to procarcinogens and their effects on the development of childhood tumors. Popular hypotheses were based upon prenatal exposure to carcinogens and the role of chemopreventive agents in DNA repair. These studies followed a classic model of carcinogen exposure and biologic response. More recent work investigating maternal diet as it relates to adult onset cancers such as breast cancer is shifting to an epigenetic model of cancer risk (Dalvai and Bystricky, 2010; Morrow and Hortobagyi, 2009).

There are a number of associations between compounds in the diet and later cancers. These are mentioned here because they are potential programming agents in development. Human studies of maternal nutrient intake and risk of acute lymphoblastic leukemia (ALL) have focused primarily on dietary compounds and combinations of foods suggested to impact DNA damage and DNA repair. Two of seven case–control studies published since 1994 used short surveys focused primarily on meat intake (Peters *et al.*, 1994; Sarasua and Savitz, 1994) and two looked only at vitamin supplement use (Shu *et al.*, 1988; Thompson *et al.*, 2001).

The remaining studies evaluated comprehensive nutrient intake during pregnancy using a food frequency questionnaire. Two of these investigations reported a reduction in risk for leukemias with greater fruit and vegetable intake (Jensen *et al.*, 2004; Petridou *et al.*, 2005). Ross *et al.* (1996) evaluated fruit and vegetables as sources of DNA topoisomerase II (DNAt2) inhibitors and reported an increase in the risk of acute myelogenous leukemia (AML) in infancy, but not ALL, with increasing fresh fruit and vegetable intake. A larger study by this group reported a reduction in risk of overall leukemia with an index of fresh fruits and vegetables and other dietary sources of DNAt2 inhibitors (Spector *et al.*, 2005). Protective effects were also reported for supplemental use of a vitamin A and D fortified cod liver oil or folate (Shu *et al.*, 1988; Thompson *et al.*, 2001). Epigenetic events have not been the focus of this research, though the findings are consistent with the hypothesis that the associations with fruit and vegetable consumption and folate supplementation are related to the role of folate as a methyl donor or sulforphane, found in cruciferous vegetables, as an histone deacetylase inhibitor. Jensen *et al.* (2004) reported an unexpected reduction in risk of ALL with high maternal consumption of protein sources (Spector *et al.*, 2005). While the authors suggest this may reflect higher levels of the antioxidant tripeptide glutathione, more recent work suggests that maternal low-protein diets may impact DNA methylation of a number of genes.

In comparison to studies on lymphoblastic leukemia, there has been a wider array of studies investigating the association between maternal dietary intake and pediatric brain tumors. Much of this work began by identifying the associations between consumption of dietary compounds that are known as neural carcinogens in animals, such as *N*-nitroso compounds and the occurrence of brain tumors. Cured meats commonly contain high amounts of *N*-nitroso compounds added during the curing process; hot dogs in particular have been implicated as a risk factor in brain tumor development. A meta-analysis published in 2004 reported on the findings of seven studies of maternal consumption of cured meats and risk of brain tumors in offspring (Huncharek and Kupelnick, 2004). This group reported a pooled 68% increase in relative risk of having a child with a brain tumor with increasing cured meat consumption. The carcinogenesis of *N*-nitroso compounds may be dramatically altered by intake of other bioactive dietary compounds such as the antioxidants in fruits and vegetables. A recent international case–control study found a reduction for risk of brain tumors in the children of mothers who consumed large amounts of yellow-orange vegetables, fresh fish, and grains (Pogoda *et al.*, 2009).

High maternal fruit and vegetable consumption during pregnancy was evaluated in a case–control study of retinoblastoma (Orjuela *et al.*, 2005). The investigators found a reduction in the risk of sporadic retinoblastoma in offspring of mothers with higher fruit and vegetable consumption and consequently, higher folate intake. These investigators argued for an adequate folate status during pregnancy to protect against increased uracil misincorporation or hypomethylation.

VII. THE PLACENTA AS A PROGRAMMING FACTOR

The placenta is the organ through which the embryo and fetus must acquire its nutrients. The size and shape of an individual's placenta are associated with increased risks for high blood pressure, heart failure, and lung cancer (Barker *et al.*, 2010). These relationships persist even in cases where birth weight is not associated with risk. The mechanisms by which placental growth powerfully affects disease risk are not known. However, a common feature of the placenta–disease risk relationship is its dependence upon maternal phenotype, typically her body mass index and/or height. The effects of these phenotypic traits can be further dependent upon the sex of the offspring. In addition, it is clear that the independent phenotypic traits of the placenta itself have predictive value. This suggests that placental thickness and dimensions are related to independent biological processes that affect fetal development.

The molecules that regulate epigenetic actions in the fetus are transferred from the mother by receptor mediated transport proteins. Essential amino acids like methionine are actively transported by several independent transporter systems (Desforges and Sibley, 2010). The components of folate metabolism and transport have been recently investigated (Solanky *et al.*, 2010). Figure 3.3 shows the potential positions of transporters and key metabolic enzymes for one carbon metabolism in the human placenta. These studies demonstrate that the folate transport system and the homocysteine metabolism system are established early in gestation. The transport of methionine and folate are dependent upon a continuous maternal supply. More research is required to understand how molecules that are important in fetal epigenetic processes are acquired by the placenta.

VIII. CONCLUSIONS

Programming is the process by which stressors in the womb lead to disease in the offspring. The programming effect is known to be a powerful process that may account for a large portion of chronic disease in populations worldwide. It also explains, in part, why diseases come and go over the ages. The inverse relationship between term birth weight and cardiovascular disease, obesity, type 2 diabetes, and

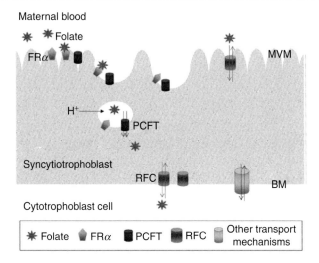

Figure 3.3. Proposed model for folate transport across human placental syncytiotrophoblast. Folate receptor alpha (FRa), localized to the microvillous membrane (MVM) surface binds 5-MTHF (folate). Colocalization of FRa and PCFT to MVM allows internalization of both transporters into an endosomal structure. Following acidification of the endosome by vacuolar proton ATPase, a favorable H^+ gradient exists allowing the H^+-coupled movement of folate by PCFT into the cytoplasm. FRa and PCFT are then recycled back to the MVM surface. RFC at the MVM surface provides an alternative folate uptake mechanism which would be favored at physiological pH. Efflux across the basement membrane, BM, does not involve FRa or PCFT. Instead, folate is transported across BM via an exchange mechanism by RFC. Other transport mechanisms localized to BM may also play a role in transporting folate across the basal plasma membrane. From Solanky *et al.* (2010) with permission.

osteoporosis is due to the developmental plasticity which allows alternate expression patterns in the embryo and fetus. These patterns of adaptation may have "predictive" survival value. Depending on the stage of gestation, the changes in gene expression may stabilize at slightly aberrant levels that lead to permanent structural changes in the developing organs. The underendowment of nephrons in the kidney is an example. One mechanism by which plasticity might bring enduring changes is the epigenetic modification of gene function. Thus, vulnerabilities for cancer may arise in the womb. Understanding the mechanisms that underlie epigenetic changes in the womb may offer preventative and therapeutic strategies in the future.

Acknowledgments

This work was supported by the National Institutes of Health and the M. Lowell Edwards Endowment.

References

Albert, M., and Peters, A. H. (2009). Genetic and epigenetic control of early mouse development. *Curr. Opin. Genet. Dev.* **19**, 113–121.

Antoniades, L., MacGregor, A. J., Andrew, T., and Spector, T. D. (2003). Association of birth weight with osteoporosis and osteoarthritis in adult twins. *Rheumatology* **42**, 791–796.

Baccarelli, A., and Bollati, V. (2009). Epigenetics and environmental chemicals. *Curr. Opin. Pediatr.* **21**, 243–251.

Baker, J., Workman, M., Bedrick, E., Frey, M. A., Hurtado, M., and Pearson, O. (2010). Brains versus brawn: An empirical test of Barker's brain sparing model. *Am. J. Hum. Biol.* **22**, 206–215.

Balch, C., Nephew, K. P., Huang, T. H., and Bapat, S. A. (2007). Epigenetic "bivalently marked" process of cancer stem cell-driven tumorigenesis. *Bioessays* **29**, 842–845.

Barker, D. J. (1990). The fetal and infant origins of adult disease. *BMJ* **301**, 1111.

Barker, D. J. P. (1998). Mothers, Babies and Health in Later Life. Churchill Livingstone, Edinburgh.

Barker, D. J. (2002). Fetal programming of coronary heart disease. *Trends Endocrinol. Metab.* **13**, 364–368.

Barker, D. J., Winter, P. D., and Osmond, C. (1989). Weight in infancy and death from ischaemic heart disease. *Lancet* **ii**, 577–580.

Barker, D. J., Osmond, C., Thornburg, K. L., Kajantie, E., Forsen, T. J., and Eriksson, J. G. (2008a). A possible link between the pubertal growth of girls and breast cancer in their daughters. *Am. J. Hum. Biol.* **20**, 127–131.

Barker, D. J., Osmond, C., Thornburg, K. L., Kajantie, E., and Eriksson, J. G. (2008b). A possible link between the pubertal growth of girls and ovarian cancer in their daughters. *Am. J. Hum. Biol.* **20**, 659–662.

Barker, D. J., Thornburg, K. L., Osmond, C., Kajantie, E., and Eriksson, J. G. (2010). The prenatal origins of lung cancer. II. The placenta. *Am. J. Hum. Biol.* **22**, 512–516.

Bernstein, B. E., Mikkelsen, T. S., Xie, X., Kamal, M., Huebert, D. J., Cuff, J., Fry, B., Meissner, A., Wernig, M., Plath, K., *et al.* (2006). A bivalent chromatin structure marks key developmental genes in embryonic stem cells. *Cell* **125**, 315–326.

Bollati, V., and Baccarelli, A. (2010). Environmental epigenetics. *Heredity* **105**, 105–112.

Brennan, K. A., Gopalakrishnan, G. S., Kurlak, L., Rhind, S. M., Kyle, C. E., Brooks, A. N., Rae, M. T., Olson, D. M., Stephenson, T., and Symonds, M. E. (2005). Impact of maternal undernutrition and fetal number on glucocorticoid, growth hormone and insulin-like growth factor receptor mRNA abundance in the ovine fetal kidney. *Reproduction* **129**, 151–159.

Bromer, J. G., Wu, J., Zhou, Y., and Taylor, H. S. (2009). Hypermethylation of homeobox A10 by in utero diethylstilbestrol exposure: An epigenetic mechanism for altered developmental programming. *Endocrinology* **150**, 3376–3382.

Bromer, J. G., Zhou, Y., Taylor, M. B., Doherty, L., and Taylor, H. S. (2010). Bisphenol-A exposure in utero leads to epigenetic alterations in the developmental programming of uterine estrogen response. *FASEB J.* (Epub ahead of print).

Burdge, G. C., Hanson, M. A., Slater-Jefferies, J. L., and Lillycrop, K. A. (2007). Epigenetic regulation of transcription: A mechanism for inducing variations in phenotype (fetal programming) by differences in nutrition during early life? *Br. J. Nutr.* **97**, 1036–1046.

Chmurzynska, A. (2010). Fetal programming: Link between early nutrition, DNA methylation, and complex diseases. *Nutr. Rev.* **68**, 87–98.

Choi, S. W., Friso, S., Dolnikowski, G. G., Bagley, P. J., Edmondson, A. N., Smith, D. E., and Mason, J. B. (2003). Biochemical and molecular aberrations in the rat colon due to folate depletion are age-specific. *J. Nutr.* **133**, 1206–1212.

Corry, G. N., Tanasijevic, B., Barry, E. R., Krueger, W., and Rasmussen, T. P. (2009). Epigenetic regulatory mechanisms during preimplantation development. *Birth Defects Res. C Embryo Today* **87**, 297–313.

Dalvai, M., and Bystricky, K. (2010). The role of histone modifications and variants in regulating gene expression in breast cancer. *J. Mammary Gland Biol. Neoplasia.* **15,** 19–33.

Dashwood, R. H., Myzak, M. C., and Ho, E. (2006). Dietary HDAC inhibitors: Time to rethink weak ligands in cancer chemoprevention? *Carcinogenesis* **27,** 344–349.

Davis, C. D., and Uthus, E. O. (2002). Dietary selenite and azadeoxycytidine treatments affect dimethylhydrazine-induced aberrant crypt formation in rat colon and DNA methylation in HT-29 cells. *J. Nutr.* **132,** 292–297.

De Cabo, S. F., Santos, J., and Fernandez-Piqueras, J. (1995). Molecular and cytological evidence of S-adenosyl-L-homocysteine as an innocuous undermethylating agent in vivo. *Cytogenet. Cell Genet.* **71,** 187–192.

Dennison, E. M., Arden, N. K., Keen, R. W., *et al.* (2001). Birthweight, vitamin D receptor genotype and the programming of osteoporosis. *Paediatr. Perinat. Epidemiol.* **15,** 211–219.

Desforges, M., and Sibley, C. P. (2010). Placental nutrient supply and fetal growth. *Int. J. Dev. Biol.* **54,** 377–390.

Dobosy, J. R., Fu, V. X., Desotelle, J. A., Srinivasan, R., Kenowski, M. L., Almassi, N., Weindruch, R., Savren, J., and Jarrardm, D. F. (2008). A methyl-deficient diet modifies histone methylation and alters Igf2 and H19 repression in the prostate. *Prostate* **68,** 1187–1195.

Dolinoy, D. C., Das, R., Weidman, J. R., and Jirtle, R. L. (2007a). Metastable epialleles, imprinting, and the fetal origins of adult diseases. *Pediatr. Res.* **61,** 30R–37R.

Dolinoy, D. C., Huang, D., and Jirtle, R. L. (2007b). Maternal nutrient supplementation counteracts bisphenol A-induced DNA hypomethylation in early development. *Proc. Natl. Acad. Sci. USA* **104,** 13056–13061.

Duthie, S. J., Grant, G., and Narayanan, S. (2000). Increased uracil misincorporation in lymphocytes from folate-deficient rats. *Br. J. Cancer* **83,** 1532–1537.

Eriksson, J. G., Lindi, V., Uusitupa, M., Forsén, T. J., Laakso, M., Osmond, C., and Barker, D. J. (2002). The effects of the Pro12Ala polymorphism of the peroxisome proliferator-activated receptor-gamma2 gene on insulin sensitivity and insulin metabolism interact with size at birth. *Diabetes* **51,** 2321–2324.

Eriksson, J. G., Thornburg, K. L., Osmond, C., Kajantie, E., and Barker, D. J. (2010). The prenatal origins of lung cancer. I. The fetus. *Am. J. Hum. Biol.* **22,** 508–511.

Fang, M. Z., Wang, Y., Ai, N., Hou, Z., Sun, Y., Lu, H., Welsh, W., and Yang, C. S. (2003). Tea polyphenol (−)-epigallocatechin-3-gallate inhibits DNA methyltransferase and reactivates methylation-silenced genes in cancer cell lines. *Cancer Res.* **63,** 7563–7570.

Fowler, B. M., Giuliano, A. R., Piyathilake, C., Nour, M., and Hatch, K. (1998). Hypomethylation in cervical tissue: Is there a correlation with folate status? *Cancer Epidemiol. Biomarkers Prev.* **7,** 901–906.

Fraga, M. F., Ballestar, E., Paz, M. F., Ropero, S., Setien, F., Ballestar, M. L., Heine-Suner, D., Cigudosa, J. C., Urioste, M., Benitez, J., *et al.* (2005). Epigenetic differences arise during the lifetime of monozygotic twins. *Proc. Natl. Acad. Sci. USA* **102,** 10604–10609.

Francis, D. D., Caldji, C., Champagne, F., Plotsky, P. M., and Meaney, M. J. (1999). The role of corticotropin-releasing factor—Norepinephrine systems in mediating the effects of early experience on the development of behavioral and endocrine responses to stress. *Biol. Psychiatry* **46,** 1153–1166.

Fulka, H., St John, J. C., Fulka, J., and Hozak, P. (2008). Chromatin in early mammalian embryos: Achieving the pluripotent state. *Differentiation* **76,** 3–14.

Gentili, S., Morrison, J. L., and McMillen, I. C. (2009). Intrauterine growth restriction and differential patterns of hepatic growth and expression of IGF1, PCK2, and HSDL1 mRNA in the sheep fetus in late gestation. *Biol. Reprod.* **80,** 1121–1127.

Ghoshal, K., Li, X., Datta, J., Bai, S., Pogribny, I., Pogribny, M., Huang, Y., Young, D., and Jacob, S. T. (2006). A folate- and methyl-deficient diet alters the expression of DNA methyltransferases and methyl CpG binding proteins involved in epigenetic gene silencing in livers of F344 rats. *J. Nutr.* **136,** 1522–1527.

Giovannucci, E. (2002). Epidemiologic studies of folate and colorectal neoplasia: A review. *J. Nutr.* **132,** 2350S–2355S.

Gluckman, P. D., and Hanson, M. A. (2004). Developmental origins of disease paradigm: A mechanistic and evolutionary perspective. *Pediatr. Res.* **56,** 311–317.

Gluckman, P. D., Hanson, M. A., Cooper, C., and Thornburg, K. L. (2008). Effect of in utero and early-life conditions on adult health and disease. *N. Engl. J. Med.* **359,** 61–73.

Gnanalingham, M. G., Mostyn, A., Dandrea, J., Yakubu, D. P., Symonds, M. E., and Stephenson, T. (2005). Ontogeny and nutritional programming of uncoupling protein-2 and glucocorticoid receptor mRNA in the ovinelung. *J. Physiol.* **565,** 159–169.

Gopalakrishnan, S., Van Emburgh, B. O., and Robertson, K. D. (2008). DNA methylation in development and human disease. *Mutat. Res.* **647,** 30–38.

Heijmans, B. T., Tobi, E. W., Lumey, L. H., and Slagboom, P. E. (2009). The epigenome: Archive of the prenatal environment. *Epigenetics* **4,** 526–531.

Herbst, A. L., Ulfelder, H., and Poskanzer, D. C. (1971). Adenocarcinoma of the vagina. Association of maternal stilbestrol therapy with tumor appearance in young women. *N. Engl. J. Med.* **284,** 878–881.

Hilakivi-Clarke, L., and de Assis, S. (2006). Fetal origins of breast cancer. *Trends Endocrinol. Metab.* **17,** 340–348.

Howell, C. Y., Bestor, T. H., Ding, F., Latham, K. E., Mertineit, C., Trasler, J. M., and Chaillet, J. R. (2001). Genomic imprinting disrupted by a maternal effect mutation in the Dnmt1 gene. *Cell* **104,** 829–838.

Huncharek, M., and Kupelnick, B. (2004). A meta-analysis of maternal cured meat consumption during pregnancy and the risk of childhood brain tumors. *Neuroepidemiology* **23,** 78–84.

Issa, J. P. (1999). Aging, DNA methylation and cancer. *Crit. Rev. Oncol. Hematol.* **32,** 31–43.

Issa, J. P. (2000). CpG-island methylation in aging and cancer. *Curr. Top. Microbiol. Immunol.* **249,** 101–118.

Jensen, C. D., Block, G., Buffler, P., Ma, X., Selvin, S., and Month, S. (2004). Maternal dietary risk factors in childhood acute lymphoblastic leukemia (United States). *Cancer Causes Control* **15,** 559–570.

Johnson, I. T., and Belshaw, N. J. (2008). Environment, diet and CpG island methylation: Epigenetic signals in gastrointestinal neoplasia. *Food Chem. Toxicol.* **46,** 1346–1359.

Jonker, S. S., Giraud, M. K., Giraud, G. D., Chattergoon, N. N., Louey, S., Davis, L. E., Faber, J. J., and Thornburg, K. L. (2010). Cardiomyocyte enlargement, proliferation and maturation during chronic fetal anaemia in sheep. *Exp. Physiol.* **95,** 131–139.

Kim, Y. I. (2005). Nutritional epigenetics: Impact of folate deficiency on DNA methylation and colon cancer susceptibility. *J. Nutr.* **13,** 2703–2709.

Langston, N. (2008). The Retreat from precaution: Regulating diethylstilbestrol (DES), endocrine disruptors, and environmental health. *Environ. Hist.* **13,** 1–26.

Leader, J. E., Wang, C., Popov, V. M., Fu, M., and Pestell, R. G. (2006). Epigenetics and the estrogen receptor. *Ann. N.Y. Acad. Sci.* **1089,** 73–87.

Li, S., Washburn, K. A., Moore, R., Uno, T., Teng, C., Newbold, R. R., McLachlan, J. A., and Negishi, M. (1997). Developmental exposure to diethylstilbestrol elicits demethylation of estrogen-responsive lactoferrin gene in mouse uterus. *Cancer Res.* **57,** 4356–4359.

Martin, G. M. (2009). Epigenetic gambling and epigenetic drift as an antagonistic pleiotropic mechanism of aging. *Aging Cell* **8,** 761–764.

Martinez, M. E., Thompson, P., Jacobs, E. T., Giovannucci, E., Jiang, R., Klimecki, W., and Alberts, D. S. (2006). Dietary factors and biomarkers involved in the methylenetetrahydrofolate reductase genotype-colorectal adenoma pathway. *Gastroenterology* **131,** 1706–1716.

Martyn, C. N., and Greenwald, S. E. (1997). Impaired synthesis of elastin in walls of aorta and large conduit arteries during early development as an initiating event in pathogenesis of systemic hypertension. *Lancet* **350,** 953–955.

McGowan, P. O., Meaney, M. J., and Szyf, M. (2008). Diet and the epigenetic (re)programming of phenotypic differences in behavior. *Brain Res.* **1237,** 12–24.

McLachlan, J. A. (2006). Commentary: Prenatal exposure to diethylstilbestrol (DES): A continuing story. *Int. J. Epidemiol.* **35,** 868–870.

McMillen, I. C., and Robinson, J. S. (2005). Developmental origins of the metabolic syndrome: Prediction, plasticity, and programming. *Physiol. Rev.* **85,** 571–633.

Miller, C., Degenhardt, K., and Sassoon, D. A. (1998). Fetal exposure to DES results in de-regulation of Wnt7a during uterine morphogenesis. *Nat. Genet.* **20,** 228–230.

Molinié, B., and Georgel, P. (2009). Genetic and epigenetic regulations of prostate cancer by genistein. *Drug News Perspect.* **22,** 247–254.

Morrow, P. K., and Hortobagyi, G. N. (2009). Management of breast cancer in the genome era. *Annu. Rev. Med.* **60,** 153–165.

Myzak, M., and Dashwood, R. H. (2006). Histone deacetylases as targets for dietary cancer preventive agents: Lessons learned with butyrate, diallyl disulfide, and sulforaphane. *Curr. Drug Targets* **7,** 443–452.

Ohm, J. E., McGarvey, K. M., Yu, X., Cheng, L., Schuebel, K. E., Cope, L., Mohammad, H. P., Chen, W., Daniel, V. C., Yu, W., *et al.* (2007). A stem cell-like chromatin pattern may predispose tumor suppressor genes to DNA hypermethylation and heritable silencing. *Nat. Genet.* **39,** 237–242.

Orjuela, M. A., Titievsky, L., Liu, X., Ramirez-Ortiz, M., Ponce-Castaneda, V., Lecona, E., Molina, E., Beaverson, K., Abramson, D. H., and Mueller, N. E. (2005). Fruit and vegetable intake during pregnancy and risk for development of sporadic retinoblastoma. *Cancer Epidemiol. Biomarkers Prev.* **14,** 1433–1440.

Peters, J. M., Preston-Martin, S., London, S. J., Bowman, J. D., Buckley, J. D., and Thomas, D. C. (1994). Processed meats and risk of childhood leukemia (California, USA). *Cancer Causes Control* **5,** 195–202.

Petridou, E., Ntouvelis, E., Dessypris, N., Terzidis, A., and Trichopoulos, D. (2005). Maternal diet and acute lymphoblastic leukemia in young children. *Cancer Epidemiol. Biomarkers Prev.* **14,** 1935–1939.

Petronis, A., Gottesman, I. I., Kan, P., Kennedy, J. L., Basile, V. S., Paterson, A. D., and Popendikyte, V. (2003). Monozygotic twins exhibit numerous epigenetic differences: Clues to twin discordance? *Schizophr. Bull.* **29,** 169–178.

Piyathilake, C. J., Johanning, G. L., Macaluso, M., Whiteside, M., Oelschlager, D. K., Heimburger, D. C., and Grizzle, W. E. (2000). Localized folate and vitamin B-12 deficiency in squamous cell lung cancer is associated with global DNA hypomethylation. *Nutr. Cancer* **37,** 99–107.

Pogoda, J. M., Preston-Martin, S., Howe, G., Lubin, F., Mueller, B. A., Holly, E. A., Filippini, G., Peris-Bonet, R., McCredie, M. R., Cordier, S., and Choi, W. (2009). An international case-control study of maternal diet during pregnancy and childhood brain tumor risk: A histology-specific an alysis by food group. *Ann. Epidemiol.* **19,** 148–160.

Pogribny, I. P., Basnakian, A. G., Miller, B. J., Lopatina, N. G., Poirier, L. A., and James, S. J. (1995). Breaks in genomic DNA and within the p53 gene are associated with hypomethylation in livers of folate/methyl-deficient rats. *Cancer Res.* **55,** 1894–1901.

Pogribny, I. P., Ross, S. A., Wise, C., Pogribna, M., Jones, E. A., Tryndyak, V. P., James, S. J., Dragan, Y. P., and Poirier, L. A. (2006). Irreversible global DNA hypomethylation as a key step in hepatocarcinogenesis induced by dietary methyl deficiency. *Mutat. Res.* **593,** 80–87.

Pufulete, M., Al-Ghnaniem, R., Leather, A. J., Appleby, P., Gout, S., Terry, C., Emery, P. W., and Sanders, T. A. (2003). Folate status, genomic DNA hypomethylation, and risk of colorectal adenoma and cancer: A case control study. *Gastroenterology* **124,** 1240–1248.

Pufulete, M., Al-Ghnaniem, R., Rennie, J. A., Appleby, P., Harris, N., Gout, S., Emery, P. W., and Sanders, T. A. (2005). Influence of folate status on genomic DNA methylation in colonic mucosa of subjects without colorectal adenoma or cancer. *Br. J. Cancer* **92,** 838–842.

Roseboom, T. J., van der Meulen, J. H., Ravelli, A. C., Osmond, C., Barker, D. J., and Bleker, O. P. (2001). Effects of prenatal exposure to the Dutch famine on adult disease in later life: An overview. *Twin Res.* **4,** 293–298.

Ross, S. A., and Milner, J. A. (2007). Epigenetic modulation and cancer: Effect of metabolic syndrome? *Am. J. Clin. Nutr.* **86,** 872–877.

Ross, J. A., Potter, J. D., Reaman, G. H., Pendergrass, T. W., and Robison, L. L. (1996). Maternal exposure to potential inhibitors of DNA topoisomerase II and infant leukemia (United States): A report from the Children's Cancer Group. *Cancer Causes Control* **7,** 581–590.

Rubin, M. M. (2007). Antenatal exposure to DES: Lessons learned…future concerns. *Obstet. Gynecol. Surv.* **62,** 548–555.

Ruden, D. M., Xiao, L., Garfinkel, M. D., and Lu, X. (2005). Hsp90 and environmental impacts on epigenetic states: A model for the trans-generational effects of diethylstibesterol on uterine development and cancer. *Hum. Mol. Genet.* **14,** R149–R155.

Ruder, E. H., Dorgan, J. F., Kranz, S., Kris-Etherton, P. M., and Hartman, T. J. (2008). Examining breast cancer growth and lifestyle risk factors: Early life, childhood, and adolescence. *Clin. Breast Cancer* **8,** 334–342.

Sarasua, S., and Savitz, D. A. (1994). Cured and broiled meat consumption in relation to childhood cancer: Denver, Colorado (United States). *Cancer Causes Control* **5,** 141–148.

Shu, X. O., Gao, Y. T., Brinton, L. A., Linet, M. S., Tu, J. T., Zheng, W., and Fraumeni, J. F., Jr. (1988). A population-based case-control study of childhood leukemia in Shanghai. *Cancer* **62,** 635–644.

Sibani, S., Melnyk, S., Pogribny, I. P., Wang, W., Hiou-Tim, F., Deng, L., Trasler, J., James, S. J., and Rozen, R. (2002). Studies of methionine cycle intermediates (SAM, SAH), DNA methylation and the impact of folate deficiency on tumor numbers in Min mice. *Carcinogenesis* **23,** 61–65.

Solanky, N., Requena Jimenez, A., D'Souza, S. W., Sibley, C. P., and Glazier, J. D. (2010). Expression of folate transporters in human placenta and implications for homocysteine metabolism. *Placenta* **31,** 134–143.

Spector, L. G., Xie, Y., Robison, L. L., Heerema, N. A., Hilden, J. M., Lange, B., Felix, C. A., Davies, S. M., Slavin, J., Potter, J. D., Blair, C. K., Reaman, G. H., *et al.* (2005). Maternal diet and infant leukemia: The DNA topoisomerase II inhibitor hypothesis: A report from the children's oncology group. *Cancer Epidemiol. Biomarkers Prev.* **14,** 651–655.

Stempak, J. M., Sohn, K. J., Chiang, E. P., Shane, B., and Kim, Y. I. (2005). Cell and stage of transformation-specific effects of folate deficiency on methionine cycle intermediates and DNA methylation in an in vitro model. *Carcinogenesis* **26,** 981–990.

Tang, W. Y., and Ho, S. M. (2007). Epigenetic reprogramming and imprinting in origins of disease. *Rev. Endocr. Metab. Disord.* **8,** 173–182.

Thompson, J. R., Gerald, P. F., Willoughby, M. L., and Armstrong, B. K. (2001). Maternal folate supplementation in pregnancy and protection against acute lymphoblastic leukaemia in childhood: A case-control study. *Lancet* **358,** 1935–1940.

Thornburg, K. L., and Louey, S. (2005). Fetal roots of cardiac disease. *Heart* **91,** 867–868.

Tobi, E. W., Lumey, L. H., Talens, R. P., Kremer, D., Putter, H., Stein, A. D., Slagboom, P. E., and Heijmans, B. T. (2009). DNA methylation differences after exposure to prenatal famine are common and timing- and sex-specific. *Hum. Mol. Genet.* **18,** 4046–4053.

Turker, M. S. (1999). The establishment and maintenance of DNA methylation patterns in mouse somatic cells. *Semin. Cancer Biol.* **9,** 329–337.

Vaissiere, T., Sawan, C., and Herceg, Z. (2008). Epigenetic interplay between histone modifications and DNA methylation in gene silencing. *Mutat. Res.* **659,** 40–48.

Wasson, G. R., McGlynn, A. P., McNulty, H., O'Reilly, S. L., McKelvey-Martin, V. J., McKerr, G., Strain, J. J., Scott, J., and Downes, C. S. (2006). Global DNA and p53 region-specific hypomethylation in human colonic cells is induced by folate depletion and reversed by folate supplementation. *J. Nutr.* **136,** 2748–2753.

Whorwood, C. B., Firth, K. M., Budge, H., and Symonds, M. E. (2001). Maternal undernutrition during early to midgestationprograms tissue-specific alterations in the expression of the glucocorticoid receptor, 11betahydroxysteroiddehydrogenase isoforms, and type 1 angiotensin ii receptor in neonatal sheep. *Endocrinology* **142,** 2854–2864.

Yamada, Y., Jackson-Grusby, L., Linhart, H., Meissner, A., Eden, A., Lin, H., and Jaenisch, R. (2005). Opposing effects of DNA hypomethylation on intestinal and liver carcinogenesis. *Proc. Natl. Acad. Sci. USA* **102,** 13580–13585.

Zandi-Nejad, K., Luyckx, V. A., and Brenner, B. M. (2006). Adult hypertension and kidney disease: The role of fetal programming. *Hypertension* **47,** 502–508.

4 Folate and One-Carbon Metabolism and Its Impact on Aberrant DNA Methylation in Cancer

Jia (Jenny) Liu*,† and Robyn Lynne Ward*,†
*Lowy Cancer Research Centre, University of New South Wales, Kensington, New South Wales, Australia
†Prince of Wales Clinical School, University of New South Wales, Kensington, New South Wales, Australia

Advances in Genetics, Vol. 71
0065-2660/10 $35.00
DOI: 10.1016/S0065-2660(10)71004-X

ABSTRACT

Folate is a methyl donor that plays an essential role in DNA synthesis and biological methylation reactions, including DNA methylation. Folate deficiency may be implicated in the development of genomic DNA hypomethylation, which is an early epigenetic event found in many cancers, particularly colorectal cancer (CRC). Numerous studies employing *in vitro* systems, animal models, and human interventional studies have tested this hypothesis. Here, we describe the role of folate as a methyl donor in the one-carbon metabolism cycle, and the consequences of cellular folate deficiency. The existing evidence on folate and its relationship to DNA methylation is discussed using CRC as an example. While there remain numerous technical challenges in this important field of research, changes to folate intake appear to be capable of modulating DNA methylation levels in the human colonic mucosa and this may potentially alter CRC risk.   © 2010, Elsevier Inc.

I. INTRODUCTION

Previous chapters have discussed the role of DNA methylation in the regulation of gene expression and the maintenance of genomic stability. It has also been noted that the profile of DNA methylation in cells is altered, perhaps permanently, by inflammation, aging, and environmental exposures. Of the environmental exposures, the water soluble B-vitamin folate is the best studied to date. This family of interconvertible coenzymes plays a key role in accepting and donating one-carbon groups in reactions such as the synthesis of purine and thymidine nucleotides and the transfer of methyl groups. This latter role is essential for a range of biological methylation reactions, including DNA methylation.

       Over the past few decades, low dietary folate and genetic polymorphisms that alter folate metabolism have been linked to diseases, such as neural tube defects (Czeizel and Dudas, 1992), Alzheimer's disease (Wang et al., 2001), Down's syndrome (James et al., 1999), and several cancers, particularly colorectal cancer (CRC; Mason and Choi, 2000b). Public health interest in the role of folate to prevent neural tube defects has culminated in the introduction of mandatory folate fortification in several countries (USA 1995, Canada 1998, Australia 2009). In terms of cancer risk, the long-term consequences of this initiative are unknown (Pfeiffer et al., 2007). Certainly folate is an essential nutrient for cancer cell proliferation, and many chemotherapy drugs, including methotrexate and 5-fluorouracil, are designed to inhibit its metabolism. Excessive folate intake increases the number of preneoplastic and neoplastic lesions in animal models and humans (Cole *et al.*, 2007; Kim, 2004c). Folate has been described as "a double edged sword" with the potential to protect against CRC but under certain circumstances to promote neoplastic progression (Kim, 2004a,b,c).

This chapter will describe the folate one-carbon metabolism pathway and its regulation in humans. With this background, we will use the large body of epidemiological and experimental data in the field of colorectal carcinogenesis to critically examine the proposition that folate status is related to aberrant DNA methylation.

II. CHEMISTRY AND METABOLISM OF FOLATE

A. Methyl donors

Methyl donors or lipotropes are a group of chemicals that directly supply or regenerate labile methyl groups. These chemicals include methionine, choline, folate, and the cofactors vitamin B_{12} and B_6. In humans, the major sources of dietary methyl groups are methionine (10 mmol of methyl groups/day), one-carbon metabolism via 5-methyltetrahydrofolate (5–10 mmol of methyl groups/day), and choline (30 mmol methyl groups/day; Niculescu and Zeisel, 2002). Of these methyl donors, folate is the most important and variable in terms of environmental exposure. Subclinical folate deficiency which predisposes to a number of diseases has a high prevalence in developed and developing populations (McDowell *et al.*, 2008). In contrast, isolated deficiencies of methionine or choline in humans are quite rare since these nutrients are ubiquitous and plentiful in many foods (Rogers, 1995).

B. Overview of folate chemistry

The synthetic form of folate (folic acid) is composed of a pteridine ring, *para*-amino benzoic acid, and glutamic acid (Fig. 4.1). Naturally, folate encompasses a family of over 150 biologically and structurally related compounds that differ in three respects. First, the pteridine ring may be reduced to give either 7,8-dihydrofolate (DHF) or the coenzymatically active 5,6,7,8-tetrahydrofolate (THF) form. Second, the reduced forms can be substituted with covalently bonded one-carbon units at the N-5 or N-10 nitrogen positions. Finally, additional glutamate residues may be conjugated to one another via γ-glutamyl bonds at the end of the molecule, forming a polyglutamate tail (Shane, 1996).

C. Folate metabolism and transport

Mammals cannot synthesize folate *de novo* and hence it is primarily acquired from the dietary intake of green vegetables (asparagus, broccoli, and spinach), legumes, oranges, and liver (predominantly in the form of polyglutamated

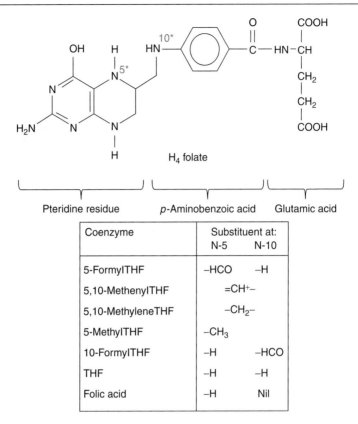

Figure 4.1. Chemical structure of tetrahydrofolate (THF) and its related folate derivatives (adapted from Lucock, 2000). THF is the metabolically active form of folate which can carry various one-carbon groups at the N-5 and N-10 positions, giving rise to numerous folate coenzymes with diverse biological functions. Abbreviations: THF, tetrahydrofolate.

derivatives). In countries which subscribe to folate fortification, folic acid is added to wheat products, breakfast cereals, and juice. Folate absorption and metabolism is a multistep process (summarized in Fig. 4.2). Folate is most efficiently absorbed in the proximal jejunum where its bioavailability depends upon the host's capacity to hydrolyze the polyglutamate chain. Hence, folic acid has a higher bioavailability than the naturally occurring reduced polyglutamated forms. After enterocyte uptake, folate is metabolized to 5-methylTHF and released into the portal circulation (Halsted, 1979, 1980) for storage in the liver and subsequent redistribution to peripheral tissues. Folate is polyglutamated within peripheral tissues to ensure its intracellular retention. Polyglutamation also enhances the affinity of folate coenzymes for several folate-dependent enzymes (Stover, 2004).

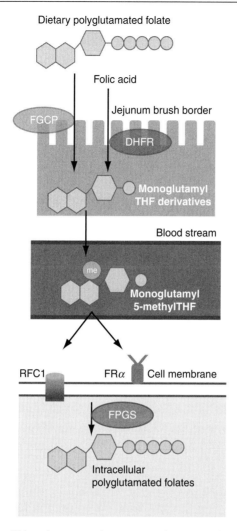

Figure 4.2. Summary of folate absorption and transport. In the jejunum, dietary polyglutamates are converted to monoglutamated folate by gamma-glutamylcarboxypeptidase activity present in the intestinal juice and/or mucosal brush border. Folate is then absorbed through a pH-dependent carrier-mediated process in the jejunum brush border. Folic acid is efficiently absorbed without the need for deconjugation of polyglutamates. Monoglutamyl folate derivatives circulates in the blood stream primarily as 5-methyltetrahydrofolate. Peripheral tissues take up folate through the folate receptor or reduced folate carrier. Folate is retained in tissues by polyglutamation. Abbreviations: FGCP, foly-poly-gamma-glutamyl-carboxypeptidase; FPGS, foly-poly-glutamate-synthase; 5-methylTHF, 5-methyltetrahydrofolate; RFC1, reduced folate carrier 1; FRα, folate receptor α.

Mammalian tissue folate contains a mixture of several polyglutamated species, primarily unsubstituted THF, 5-methylTHF, and several formyl-substituted folates. The folate concentration in tissues depends on the expression of folate receptors (FRs) and exporters (Ross et al., 1994), and also the intracellular concentration of folate-binding proteins which protect folate from catabolism (Suh et al., 2001). For these reasons, there is wide variability in the levels of folate in different tissues. In rodents, tissue folate concentration is highest in the liver, and lowest in skeletal muscle and erythrocytes (Richardson et al., 1979). Localized folate deficiency can also occur in certain tissues that have a high cellular turnover rate (Eisenga et al., 1992; Gregory and Scott, 1996; Varela-Moreiras and Selhub, 1992). Colonic epithelial cells in particular have a very short lifespan (3–5 days; Potten and Allen, 1977), and their rapid renewal confers high folate demands. Due to the presence of carrier-mediated folate transport mechanisms, human colonocytes are capable of absorbing folate that is synthesized by intestinal microflora (Dudeja et al., 1997).

D. Cellular folate transport

Mammalian cellular folate transport occurs primarily via two pathways: membrane carrier mediated and folate-binding protein mediated (summarized in Table 4.1). The reduced folate carrier (RFC) is a transmembrane protein which uses a bidirectional anion-exchange mechanism to transfer hydrophilic folates across the cell membrane (Kaufman et al., 2004). It is responsible for mediating uptake of serum 5-methylTHF across most tissues in the body (Kane and Waxman, 1989). In contrast, FRs bind folate with high affinity and mediate cellular uptake by an endocytotic mechanism. FRs exist in three isoforms which have tissue-specific expression (Antony, 1996). FRα and FRβ are membrane bound receptors, while FRγ is nonmembrane bound and found only in hematopoietic tissues (Shen et al., 1995). Differential expression of FRs, particularly FRα may explain some of the tissue-specific differences in folate uptake. It has been shown that FRα is highly upregulated in several malignant tumors, including ovarian, endometrial, kidney, lung, mesothelioma, brain, breast, and myeloid leukemia (Parker et al., 2005). Upregulated FRα is also associated with higher grade, poor prognosis, and treatment resistance in ovarian tumors (Toffoli et al., 1997).

III. BIOCHEMISTRY OF CELLULAR FOLATE-MEDIATED ONE-CARBON METABOLISM

A. Biochemical role of folate coenzymes

Folate-mediated one-carbon metabolism describes a complex biochemical network of intracellular one-carbon transfer reactions. The function of these reactions is to pick up carbons from amino acids (primarily serine, but also glycine

Table 4.1. Key Features of Cellular Folate Transport Systems

	Reduced folate carrier	Folate receptor
Structure	65 kDa transmembrane protein comprising 591 amino acid residues	Folate receptors α and β are attached to membranes by a glycosyl-phosphatidyl-inositol (GPI) anchor
Mechanism	Bidirectional anion-exchange mechanism	Unidirectional endocytotic mechanism
Tissue expression	Expressed in most mammalian cells	Expression is highly tissue specific: FRα: genitourinary organs, choroid; FRβ: plexus and type I/II pneumocytes; placenta and haematopoetic tissues; FRγ: haematopoeitic tissues only.
Function	The primary transport route for reduced folates at physiological concentrations. Mediates uptake of circulating 5-methylTHF.	Tissue specific; for example, involved in folate reabsorption across glomerular membrane, transplacental transport, and folate transfer from plasma to the cerebral spinal fluid.
Tumor expression	Downregulated compared to normal tissues. May confer methotrexate resistance.	Upregulated in malignant tumors compared to normal tissues

and methionine) and transfer them as methyl groups for nucleotide synthesis and methylation reactions (Nijhout *et al.*, 2008; Wagner, 1995). Within the cell, folate exists in several functionally discrete, but metabolically interconnected "pools." A simplified version of the folate metabolism cycle that illustrates the key functional pools of folate coenzymes is shown in Fig. 4.3.

1. Nucleotide biosynthesis

The folates 5,10-methyleneTHF, 5,10-methenylTHF, and 10-formylTHF exist in metabolic equilibrium in the cell and are required for *de novo* deoxynucleotide triphosphate (dNTP) synthesis (Fig. 4.3, pools 1 and 2). In the synthesis of purines, 10-formylTHF donates one-carbon units to carbon atoms 2 and 8 of the purine ring. In the synthesis of pyrimidines, 5,10-methyleneTHF methylates deoxyuridylate monophosphate (dUMP) to form deoxythymidylate monophosphate (dTMP), a reaction that is rate limiting in nucleotide synthesis (Fukushima *et al.*, 2003). Given that dNTPs have an extremely high turnover, the fidelity of DNA replication and repair is crucially reliant upon maintaining a correct balance of dNTPs (Lucock, 2000).

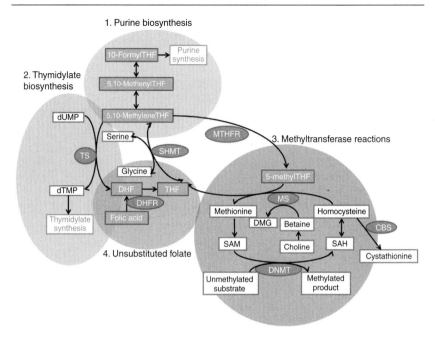

Figure 4.3. Folate-mediated one-carbon metabolism. A simplified diagram of the folate-mediated one-carbon metabolism cycle illustrating the different functional pools of folate is shown here. These functional pools are (1) purine synthesis, (2) thymidylate synthesis, (3) homocysteine remethylation, and (4) unsubstituted folates. Normal flux of one-carbons through cycles 1 and 2 is essential for the *de novo* synthesis of nucleotides required for DNA synthesis and repair. In contrast, flux of one-carbons through cycle 3 is required for numerous cellular methylation reactions of which DNA methylation is a key example. The balance of folate between these two pathways is regulated by the enzymes MTHFR and SHMT. Folate coenzymes are shown in shaded boxes. Other metabolites are shown in unshaded boxes. Several key enzymes of this cycle are shown in shaded ovals. Abbreviations: THF, tetrahydrofolate; DHF, dihydrofolate; dUMP, deoxyuridylate; dTMP, deoxythymidylate; SAM, S-adenosylmethionine; SAH, S-adenosyl-homocysteine; DMG, dimethylglycine; TS, thymidylate synthase; DHFR, dihydrofolate reductase; SHMT, serine hydroxymethyltransferase; MTHFR, 5,10-methylenetetrahydrofolate reductase; MS, methionine synthase; DNMT, DNA methyltransferase; CBS, cystathionine beta-synthase.

2. Methyltransferase reactions

5-methylTHF donates methyl groups for one-carbon reactions to regenerate methionine from homocysteine (Fig. 4.3, pool 3). This irreversible reaction is catalyzed by 5,10-methylenetetrahydrofolate reductase (MTHFR) which reduces 5,10-methyleneTHF to 5-methylTHF, thus committing methyl groups for methylation reactions. The synthesis of S-adenosylmethionine (SAM) from methionine provides the substrate for the cellular methyltransferases which methylate DNA, RNA, lipids, and proteins (Rampersaud *et al.*, 2000). DNA methyltransferases (DNMTs), for example, methylate cytosine residues in CpG dinucleotides, forming 5-methylcytosine. After donating its methyl group, SAM is converted to S-adenosylhomocysteine (SAH), a competitive inhibitor of numerous methyltransferases. Hence, the ratio of SAM to SAH (SAM:SAH) denotes the methylation capacity of a cell. SAH is hydrolyzed to homocysteine and adenosine by SAH hydrolase. Homocysteine is an important branch point in the remethylation cycle. It can either be remethylated by methionine synthase to methionine, or catabolized via the transsulfuration pathway to cysteine and cystathionine. In certain tissues with an active betaine pathway, such as the liver and kidney, homocysteine can also be remethylated using betaine to form dimethyl gycline (Ryan and Weir, 2001). Accumulation of homocysteine is associated with a number of diseases, hence the presence of numerous pathways which facilitate its removal (Wald *et al.*, 2002).

3. Unsubstituted folate

THF and DHF are unsubstituted folate species which do not play a direct metabolic role, but rather receive one-carbon moieties and transfer them toward either methylation-dependent or DNA-synthesis-dependent pathways (Fig. 4.3, pool 4). DHF reductase is a critical enzyme that is responsible for reducing folic acid and DHF into the metabolically active THF. Methotrexate is a competitive and irreversible inhibitor of this enzyme, and thus the presence of this drug inhibits both nucleotide synthesis and remethylation pathways.

Given its metabolic importance, the folate metabolism cycle is tightly controlled through feedback regulation and end product inhibition. This regulation is necessary to ensure the efficient utilization of dietary folate and maintain an appropriate balance of intracellular folate pools. The details of folate cycle regulation have been thoroughly studied with the aid of complex mathematical models and are described elsewhere (Nijhout *et al.*, 2004, 2006; Ulrich, 2005).

B. Consequences of folate deficiency

Given the importance of folate in DNA synthesis, repair, and methylation, maintaining adequate intracellular folate is intimately related to ensuring normal cell proliferation. There are several mechanisms by which folate deficiency can promote carcinogenesis (Fig. 4.4):

1. Impaired thymidylate synthesis: leads to the misincorporation of uracil into DNA causing DNA instability and increased mutagenesis.

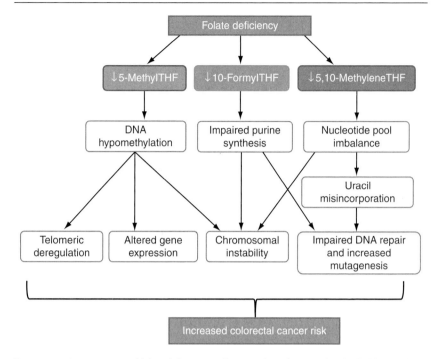

Figure 4.4. Consequences of folate deficiency and potential mechanisms by which folate modifies colorectal cancer risk.

2. Impaired purine synthesis: restricts the cell's capacity to synthesize and repair DNA.
3. Reduced homocysteine remethylation: can inhibit DNMTs and limit SAM availability thus resulting in DNA hypomethylation.

1. Impaired DNA synthesis/repair

The misincorporation of dUTP into DNA from impaired thymidylate synthesis creates a temporary strand nick and if two nicks appear transversely within 12 bases of each other, a strand breakage can occur (Quinlivan *et al.*, 2005). Uracil misincorporation and double strand breaks have been observed in cultured tumor cells, and blood and tissues from rodents and humans with low folate status (Duthie and Hawdon, 1998; Duthie *et al.*, 2000a, 2002). Folate deficiency also impairs *de novo* purine synthesis. It has been demonstrated that folate deficiency due to impaired dNTP synthesis is causative of chromosomal instability *in vitro*. When cells were cultured in folate-deplete media supplemented with hypoxanthine (a purine precursor which bypasses the need for folate-dependent purine synthesis), chromosomal damage was significantly reduced compared to cells cultured in folate-deplete media alone (Libbus et al., 1990).

2. Impaired homocysteine remethylation

A major consequence of folate deficiency is a reduction in the availability of one-carbon groups required for methylation reactions, which in turn causes accumulation of homocysteine. Homocysteinaemia is associated with vascular disease, neural tube defects and Alzheimer's disease, although the precise mechanism and causal association has yet to be determined (Wilcken and Wilcken, 2001). Elevated homocysteine also leads to an accumulation of SAH, which actively inhibits SAM-dependent methyltransferases, including DNMTs (Yi et al., 2000). Both a deficiency of SAM and an accumulation of SAH may cause DNA hypomethylation. The evidence for an association between folate status, DNA hypomethylation, and CRC will be further discussed in Sections IV and V.

C. Folate metabolism and requirements during pregnancy

DNA methylation undergoes extensive reprogramming during early gestation, and adequate maternal folate intake during this critical time period has important consequences on the offspring's epigenome. Genome-wide demethylation occurs during gametogenesis in the primordial germ cells. At fertilization, both parental genomes undergo further demethylation which is thought to restore totoipotency of the fertilized ovum (Waterland and Michels, 2007). Lineage-specific reestablishment of DNA methylation occurs following implantation. This process is critical for determining selective gene expression and hence the developmental fate of differentiating tissues (Morgan et al., 2005). Tissue-specific DNA methylation patterns must then be maintained over successive rounds of rapid cellular proliferation during fetal and postnatal development (Waterland and Jirtle, 2004).

An adequate supply of folate and other methyl donors is critical for the early embryo, given the rapid cellular division and dynamic changes in DNA methylation occurring during this time. Folate demands of the embryo and placental tissues must be obtained entirely from the dietary intake of the mother, whose requirements is 5–10-fold higher than that of nonpregnant women. Circulating maternal 5-methylTHF is captured by placental FRs that are located densely on the maternally facing chorionic surface (Henderson et al., 1995). The resulting intervillous blood folate concentration (which is threefold higher than maternal blood folate concentration) allows for folate to be transferred to the fetal circulation along a downhill concentration gradient. This ensures unidirectional transplacental maternal-fetal transport and allows the fetus ready access to maternal folate stores (Henderson *et al.*, 1995).

While folate remains important *in utero* and in the postnatal period, the offspring is most sensitive to deficiency in the first 4 weeks of life. During the third week of embryogenesis (embryonic day 20–24), the embryonic neural tube and neural crest cells have a doubling time of 5 h. Small delays in remethylation of genes during the neuralation phase could have major consequences since the expression of genes involved in neural tube closure are following each other in a cascade event (Antony, 2007). *In utero* folate deficiency caused by knockout of the murine FR *Folbp1* in mice resulted in severe neural tube defects and embryonic lethality (Piedrahita *et al.*, 1999). However, these offspring could be rescued with very large doses of 5-formylTHF (25 mg/kg; Piedrahita *et al.*, 1999), suggesting that functional *Folbp1* is essential for intrauterine viability by supplying the proper amount of folate to tissues at critical times during embryogenesis. Furthermore, *Folbp1* is not the only murine FR, and uptake of supplemental folate by the *Folbp1−/−* mice occurred via alternative folate membrane transporters. Several robust randomized control studies have confirmed that periconceptional folate supplementation may prevent the recurrence and occurrence of neural tube defects in humans (Czeizel and Dudas, 1992; Czeizel *et al.*, 1994). However, patients with neural tube defects and their mothers tend not to have significant tissue folate depletion, but only a mild elevation in homocysteine (Van der put, 1997). Consequently, it is believed that folate supplementation does not correct a simple nutritional deficiency in many cases, but rather overcomes an underlying metabolic defect in folate metabolism (Lucock *et al.*, 1998; Mills *et al.*, 1996). While many common polymorphisms in folate metabolism enzymes have been studied for their association with neural tube defects, only the MTHFR C677T has been associated with an increased risk (Lucock *et al.*, 2001). A 50% increased risk of neural tube defects was found for the maternal TT genotype, and an 80% increased risk was found for the fetal TT genotype (Blom *et al.*, 2006).

Aberrant DNA methylation and premature "epigenetic aging" (Waterland and Michels, 2007) due to malnutrition *in utero* has been implicated in the pathogenesis of a number of diseases including diabetes, cardiovascular and neurological conditions, and cancer (McKay *et al.*, 2004). Feeding a protein-restricted diet to rats during pregnancy induced hypomethylation of the peroxisomal proliferator-activated receptor and glucocorticoid receptor promoters and increased expression of these genes in the liver of weaned offspring (Lillycrop *et al.*, 2005) suggesting that stable changes to gene expression can be induced by changes to maternal intake during critical periods of exposure. This protein-restricted diet was associated with downregulation of *DNMT1* expression, which could be a possible mechanism for the induction of targeted gene-specific promoter hypomethylation. Feeding these rodents with fivefold higher folic acid levels prevented both DNA hypomethylation (Lillycrop *et al.*, 2005) as well as *DNMT1* repression (Lillycrop *et al.*, 2007). Recently, animal studies have also found that folate supplementation *in utero* was also associated with reduced risk of breast cancer (Sie *et al.*, 2009) and azoxymethane-induced CRC in the offspring (published in abstract form only). While altered susceptibility to disease may reflect the early life environment, it is likely that disease susceptibility may still be modified by environmental exposures throughout the life-course.

D. Factors that alter folate status or folate metabolism

Folate status is modulated by lifestyle factors, comorbidities and medications, and genetic variants in the folate metabolism enzymes (Table 4.2). In the USA, folate fortification (using synthetic folic acid) resulted in an estimated population-wide increase of almost 60% in red cell folate and a threefold increase in serum folate (McDowell *et al.*, 2008). This intervention is not without its detractors. Smith *et al.* (2008) noted that folic acid may interfere with the metabolism of naturally occurring reduced folates in foods and a number of investigators have also raised concerns about the potentially harmful effects of high folate status on a patient's response to antifolate drugs (used for the treatment of cancer, autoimmune diseases, and some infections), the possibility of accelerated progression of preneoplastic lesions (e.g., adenomas; Cole *et al.*, 2007), and masking the effects of vitamin B_{12} deficiency (Morris *et al.*, 2007).

In contrast, inadequate dietary folate intake is the most common cause of mild, subclinical folate deficiency (Thuesen *et al.*, 2009). The likelihood of folate deficiency is increased when marginal intake is compounded by other

Table 4.2. Factors that Increase or Decrease Folate Status

Factors	Folate status enhanced	Folate status diminished
Dietary	Folate fortification of foods Intake of green vegetables (e.g., spinach), liver, oranges	Dietary deficiency of folate and related B-vitamins
Supplements	Multivitamin or B-vitamin supplement use	Antifolate medications, for example, methotrexate High (>100 grams/day) alcohol consumption
Intrinsic	Polymorphisms that enhance folate uptake or metabolism	Polymorphisms that impair folate uptake or metabolism Comorbidities that impair folate absorption or increase folate losses, for example, malabsorption

exposures, particularly high alcohol intake. Alcohol impairs intestinal folate absorption, increases folate catabolism, blocks the release of folate from hepatocytes and inhibits critical folate-metabolizing enzymes (Trimble et al., 1993). Acetaldehyde, a mutagenic metabolite of alcohol, catabolizes folate, forms DNA adducts, and interferes with DNA repair (Poschl et al., 2004). A range of medications (methotrexate, 5-flurouracil, sulfasalazine trimethoprim) and diseases (malabsorption, chronic renal failure on hemodialysis, haemolytic anemia) can also perturb folate status.

There are close interrelations between folate and other B-vitamins in the folate cycle, particularly vitamins B_6 and B_{12}. Methionine synthase is a B_{12}-dependent enzyme, and deficiency of vitamin B_{12} can impair the remethylation of homocysteine, thus accumulating cellular folate as 5-methylTHF, a process termed the methyl trap hypothesis (Wagner et al., 1985). This depletes one-carbon donors from the nucleotide synthesis pathway. Malabsorption or inadequate dietary intake (e.g., in vegans) may lead to vitamin B_{12} deficiency.

Finally, several folate metabolism enzymes are highly polymorphic within and between populations. Common single nucleotide polymorphisms (SNPs) have been found to interact with dietary folate intake and alter tissue-specific folate levels as well as systemic markers of folate status. For example, a common polymorphism of MTHFR c.667C > T results in a thermolabile enzyme with an activity that is 65% of the wild-type activity in heterozygotes (CT), and 30% of wild-type activity in homozygotes (TT; Kim, 2000). In Caucasian populations, up to 50% CT prevalence and 8–20% TT prevalence have been reported, although this varies with ethnicity (Sharp and Little, 2004). TT individuals have significantly lower red cell folate and serum folate and higher homocysteine than CC individuals (Molloy et al., 1997). Numerous studies have found that the TT genotype is associated with a moderately reduced risk of CRC, although the evidence is not entirely consistent (Sharp and Little, 2004). Table 4.3 summarizes the most commonly studied polymorphisms and their influences on folate metabolism, DNA methylation, and CRC risk.

E. Challenges in folate status assessment

Folate status is usually assessed with dietary questionnaires and blood folate markers although tissue-specific folate measurements have been performed in a limited number of studies. Accurate assessment of folate status is difficult for many reasons and this confers limitations to each of the different methods used (Table 4.4). Dietary folate assessment relies mainly on self-reported food frequency questionnaires, which requires subjects to provide information about their "usual" food intake over a period of time. Their dietary folate intake is then calculated using the estimated folate content in their food. Unfortunately, folate bioavailability varies immensely depending upon the method of food

Table 4.3. Common Polymorphisms in Folate Metabolism Enzymes and Their Consequences.

SNP	Description	Effect on folate pathway	Effect on disease	References
MTHFR C677T	Alanine–valine conversion	Thermolabile enzyme with reduced catalytic efficiency. Redistributes methyl groups away from homocysteine remethylation.	TT individuals associated with reduced CRC risk in many studies, but protective effect is reversed in individuals with low folate or high alcohol intake. Low folate status and TT genotype is associated with increased risk of NTDs and vascular disease.	Friso et al. (2002), Boyapati et al. (2004)
MTHFR A1298C	Alanine–glutamine conversion	Reduced enzyme efficiency, but to lesser extent than C677T. Unlike C677T, A1298C is not associated with lower blood folate or higher homocysteine.	Moderately reduced risk of CRC in CCs compared to AAs.	Chen et al. (1999), Le Marchand et al. (2002)
MTR A2756G	Aspartic acid–glycine conversion	Functional effect not well established. Several studies suggest it is activating polymorphism (lower homocysteine concentration with the G allele).	One CRC and one adenoma study found lower risk in GGs. Tumors in GG subjects showed lower number of hypermethylated CpG islands.	Ma et al. (1997), Chen et al. (1999), Paz et al. (2002)
MTRR A66G	Isoleucine–methionine conversion	AA subjects had elevated homocysteine in some studies but no differences in blood folate	Studies have found variant allele is associated with increased risk of NTDs and adult lymphoblastic leukemia but reduced risk of colorectal adenoma recurrence.	Gaughan et al. (2001), Hubner et al. (2006)
CBS 844ins68	68 Base insert at exon 8 coding region	68 Base insert is associated with lower homocysteine levels in four studies	The insert was associated with reduced risk of proximal CRCs in one study. In transitional cell carcinomas, the insert was not associated with altered cancer risk or DNA hypomethylation.	Wang et al. (1999)

(Continues)

Table 4.3. (*Continued*)

SNP	Description	Effect on folate pathway	Effect on disease	References
TYMS 1494del	6 base pair deletion at 3'UTR	It results in low TYMS mRNA stability and low TS expression compared to WT allele. Functional consequences uncertain.	A few studies have found an association with CRC/adenoma.	Sharp and Little (2004)
TYMS Exon 1 3R→2R	5'UTR contains 28 base pair tandem repeats (2 repeats or 3 repeats most common).	*In vitro*, 3R is associated with two- to fourfold greater TS expression. The 3R is associated with lower plasma folate and higher homocysteine in one cohort study.	Reduced CRC/adenoma risk and better survival after CRC diagnosis in 3R/3R compared to 2R/2R; especially in those with high folate or TT genotype of MTHFR C677TT. Polymorphism may affect response to chemotherapy but results are conflicting.	Trinh *et al.* (2002)

Abbreviations: NTD, neural tube defects; CRC, colorectal cancer; MTHFR, 5,10-methylenetetrahydrofolate reductase; MTR, methionine synthase; MTRR, methionine synthase reductase; CBS, cystathionine β-synthase; TYMS, thymidylate synthase.

Table 4.4. Comparison of Different Methods of Folate Status Assessment

Method	Strengths	Limitations
Dietary questionnaires	• Inexpensive • Easy to administer in large-scale population-based studies	• Recall bias • Inaccurate estimates of food folate composition • Folate intake may be confounded by other factors, for example, alcohol consumption, medications, diseases
Plasma folate	• Cheapest of all blood folate assays	• Value fluctuates on a daily basis due to recent diet or alcohol consumption
Red cell folate	• Gives a good indication of systemic longer term folate stores	• Some interassay variability have been reported depending on method of measurements • Does not necessarily reflect local folate levels in tissue of interest
Plasma homocysteine	• Elevated homocysteine is a good inverse measure of reduced one-carbon groups for the methylation pathway	• May be confounded by deficiencies in other B-vitamins or the presence of comorbidities that elevate homocysteine (e.g., heart disease, renal failure)
Tissue folate	• Direct measure of folate levels in the tissue of interest; able to detect localized tissue folate deficiencies	• Difficult to acquire these samples in large population-based studies • Tissue folate assays are time-consuming and labor intensive (no automated assays available)

preparation, storage, and cooking since folate is highly susceptible to degradation (Arcot and Shrestha, 2005). Recall of recent diet in research subjects may not be representative of their dietary patterns during the aetiologically relevant period of exposure (Bird *et al.*, 1995).

Plasma folate, red cell folate, and plasma homocysteine are common blood folate markers used to estimate systemic folate stores. Human plasma mainly comprises monoglutamate 5-methylTHF, levels of which fluctuate on a daily basis thus conferring plasma folate as a short-term indicator of folate status (Halsted *et al.*, 2002). In contrast, red cell folate is less sensitive to systemic fluctuations as it reflects folate stores at the time of erythropoiesis. Red cell folate thus serves as a more useful biomarker of long-term folate status, only decreasing after several months of low intake (Bailey, 1990). Plasma homocysteine is a sensitive maker of methyl donor deficiency (Selhub and Miller, 1992), but it is also nonspecifically elevated by comorbidities and vitamin B_{12} and B_6 deficiency, which serve as cofactors for its metabolism (Hortin, 2006).

Only on a few occasions have attempts been made to characterize folate concentration in the tissue of interest. When studies did measure and compare human colonic folate levels with blood folate markers, a direct correlation was not always found (Kim, 1999; Meenan *et al.*, 1997). This is probably because localized folate deficiency occurs in the colonic tissue independent of systemic folate levels (Weir and Scott, 1998) and this has been hypothesized to promote CRC especially in patients with the MTHFR C677T polymorphism (Brockton, 2006). Hence, it is important for studies to characterize colonic mucosal folate levels specifically rather than assuming that blood folate markers adequately reflect folate levels in the colonic mucosa. However, accurate measurement of tissue-specific folate is difficult because of the inherent low quantities found in human tissues (pmol/g), the multiple coenzyme forms of folate and the complex tissue matrix from which folate must be extracted prior to analysis. Previous research which characterized human colonic tissue folate used the microbiological assay (Kim *et al.*, 1998, 2001a,b) which has limited specificity and is unable to characterize the distribution of different folate coenzymes (Quinlivan *et al.*, 2006). Tissue folate distribution is well characterized in rodents but not humans. In mice, plasma folate is $>90\%$ 5-methylTHF. In contrast, mice liver comprises 75% THF, 10% formyl folate, and only 10% 5-methylTHF, while mice brain comprises 50% THF, 35% 5-methylTHF, and 2% formyl folate (Ghandour *et al.*, 2004). Tissue-specific changes in the intracellular distribution of folate coenzymes will influence the rates of folate-dependent reactions and it is thus important to be able to measure these variations.

IV. CHARACTERISTICS OF DNA HYPOMETHYLATION IN CANCER

In normal tissues, 3.5–4% of all cytosines in the genome are methylated (Ehrlich *et al.*, 1982). Most of these 5-methylcytosines are located within repeat sequences that constitute at least 50% of the human genome. Repeat sequences are primarily found at centromeric, pericentromeric, and subtelomeric regions as well as in intergenic regions (e.g., LINEs, SINEs, and Alu repeats). Maintaining adequate levels of DNA methylation in these areas is critical for genomic stability (Gaudet *et al.*, 2003), for the regulation of telomere length (Gonzalo *et al.*, 2006) and in the repression of transposable elements and imprinted genes (Schofield *et al.*, 2001).

In cancer, DNA hypomethylation is a relative term which denotes a reduction of 5-methylcytosine across the genome compared to the corresponding normal tissue (Ehrlich, 2002). There is no arbitrary value which defines "DNA hypomethylation" because of interindividual and tissue-specific differences in global DNA methylation. Generally, malignant tumors have been found to contain 20–60% less 5-methylcytosine compared to their corresponding normal tissue (Esteller, 2005; Wilson *et al.*, 2007). However, within and between cancer types, the absolute and relative level of hypomethylation varies continuously compared to matched normal tissue (Molloy, 2009). Large decreases of genomic DNA methylation in cancer cells are ascribed primarily to hypomethylation of CpG-rich noncoding regions of the genome (Hoffmann and Schulz, 2005). These sites include the pericentromeric regions, subtelomeric regions, transposable elements, and at specific genes. Figure 4.5 summarizes how hypomethylation at each of these regions has specific consequences that contribute to carcinogenesis.

V. FOLATE DEFICIENCY, DNA HYPOMETHYLATION, AND CRC: A CRITICAL REVIEW

A common question raised by many researchers in this field is whether isolated folate deficiency alone is sufficient to cause DNA hypomethylation and the induction of CRC. A large number of *in vitro*, animal and human observational or interventional studies have been undertaken to determine whether folate is involved in colorectal carcinogenesis through an effect on DNA methylation. While a significant and prolonged depletion of folate results in DNA hypomethylation in the colorectum in some models, this finding has not been consistently replicated and is unlikely to be of physiological relevance to humans. In several comprehensive reviews, Young-in Kim highlighted the inconsistencies in this body of literature, and concluded that the effects of folate deficiency are highly gene and site specific and depend on the cell type, target organ, and regimen of folate depletion (Kim, 2004a,b,c, 2005). Here, we propose that findings from these studies indicate that a deficiency of dietary folate alone is not sufficient to modulate colonic mucosal DNA methylation. Other adverse events such as a deficiency of alternative methyl donors or a polymorphism in a key folate metabolism enzyme may be required as a "second hit" before the effects of folate bioavailability manifest in the DNA. Aberrant DNA methylation, together with further genetic and epigenetic "hits," may subsequently lead to the development of adenomas and eventually, cancer. This hypothesis is illustrated in Fig. 4.6. In Section V, a critical review of *in vitro*, animal, and human studies on the relationship between folate and DNA methylation will be presented, followed by a summary of the gaps remaining in this field and future directions for research.

A. *In vitro* studies

A small number of studies have examined the effects of folate supplementation or depletion on cell behavior and DNA methylation using *in vitro* cultures of pure cell populations. Three *in vitro* studies used nontransformed cell lines specifically NIH/3 T3 mouse fibroblast cells, CHO-K1 Chinese hamster ovary cells

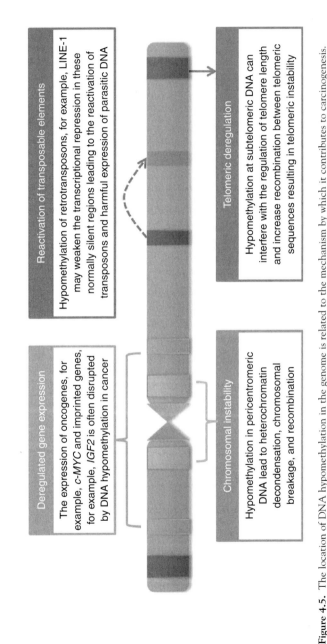

Reactivation of transposable elements

Hypomethylation of retrotransposons, for example, LINE-1 may weaken the transcriptional repression in these normally silent regions leading to the reactivation of transposons and harmful expression of parasitic DNA

Telomeric deregulation

Hypomethylation at subtelomeric DNA can interfere with the regulation of telomere length and increase recombination between telomeric sequences resulting in telomeric instability

Deregulated gene expression

The expression of oncogenes, for example, *c-MYC* and imprinted genes, for example, *IGF2* is often disrupted by DNA hypomethylation in cancer

Chromosomal instability

Hypomethylation in pericentromeric DNA lead to heterochromatin decondensation, chromosomal breakage, and recombination

Figure 4.5. The location of DNA hypomethylation in the genome is related to the mechanism by which it contributes to carcinogenesis.

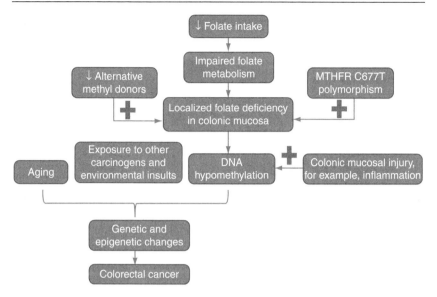

Figure 4.6. A model of how dietary folate deficiency together with alterations in tissue-specific folate metabolism is required to alter colonic mucosal DNA methylation. The effect of reduced dietary folate intake is aggravated by impaired tissue folate uptake and/or metabolism. This, together with a deficiency of the other methyl donors, or a polymorphism in key folate metabolism enzymes, for example, MTHFR C677T can result in localized folate deficiency in the colonic mucosa. DNA hypomethylation may then result, particularly if the colonic mucosa is already damaged from local irritants and inflammatory processes. Hypomethylation itself, however, is not adequate to drive CRC progression unless combined with several other environmental insults and molecular changes associated with the aging process.

(Stempak *et al.*, 2005), immortalized normal colonocytes (HCEC; Duthie *et al.*, 2000a,b,c), and the normal colonic mucosal cell lines NCM356 and NCM460 (Crott *et al.*, 2008). Two others used cancer cell lines (Sohn *et al.*, 2009; Wasson *et al.*, 2006). In most of these studies, cells were cultured in either standard media containing 2.3–3 μM folic acid (which is 200-fold higher than physiological levels of folate found in human serum) or folate-deplete media containing 0–0.6 nM folic acid (0–5% of human serum concentration). Under these extreme changes in media folate concentration, cell growth and intracellular folate content was significantly lowered (>80% reduction), although the extent of growth restriction varied between different cell lines (Stempak *et al.*, 2005; Wasson *et al.*, 2006).

In two of the above three studies that used nontransformed cells, global DNA methylation was significantly decreased (by 18–25%, $p < 0.05$) following media folate depletion for 12–14 days (Duthie *et al.*, 2000a,b,c;

Stempak *et al.*, 2005). Stempak *et al.* (2005) did not find significant hypomethylation in the CRC cell lines HCT116 and Caco2 subjected to the same folate-depletion conditions as NIH/3 T3 and CHO-K1. These results suggest that hypomethylation of nontransformed cells occurs more readily in response to media folate depletion compared to transformed cells (Stempak *et al.*, 2005). Nevertheless, in one study using nontransformed cells, Crott *et al.* (2008) did not observe any changes in DNA methylation in the normal colonic cells NCM356, NCM460, and HCEC after 5 weeks of culture in media containing 25–150 nM folic acid (Crott *et al.*, 2008). The negative result from this study may be because the level of folate in the media was still above physiological levels found in human serum. Alternatively, the high level of methionine in the media may have provided sufficient one-carbon units for DNA methylation without the need for folate-mediated remethylation of homocysteine (Crott *et al.*, 2001).

Of the studies that used cancer cell lines, Wasson *et al.* observed a significant decrease in DNA methylation after 14 days of culture in folate-free media (Wasson *et al.*, 2006). However, both Sohn *et al.* (2009) and Stempak *et al.* (2005) did not detect any significant changes in their cancer cell lines. These findings collectively suggest that the ability of media folate depletion to induce genomic DNA hypomethylation depends upon three factors: the magnitude of depletion, the duration of depletion, and the stage of transformation of the cell line. Studies that detected a significant level of genomic hypomethylation tended to use a greater magnitude of folate depletion, longer duration of culture under folate deficiency, and untransformed normal epithelial cell lines rather than cancer cell lines (Stempak *et al.*, 2005; Wasson *et al.*, 2006). In addition to the hypomethylated baseline state of cancer cell lines, prolonged time in culture creates further changes in DNA methylation patterns that can make them even less representative of normal human tissues and primary tumors (Baylin and Bestor, 2002). Moreover, cancer cells have upregulated FRs (Parker *et al.*, 2005), allowing them to better compensate for extracellular folate deficiency.

The effect of the MTHFR C677T polymorphism on genomic DNA methylation and folate metabolism has also been modeled *in vitro* using human HCT116 colon and MDA-MB-435 breast cancer cells stably transfected with the wild-type or mutant 677T MTHFR cDNAs (Sohn *et al.*, 2009). HCT116 cells with the mutant 677T MTHFR had a significantly lower SAM:SAH and increased homocysteine and uracil misincorporation. These changes are consistent with a lower MTHFR efficiency thus diverting folate away from DNA methylation and toward DNA synthesis. However, global DNA methylation in these cells was significantly decreased only at the lowest media folate concentration of 5 nM. In contrast, mutant MDA-MB-435 cells had increased SAM:SAH, decreased homocysteine levels, and increased uracil misincorporation, effects that were opposite to those observed in HCT116. Paradoxically, mutant MDA-MB-435 cells were hypomethylated even when cultured in folate-sufficient media (Sohn et al., 2009). These results suggest that folate metabolism is highly cell specific and tissue specific but the implication of this study on primary tissues is limited due to the use of transformed cell lines in this study.

In summary, findings from cell culture experiments suggest that folate deficiency influences global DNA methylation in nontransformed cells only when they are cultured for a sustained period of time in a severely folate-deplete media. Due to technical difficulties of culturing primary colonic mucosal cells, it has not been determined whether these results can be replicated in primary cultures. The slow and fickle growth rates of colonic mucosal cells would also make measurements of cellular folate and DNA methylation at various time points more challenging.

Most *in vitro* studies used measures of global DNA methylation as the experimental readout. Knowledge of the specific genes and DNA regions that are most susceptible to aberrant DNA methylation as a result of folate deficiency is necessary. For example, in human nasopharyngeal carcinoma cells cultured in folate-deplete media, hypermethylation of *H-cadherin* was associated with down-regulation of *H-cadherin* expression (Jhaveri et al., 2001). It appears that altera-tions in folate supply may target DNA methylation at certain genes rather than at a genome-wide level in some cells. Further research is required to clarify the downstream consequences of DNA methylation changes on gene expression and genomic stability, and their implications on cancer progression.

B. Animal studies

Studies using animal models have provided important insights into the effects of early life nutrition on DNA methylation of the offspring. In addition, a large number of rodent feeding studies have also sought to determine the influence of dietary folate and methyl donor intake on tissue folate reserves and tissue-specific DNA methylation patterns.

As discussed in Section III.C, *in utero* deficiencies of folate and other nutrients have been shown to permanently affect the epigenotype and phenotype of the offspring in later life (Waterland and Jirtle, 2004). This is classically illustrated by the honeybee and agouti mouse models. In the honeybee model, feeding female larvae copious amounts of royal jelly will produce queen bees. The queens are genetically identical to worker bees which develop from larvae that have not been fed royal jelly (Lue and Dizon, 1967). It has been shown that DNA methylation plays a critical role in the differential gene expression and subsequent phenotype found in fertile queens compared to sterile workers (Kucharski et al., 2008). This model emphasizes that early life nutritional exposure affects the phenotype of the offspring through changes in DNA meth-ylation patterns.

The agouti viable yellow (A^{vy}) mouse model also illustrates how early life methyl donor exposure affects offspring phenotype and health. The A^{vy} arises from a spontaneous insertion of a retrotransposon upstream to the gene. A^{vy} is a metastable epiallele, which is variably expressed in genetically identical offspring due to differences in DNA methylation at its promoter (Wolff et al., 1998). Methylation of the retrotransposon silences this gene producing lean mice with brown fur, while hypomethylation of the retrotransposon activates the gene producing obese mice with yellow fur, and an increased risk of metabolic disease and cancer (Whitelaw and Martin, 2001). Wolff et al. (1998) demonstrated that maternal methyl donor supplementation significantly increased DNA methylation in the A^{vy} allele and the percentage of lean offspring with brown fur. Although metastable epialleles have not been implicated in human disease, these studies demonstrate that dietary methyl donor intake may alter epigenetics readily and permanently during certain periods of development.

Other investigators have noticed that combined depletion of the methyl donors folate, choline, methionine, and vitamin B_{12} consistently induced rapid and substantial DNA hypomethylation in the liver of adult rodents (summarized in Table 4.5; Pogribny et al., 2004; Wainfan and Poirier, 1992; Wilson et al., 1984). Changes in hepatic genomic and gene-specific DNA methylation at c-myc, c-fos, and c-Ha-Ras were accompanied by concomitant alterations in gene expression of these protooncogenes (Dizik et al., 1991; Wainfan and Poirier, 1992). Spontaneous hepatocellular carcinoma has also occurred following methyl donor deficiency, thus confirming that methyl-deficient diets are cancer promoting in the liver (Wilson et al., 1984). Curiously, similar methyl-deficient diets failed to produce measurable changes in DNA methylation in the rat colon (Duthie et al., 2000b), spleen, thymus, kidney, and pancreas (Pogribny et al., 2004). Why the hepatocyte DNA methylation pathway is metabolically more sensitive toward methyl donor deficiency compared to peripheral tissues such as the colon is not fully understood.

The relationship between isolated folate deficiency and DNA hypomethylation in the liver or colon of rodents is far less consistent (Table 4.5). Only one study detected significant global hypomethylation after feeding rats a severely folate-deplete diet for 4 weeks (Balaghi et al., 1993), but this result has not been subsequently confirmed. In the rodent colon, only 2 of 10 studies observed colonic DNA hypomethylation following isolated folate deficiency. Linhart et al. (2009) fed wild-type mice and mice deficient in DNA uracil glycosylase a folate-deficient diet for 8 months, which was far longer than most other rodent studies. While colonic epithelial DNA hypomethylation occurred in the folate-deficient mice, hypomethylation alone or combined with uracil glycosylase deficiency did not increase tumor burden in these animals. In the second rodent colon study, dietary folate deficiency resulted in hypomethylation in the colons of older (18-month-old) but not the younger (4-month-old) mice.

Diet	Target organ	Stud-ies (n)	Duration of deficiency	Effect on global DNA methylation	Effect on site-specific DNA methylation	Cancer promoting?
Combined methyl donor depletion	Liver	9[a]	1–36 Weeks	7 studies found significant global DNA hypomethylation (15–60% decrease) following severe combined methyl donor depletion.	Hypomethylation of c-myc, c-fos, and c-Ha-Ras protooncogenes confirmed in three studies (Christman et al., 1993; Dizik et al., 1991; Wainfan and Poirier, 1992)	Yes
Isolated folate deficiency	Liver	5[b]	4–25 Weeks	Only one study found a significant decrease in global DNA methylation (20% decrease; $p = 0.032$) following 4 weeks of severe dietary folate deficiency (Balaghi et al., 1993).	Kim et al., 1997 detected exon 6–7 hypomethylation in p53 (40% decrease, $p = 0.002$) following 6 weeks of severe dietary folate deficiency.	No
Combined methyl donor depletion	Colon	1[c]	10 Weeks	DNA methylation was unaffected either by folate and/or methionine and choline depletion in isolated colonocytes.	Not studied.	No
Isolated folate deficiency	Colon	10[d]	5–32 Weeks	Two studies found a significant but modest decrease in global DNA methylation (6–7% decrease) after prolonged folate deficiency (Linhart et al., 2009) or in older animals (Keyes et al., 2007).	Exon 8 p53 hypomethylation (25% decrease, $p = 0.038$) after 20 weeks of mild folate depletion (Kim et al., 1996), but not following 5 weeks of severe folate depletion (Sohn et al., 2003).	No

Global and/or gene-specific changes in DNA methylation following combined methyl donor depletion has been found in the liver of almost all studies but not the colon of rodents. Isolated folate deficiency did not produce similar changes in DNA methylation suggesting that rodents are able to utilize alternative sources of methyl donors to compensate for folate deficiency in the colon. Studies included in this table are a survey of the literature using PubMed database and are not intended to be exhaustive.

[a]Wainfan et al. (1989), Dizik et al. (1991), Wainfan and Poirier (1992), Zapisek et al. (1992), Balaghi et al. (1993), Christman et al. (1993), Pogribny et al. (1995, 2004), Choi et al. (2005).

[b]Balaghi et al. (1993), Kim et al. (1995), Le Leu et al. (2000a,b), Song et al. (2000), Kim (2004a,b,c).

[c]Duthie et al. (2000a).

[d]Kim et al. (1995, 1996), Duthie et al. (2000a), Le Leu et al. (2000a,b), Choi et al. (2003), Davis and Uthus (2003), Sohn et al. (2003), Keyes et al. (2007), Linhart et al. (2009).

This study corroborates previous findings that older rodents have significantly lower colonic folate concentration (Choi *et al.*, 2003) and tissue DNA methylation compared to younger rodents (Richardson, 2003; Vanyushin *et al.*, 1973). Several researchers have also noted that exon-specific hypomethylation at p53 occurs in both the liver and colon following severe folate deficiency in rodents (Kim *et al.*, 1996, 1997). These changes were accompanied by DNA strand breaks at these sites. However, gene-specific DNA methylation changes in animal models have not yet been systematically profiled.

Overall, animal models have highlighted that

1. Effects of methyl deficiency on DNA methylation are tissue specific. The rodent liver but not the colon is responsive to dietary methyl deficiency.
2. Isolated folate deficiency is not sufficient in most cases to promote DNA hypomethylation or carcinogenesis in rodents.
3. Certain time periods of exposure, for example, *in utero* and in the elderly confer increased susceptibility of the epigenome to folate deficiency.
4. The effects of folate deficiency on DNA methylation may be gene specific rather than genome wide.

These conclusions support the hypothesis that folate deficiency alone does not result in genome-wide DNA hypomethylation and CRC unless combined with other methyl deficiencies, molecular changes, or adverse environmental exposures.

C. Human studies

A growing number of observational and interventional studies in humans have investigated the relationship between folate and genomic DNA methylation. Blood folate markers predicted colonic mucosal DNA methylation in both healthy subjects (Pufulete *et al.*, 2005b) and patients with CRC or colorectal adenoma (Pufulete *et al.*, 2003, 2005b). The relationship between folate metabolism and tissue DNA methylation is further reinforced by the consistent effects of the MTHFR C677T polymorphism on folate distribution and DNA methylation. In 2002, Stern *et al.* first reported that the MTHFR C677T polymorphism was associated with altered DNA methylation in healthy adults. TT subjects had significantly lower genomic leukocyte DNA methylation compared to CC subjects, and a direct relationship between red cell folate status and leukocyte genomic methylation was found only in TT subjects (Stern *et al.*, 2000). These interesting results were further validated by several other research groups with larger sample sizes and diverse study populations (Axume *et al.*, 2007; Castro *et al.*, 2004; Friso *et al.*, 2002). The T allele of the MTHFR C677T is also associated with DNA hypomethylation in the corresponding normal tissues of resected lung, breast, and CRC specimens in one study (Paz *et al.*, 2002). However, the role of DNA hypomethylation associated with the MTHFR C677T polymorphisms in CRC susceptibility has yet to be determined.

Two small case–control studies examined the relationship between folate metabolism, DNA methylation, and the risk of developing CRC or colorectal adenomas. In a study of 35 patients with adenomas, 28 patients with CRC, and 76 healthy controls recruited at colonoscopy (Pufulete *et al.*, 2003), low folate status (determined by erythrocyte folate, serum folate, and dietary folate intake combined into a single "folate status score") was significantly associated with CRC but not adenomas. Furthermore, DNA hypomethylation in both the blood and the colonic mucosa was associated with a significantly elevated risk for adenoma and a nonsignificant elevated risk for CRC (Pufulete *et al.*, 2003). The extent of hypomethylation was similar in adenoma and cancer patients. This reinforces the theory that hypomethylation is a very early lesion in CRC progression. However, it was not possible to determine whether folate deficiency was a cause for hypomethylation or colonic tumors in this study. Confounding factors such as higher alcohol consumption and older age in the cases compared to the healthy controls may have also influenced DNA methylation independently of folate status. In the second study, DNA methylation and folate intake of 115 female colorectal adenoma patients were compared with 115 healthy women undergoing colonoscopy (Lim *et al.*, 2008). Leukocyte DNA hypomethylation was again predictive of colorectal adenoma. Furthermore, genomic methylation showed a stronger inverse association with adenoma amongst those with low folate intake (Lim *et al.*, 2008). These studies indicate that a systemic genomic hypomethylation associated with folate deficiency may be a potential etiological factor for an early stage of colorectal adenoma.

Several randomized control trials and dietary intervention studies have investigated whether folate supplementation or depletion modifies DNA methylation in colonic and/or leukocyte DNA (Table 4.6). High-dose folate supplementation in CRC and colorectal adenoma patients significantly increased rectal mucosa DNA methylation compared to placebo (Cravo *et al.*, 1994). However, four subsequent randomized control trials of low dose folate supplementation in adenoma patients did not report any convincing increases in colonic mucosal DNA methylation. Limitations of these studies include their small sample size, short duration of intervention, and inability to exclude confounding factors such as concurrent folate intake from folate-fortified foods (Kim, 2005). When healthy postmenopausal women were housed in a metabolic unit to ensure greater control of dietary confounding factors, moderate folate depletion resulted in a significant reduction in leukocyte global DNA methylation (Jacob *et al.*, 1998; Rampersaud *et al.*, 2000). Whether such changes are also observed in the human colon following folate deficiency has yet to be determined.

Table 4.6. The Influence of Folate Supplementation (S) and Deficiency (D) on DNA Methylation in Human Intervention Studies

Subjects (n)	n	Folate intervention S/D (μg/day)	Duration	DNA	Result	Reference
CRC and colorectal adenoma patients	22	S: 10 mg/day	6 Months	Genomic; rectal mucosa	• 93% increase in patients receiving folic acid compared to placebo ($p < 0.002$).	Cravo et al. (1994)
Colorectal adenoma patients	20	S: 5 mg/day	3 Months	Genomic; colonic mucosa	• Significant increase in 7/20 patients • 37% increase in patients with only 1 polyp ($p = 0.05$) • No significant change in patients with > 1 polyp ($p > 0.05$)	Cravo et al. (1998)
Colorectal adenoma patients	20	S: 5 mg/day	1 Year	Genomic; colonic mucosa	• 57% increase in supplemented compared to placebo group at 6 months ($p = 0.001$) • Significant increase in methylation in both supplemented and placebo groups compared to baseline levels. • No significant difference between supplemented and placebo at 1 year.	Kim et al. (2001a,b)
Colorectal adenoma patients	31	S: 400μg/day	10 Weeks	Genomic; colonic mucosa and leukocyte	• 31% increase in leukocyte DNA ($p = 0.05$) of patients receiving folic acid • No significant change in colonic mucosal DNA in supplemented or placebo groups.	Pufulete et al. (2005a,b)
Colorectal adenoma patients	388	S: 1 mg/day	3 Years	LINE-1 methylation; colonic mucosa	• No significant difference in colonic LINE-1 methylation in supplemented compared to placebo.	Figueiredo et al. (2009)

Healthy adults	63	S: 700 μg folate + 7 μg B$_{12}$ then 2 mg folate + 20 μg B$_{12}$.	6 Months	Genomic; leukocyte	• No significant difference in supplemented compared to placebo group	Fenech et al. (1998)
Postmenopausal women	8	D: 56–111 μg/day	9 Weeks	Genomic; leukocyte	• 120% decrease compared to baseline ($p < 0.05$)	Jacob et al. (1998)
Postmenopausal women	33	D: 118 μg/day	7 Weeks	Genomic; leukocyte	• 10% decrease ($p = 0.0012$) compared to baseline	Rampersaud et al. (2000)

Five randomized control trials of folic acid supplementation in colorectal cancer (CRC) and colorectal adenoma patients have been performed. Four out of five of these studies showed a significant improvement in DNA methylation following the intervention in at least one patient subgroup or timepoint. Three folate intervention studies have been performed in healthy individuals. Short term folate depletion but not supplementation altered leukocyte DNA methylation levels in these studies.

The relationship between folate status and site-specific changes in DNA methylation has been poorly characterized in human studies. High-dose folate supplementation in patients homozygous for the variant MTHFR C677T allele lead to increased gene-specific methylation of *MGMT*, an important DNA repair gene (van den Donk *et al.*, 2007). However, several epidemiological studies have also reported that dietary folate deficiency was associated with aberrant promoter hypermethylation (van den Donk *et al.*, 2007; van Engeland *et al.*, 2003), while others found no significant relationships (Curtin *et al.*, 2007; Slattery *et al.*, 2007). Further studies that investigate how site-specific methylation levels are associated with polymorphisms of folate metabolism enzymes and folate status are required.

In human observational and case–control studies, a weak relationship between blood folate markers, colonic mucosal DNA methylation, and colorectal adenoma has been found. Moderate folate depletion in a highly controlled environment has also resulted in significant changes in leukocyte DNA methylation in healthy individuals. However, results from a number of preliminary studies are inconsistent on whether folate supplementation can independently modify colonic mucosal DNA methylation levels in healthy patients and those with colorectal pathology. Given the small sample size in most of these studies, it was not possible to control for intake of alternative methyl donors or polymorphisms in folate metabolism enzymes, two arms in the hypothesis flowchart (Fig. 4.6), which can also influence colonic mucosal folate and DNA methylation. Only one study confirmed that dietary folate intervention actually modified colonic mucosal folate stores (Kim, 2001). In most, blood folate markers were used to estimate systemic folate levels, which is not always reflective of tissue-specific exposures as previously discussed. More studies are required to investigate the effect of tissue-specific folate status and folate polymorphisms on colonic mucosal DNA methylation.

D. Species-specific differences in folate metabolism

It is logistically difficult to folate-deplete large numbers of human subjects and monitor colonic DNA methylation changes. However, in the human folate-depletion studies conducted, a significant effect on leukocyte DNA methylation was seen. In contrast, a large number of animal studies failed to induce changes in DNA methylation following dietary folate deficiency. There are several important species-specific differences in folate metabolism which begs us to consider whether rodents serve as appropriate animal models when investigating folate-related diseases.

A. *Rodents have an upregulated betaine pathway.* Betaine is an alternative methyl donor in the remethylation of homocysteine through the action of the enzyme betaine homocysteine methyltransferase (BHMT). MTHFR+/− and MTHFR −/− mice supplemented with betaine show significantly reduced plasma homocysteine and increased global DNA methylation (Schwahn *et al.*,

2004a,b). Upregulation of *BHMT* also occurs when rodents are fed a methyl-deficient diet (Park and Garrow, 1999), suggesting that betaine acts as a surrogate methyl donor. However, *BHMT* is primarily expressed in the liver and kidney (Delgado-Reyes *et al.*, 2001) and it is not known whether the betaine-dependent pathway functions in the colon (Knock *et al.*, 2008). The betaine pathway is also known to be more active in rodents compared to humans (Mason and Choi, 2000a,b). This affects the suitability of rodents as animal models to study folate-related diseases in humans.

B. *Rodents have higher tissue folate stores, and depend less on the transsulfuration pathway to remove excess homocysteine.* Another critical difference between rodents and humans is that systemic and tissue folate concentrations are consistently higher in rodents (Kim *et al.*, 1995) compared to humans (Kim *et al.*, 2001a,b; Likogianni *et al.*, 2006; Meenan *et al.*, 1997). Rodents can thus remove homocysteine more efficiently via remethylation. In contrast, humans would rely more heavily upon the transsulfuration pathway in situations of homocysteine overload (which removes homocysteine by degrading it into cysteine and cystathionine). Consequently, the remethylation/transulfuration ratio is higher in rodents than humans (Likogianni *et al.*, 2006). This metabolic difference is of particular importance in tissues lacking the transsulfuration pathway, such as the colonic mucosa (Finkelstein, 2000). Indeed, reports of rat colonic SAH levels (1.7 ± 0.5 nmol/g; Kim *et al.*, 1995) are over threefold lower than that of human colonic SAH (5.3 ± 0.8 nmol/g; Alonso-Aperte *et al.*, 2008), suggesting the rat colon is better able to remove SAH and homocysteine compared to the human colon. Consequently, high colonic folate stores in rodents constitute a high "tolerance threshold" which is capable of buffering the rodent colon from short-term dietary folate deficiencies modeled in most animal studies.

C. *Rodents have much higher DHFR activity compared to humans.* DHFR is a key enzyme which is responsible for metabolizing folic acid into the metabolically active THF. Accurate HPLC techniques have confirmed that human liver DHFR activity is over 70-fold lower than that of rat liver (Bailey and Ayling, 2009). Hence, the metabolism of synthetic folic acid in the systemic circulation of humans is much slower than that in rodents. Humans are thus much less efficient in utilizing supplemental folic acid to increase tissue folate stores compared to rodents.

VI. CONCLUSION

Folate is an essential cofactor required for the transfer of one-carbon groups in several important metabolic processes including DNA methylation. Changes in folate status may directly influence DNA methylation patterns in some tissues and as a consequence predispose to the development of cancer. The relationship between folate and aberrant DNA methylation in CRC has been extensively studied using a number of experimental approaches, as reviewed above. However, these studies have provided conflicting and inconclusive evidence on whether folate status is a direct and independent modulator of DNA methylation patterns. As proposed in Fig. 4.6, dietary folate deficiency alone may not be sufficient to induce localized folate deficiency in the colonic mucosa and subsequently colonic DNA hypomethylation. Polymorphic variants in important folate enzymes that hinder the folate metabolism or concurrent deficiencies of alternative methyl donors may also play an important role in determining interindividual susceptibility of the methylation machinery to reduced folate inputs. There is also little evidence to currently suggest that folate deficiency alone leads to CRC. It is likely that other sporadic and/or inherited genetic and epigenetic "hits" are required in concert with localized folate deficiency to give rise to colorectal neoplasms.

There is a need for more appropriate and physiologically relevant *in vitro* models in this research field. Ideally, this would be primary cultures of nontransformed colonic epithelial cells. This would ensure that cancer-associated changes in DNA methylation patterns and changes in cell behavior associated with multiple passages do not confound experimental results. By using primary cultures, it is likely that the magnitude of folate depletion and/or supplementation used can also be within more physiological realms. Previous studies using cell lines required over 1000-fold changes in media folate levels before a change in DNA methylation levels was observed (Duthie *et al.*, 2000a,b,c; Stempak *et al.*, 2005).

Animal models are useful as they allow precise control of dietary folate intake in a highly inbred population, thereby minimizing genetic and environmental confounding factors that could affect folate metabolism or the DNA methylation machinery. However, the folate metabolism cycle in rodents differs in many regards compared to humans and any extrapolation of findings from animal studies to humans must be done with extreme caution. Indeed, severe combined methyl donor depletion in rodents was necessary to modulate tissue-specific DNA methylation levels (Wilson *et al.*, 1984). The colonic mucosa was also found to be particularly resistant to dietary folate supplementation or depletion (Sohn *et al.*, 2003), a result also seen in preliminary folate intervention studies in humans. At present, there are a vast number of epidemiological studies on the relationship between folate status and CRC risk. Given the difficulty of accurately determining dietary folate intake (Kristal and Potter, 2006), it would be desirable to incorporate accurate measures of folate status and DNA methylation levels in these studies.

There are also several technical obstacles in this research field, both in the measurement of folate and the characterization of DNA methylation patterns. Most studies have characterized total cellular or tissue folate rather than the concentration of the individual folate coenzymes. However, it may not be total folate levels, but rather the distribution and balance of the different folate coenzymes which determine the relative efficiency of DNA synthesis and DNA methylation reactions (Reed *et al.*, 2006). Using liquid chromatography tandem mass spectrometry, Friso *et al.* (2002) have shown that MTHFR TT subjects have a relative deficiency of red cell 5-methylTHF and elevated red cell formylfolates. This redistribution of folate coenzymes away from the methylation pathway is consistent with the DNA hypomethylation found in the peripheral blood of these subjects. More studies incorporating HPLC-based methods of folate separation and detection are warranted.

As animal and *in vitro* studies have indicated, there are tissue-specific differences in folate metabolism and DNA methylation patterns. Changes in DNA methylation in peripheral leukocytes may not necessarily reflect changes in the colonic mucosa, due to differences in cellular turnover and regulation of DNA methylation. Hence, there is a need for more human observational and interventional studies that interrogate the influence of folate on methylation in the colorectum specifically (Brockton, 2006). Furthermore, no studies have systematically interrogated the specific patterns of global DNA hypomethylation induced by folate deficiency. Where does the hypomethylation occur in the genome and what are the direct consequences of hypomethylation on genomic stability and gene expression in those areas? How does folate affect methylation patterning across the entire genome? Traditionally studies have used candidate gene approaches and expression microarrays. As the cost of next-generation sequencing technology falls, researchers will increasingly be able to systematically interrogate DNA methylation patterns across the genome and begin to define its relationship to folate status.

Acknowledgments

J. Liu is supported by an Australian Postgraduate Award and the NSW Cancer Institute's Research Scholar Award. The authors declare no conflicts of interest.

References

Alonso-Aperte, E., Gonzalez, M. P., Poo-Prieto, R., and Varela-Moreiras, G. (2008). Folate status and S-adenosylmethionine/S-adenosylhomocysteine ratio in colorectal adenocarcinoma in humans. *Eur. J. Clin. Nutr.* **62**(2), 295–298.

Antony, A. C. (1996). Folate receptors. *Annu. Rev. Nutr.* **16**, 501–521.

Antony, A. C. (2007). In utero physiology: Role of folic acid in nutrient delivery and fetal development. *Am. J. Clin. Nutr.* **85**(2), 598S–603S.

Arcot, J., and Shrestha, A. (2005). Folate: Methods of analysis. *Trends Food Sci. Technol.* **16**, 253–266.

Axume, J., Smith, S. S., Pogribny, I. P., Moriarty, D. J., and Caudill, M. A. (2007). The MTHFR 677TT genotype and folate intake interact to lower global leukocyte DNA methylation in young Mexican American women. *Nutr. Res.* **27**(1), 1365–1367.

Bailey, L. B. (1990). Folate status assessment. *J. Nutr.* **120**(Suppl. 11), 1508–1511.

Bailey, S. W., and Ayling, J. E. (2009). The extremely slow and variable activity of dihydrofolate reductase in human liver and its implications for high folic acid intake. *Proc. Natl. Acad. Sci. USA* **106**(36), 15424–15429.

Balaghi, M., Horne, D. W., and Wagner, C. (1993). Hepatic one-carbon metabolism in early folate deficiency in rats. *Biochem. J.* **291**(Pt 1), 145–149.

Baylin, S., and Bestor, T. H. (2002). Altered methylation patterns in cancer cell genomes: Cause or consequence? *Cancer Cell* **1**(4), 299–305.

Bird, C. L., Swendseid, M. E., Witte, J. S., Shikany, J. M., Hunt, I. F., Frankl, H. D., Lee, E. R., Longnecker, M. P., and Haile, R. W. (1995). Red cell and plasma folate, folate consumption, and the risk of colorectal adenomatous polyps. *Cancer Epidemiol. Biomarkers Prev.* **4**(7), 709–714.

Blom, H. J., Shaw, G. M., den Heijer, M., and Finnell, R. H. (2006). Neural tube defects and folate: Case far from closed. *Nat. Rev. Neurosci.* **7**(9), 724–731.

Boyapati, S. M., Bostick, R. M., McGlynn, K. A., Fina, M. F., Roufail, W. M., Geisinger, K. R., Hebert, J. R., Coker, A., and Wargovich, M. (2004). Folate intake, MTHFR C677T polymorphism, alcohol consumption, and risk for sporadic colorectal adenoma (United States). *Cancer Causes Control* **15**(5), 493–501.

Brockton, N. T. (2006). Localized depletion: The key to colorectal cancer risk mediated by MTHFR genotype and folate? *Cancer Causes Control* **17**(8), 1005–1016.

Castro, R., Rivera, I., Ravasco, P., Camilo, M. E., Jakobs, C., Blom, H. J., and de Almeida, I. T. (2004). 5,10-methylenetetrahydrofolate reductase (MTHFR) 677C→T and 1298A→C mutations are associated with DNA hypomethylation. *J. Med. Genet.* **41**(6), 454–458.

Chen, J., Giovannucci, E. L., and Hunter, D. J. (1999). MTHFR polymorphism, methyl-replete diets and the risk of colorectal carcinoma and adenoma among U.S. men and women: An example of gene-environment interactions in colorectal tumorigenesis. *J. Nutr.* **129**(2 S Suppl.), 560S–564S.

Choi, S. W., Friso, S., Dolnikowski, G. G., Bagley, P. J., Edmondson, A. N., Smith, D. E., and Mason, J. B. (2003). Biochemical and molecular aberrations in the rat colon due to folate depletion are age-specific. *J. Nutr.* **133**(4), 1206–1212.

Choi, S. W., Friso, S., Keyes, M. K., and Mason, J. B. (2005). Folate supplementation increases genomic DNA methylation in the liver of elder rats. *Br. J. Nutr.* **93**(1), 31–35.

Christman, J. K., Sheikhnejad, G., Dizik, M., Abileah, S., and Wainfan, E. (1993). Reversibility of changes in nucleic acid methylation and gene expression induced in rat liver by severe dietary methyl deficiency. *Carcinogenesis* **14**(4), 551–557.

Cole, B. F., Baron, J. A., Sandler, R. S., Haile, R. W., Ahnen, D. J., Bresalier, R. S., McKeown-Eyssen, G., Summers, R. W., Rothstein, R. I., Burke, C. A., *et al.* (2007). Folic acid for the prevention of colorectal adenomas: A randomized clinical trial. *JAMA* **297**(21), 2351–2359.

Cravo, M., Fidalgo, P., Pereira, A., *et al.* (1994). DNA methylation as an intermediate biomarker in colorectal cancer: Modulation by folic acid supplementation. *Eur. J. Cancer Prev.* **3**, 473–479.

Cravo, M. M. J., Dayal, Y., *et al.* (1998). Effect of foalte supplemention on DNA methylation of rectal mucosa in patients with colonic adenomas: Correlation with nutrient intake. *Clin. Nutr.* **17**, 45–49.

Crott, J. W., Mashiyama, S. T., Ames, B. N., and Fenech, M. F. (2001). Methylenetetrahydrofolate reductase C677T polymorphism does not alter folic acid deficiency-induced uracil incorporation into primary human lymphocyte DNA in vitro. *Carcinogenesis* **22**(7), 1019–1025.

Crott, J. W., Liu, Z., Keyes, M. K., Choi, S. W., Jang, H., Moyer, M. P., and Mason, J. B. (2008). Moderate folate depletion modulates the expression of selected genes involved in cell cycle, intracellular signaling and folate uptake in human colonic epithelial cell lines. *J. Nutr. Biochem.* **19**(5), 328–335.

Curtin, K., Slattery, M. L., Ulrich, C. M., Bigler, J., Levin, T. R., Wolff, R. K., Albertsen, H., Potter, J. D., and Samowitz, W. S. (2007). Genetic polymorphisms in one-carbon metabolism: Associations with CpG island methylator phenotype (CIMP) in colon cancer and the modifying effects of diet. *Carcinogenesis* **28**(8), 1672–1679.

Czeizel, A. E., and Dudas, I. (1992). Prevention of the first occurrence of neural-tube defects by periconceptional vitamin supplementation. *N. Engl. J. Med.* **327**(26), 1832–1835.

Czeizel, A. E., Dudas, I., and Metneki, J. (1994). Pregnancy outcomes in a randomised controlled trial of periconceptional multivitamin supplementation. Final report. *Arch. Gynecol. Obstet.* **255**(3), 131–139.

Davis, C. D., and Uthus, E. O. (2003). Dietary folate and selenium affect dimethylhydrazine-induced aberrant crypt formation, global DNA methylation and one-carbon metabolism in rats. *J. Nutr.* **133**(9), 2907–2914.

Delgado-Reyes, C. V., Wallig, M. A., and Garrow, T. A. (2001). Immunohistochemical detection of betaine-homocysteine S-methyltransferase in human, pig, and rat liver and kidney. *Arch. Biochem. Biophys.* **393**(1), 184–186.

Dizik, M., Christman, J. K., and Wainfan, E. (1991). Alterations in expression and methylation of specific genes in livers of rats fed a cancer promoting methyl-deficient diet. *Carcinogenesis* **12**(7), 1307–1312.

Dudeja, P. K., Torania, S. A., and Said, H. M. (1997). Evidence for the existence of a carrier-mediated folate uptake mechanism in human colonic luminal membranes. *Am. J. Physiol.* **272**(6 Pt 1), G1408–G1415.

Duthie, S. J., and Hawdon, A. (1998). DNA instability (strand breakage, uracil misincorporation, and defective repair) is increased by folic acid depletion in human lymphocytes in vitro. *FASEB J.* **12**(14), 1491–1497.

Duthie, S. J., Grant, G., and Narayanan, S. (2000a). Increased uracil misincorporation in lymphocytes from folate-deficient rats. *Br. J. Cancer* **83**(11), 1532–1537.

Duthie, S. J., Narayanan, S., Blum, S., Pirie, L., and Brand, G. M. (2000b). Folate deficiency in vitro induces uracil misincorporation and DNA hypomethylation and inhibits DNA excision repair in immortalized normal human colon epithelial cells. *Nutr. Cancer* **37**(2), 245–251.

Duthie, S. J., Narayanan, S., Brand, G. M., and Grant, G. (2000c). DNA stability and genomic methylation status in colonocytes isolated from methyl-donor-deficient rats. *Eur. J. Nutr.* **39**(3), 106–111.

Duthie, S. J., Narayanan, S., Brand, G. M., Pirie, L., and Grant, G. (2002). Impact of folate deficiency on DNA stability. *J. Nutr.* **132**(8 Suppl), 2444S–2449S.

Ehrlich, M. (2002). DNA methylation in cancer: Too much, but also too little. *Oncogene* **21**(35), 5400–5413.

Ehrlich, M., Gama-Sosa, M. A., Huang, L. H., Midgett, R. M., Kuo, K. C., McCune, R. A., and Gehrke, C. (1982). Amount and distribution of 5-methylcytosine in human DNA from different types of tissues of cells. *Nucleic Acids Res.* **10**(8), 2709–2721.

Eisenga, B. H., Collins, T. D., and McMartin, K. E. (1992). Incorporation of 3H-label from folic acid is tissue-dependent in folate-deficient rats. *J. Nutr.* **122**(4), 977–985.

Esteller, M. (2005). Aberrant DNA methylation as a cancer-inducing mechanism. *Annu. Rev. Pharmacol. Toxicol.* **45**, 629–656.

Fenech, M., Aitken, C., and Rinaldi, J. (1998). Folate, vitamin B12, homocysteine status and DNA damage in young Australian adults. *Carcinogenesis* **19**(7), 1163–1171.

Figueiredo, J. C., Grau, M. V., Wallace, K., Levine, A. J., Shen, L. L., Hamdan, R., Chen, X. L., Bresalier, R. S., McKeown-Eyssen, G., Haile, R. W., et al. (2009). Global DNA hypomethyation (LINE-1) in the normal colon and lifestyle characteristics and dietary and genetic factors. Cancer Epidemiol. Biomark. Prev. 18(4), 1041–1049.

Finkelstein, J. D. (2000). Pathways and regulation of homocysteine metabolism in mammals. Semin. Thromb. Hemost. 26(3), 219–225.

Friso, S., Choi, S. W., Girelli, D., Mason, J. B., Dolnikowski, G. G., Bagley, P. J., Olivieri, O., Jacques, P. F., Rosenberg, I. H., Corrocher, R., et al. (2002). A common mutation in the 5, 10-methylenetetrahydrofolate reductase gene affects genomic DNA methylation through an interaction with folate status. Proc. Natl. Acad. Sci. USA 99(8), 5606–5611.

Fukushima, M., Morita, M., Ikeda, K., and Nagayama, S. (2003). Population study of expression of thymidylate synthase and dihydropyrimidine dehydrogenase in patients with solid tumors. Int. J. Mol. Med. 12(6), 839–844.

Gaudet, F., Hodgson, J. G., Eden, A., Jackson-Grusby, L., Dausman, J., Gray, J. W., Leonhardt, H., and Jaenisch, R. (2003). Induction of tumors in mice by genomic hypomethylation. Science 300 (5618), 489–492.

Gaughan, D. J., Kluijtmans, L. A., Barbaux, S., McMaster, D., Young, I. S., Yarnell, J. W., Evans, A., and Whitehead, A. S. (2001). The methionine synthase reductase (MTRR) A66G polymorphism is a novel genetic determinant of plasma homocysteine concentrations. Atherosclerosis 157(2), 451–456.

Ghandour, H., Chen, Z., Selhub, J., and Rozen, R. (2004). Mice deficient in methylenetetrahydrofolate reductase exhibit tissue-specific distribution of folates. J. Nutr. 134(11), 2975–2978.

Gonzalo, S., Jaco, I., Fraga, M. F., Chen, T., Li, E., Esteller, M., and Blasco, M. A. (2006). DNA methyltransferases control telomere length and telomere recombination in mammalian cells. Nat. Cell Biol. 8(4), 416–424.

Gregory, J. F., 3 rd, and Scott, K. C. (1996). Modeling of folate metabolism. Adv. Food Nutr. Res. 40, 81–93.

Halsted, C. H. (1979). The intestinal absorption of folates. Am. J. Clin. Nutr. 32(4), 846–855.

Halsted, C. H. (1980). Intestinal absorption and malabsorption of folates. Annu. Rev. Med. 31, 79–87.

Halsted, C. H., Villanueva, J. A., Devlin, A. M., and Chandler, C. J. (2002). Metabolic interactions of alcohol and folate. J. Nutr. 132(8 Suppl), 2367S–2372S.

Henderson, G. I., Perez, T., Schenker, S., Mackins, J., and Antony, A. C. (1995). Maternal-to-fetal transfer of 5-methyltetrahydrofolate by the perfused human placental cotyledon: Evidence for a concentrative role by placental folate receptors in fetal folate delivery. J. Lab. Clin. Med. 126(2), 184–203.

Hoffmann, M. J., and Schulz, W. A. (2005). Causes and consequences of DNA hypomethyation in human cancer. Biochem. Cell Biol. 83(3), 296–321.

Hortin, G. L. (2006). Homocysteine: Clinical significance and laboratory measurement. Lab. Med. 37(9), 551–553.

Hubner, R. A., Muir, K. R., Liu, J. F., Sellick, G. S., Logan, R. F., Grainge, M., Armitage, N., Chau, I., and Houlston, R. S. (2006). Folate metabolism polymorphisms influence risk of colorectal adenoma recurrence. Cancer Epidemiol. Biomarkers Prev. 15(9), 1607–1613.

Jacob, R. A., Gretz, D. M., Taylor, P. C., James, S. J., Pogribny, I. P., Miller, B. J., Henning, S. M., and Swendseid, M. E. (1998). Moderate folate depletion increases plasma homocysteine and decreases lymphocyte DNA methylation in postmenopausal women. J. Nutr. 128(7), 1204–1212.

James, S. J., Pogribna, M., Pogribny, I. P., Melnyk, S., Hine, R. J., Gibson, J. B., Yi, P., Tafoya, D. L., Swenson, D. H., Wilson, V. L., et al. (1999). Abnormal folate metabolism and mutation in the methylenetetrahydrofolate reductase gene may be maternal risk factors for Down syndrome. Am. J. Clin. Nutr. 70(4), 495–501.

Jhaveri, M. S., Wagner, C., and Trepel, J. B. (2001). Impact of extracellular folate levels on global gene expression. *Mol. Pharmacol.* **60**(6), 1288–1295.

Kane, M. A., and Waxman, S. (1989). Role of folate binding proteins in folate metabolism. *Lab. Invest.* **60**(6), 737–746.

Kaufman, Y., Drori, S., Cole, P. D., Kamen, B. A., Sirota, J., Ifergan, I., Arush, M. W., Elhasid, R., Sahar, D., Kaspers, G. J., *et al.* (2004). Reduced folate carrier mutations are not the mechanism underlying methotrexate resistance in childhood acute lymphoblastic leukemia. *Cancer* **100**(4), 773–782.

Keyes, M. K., Jang, H., Mason, J. B., Liu, Z., Crott, J. W., Smith, D. E., Friso, S., and Choi, S. W. (2007). Older age and dietary folate are determinants of genomic and p16-specific DNA methylation in mouse colon. *J. Nutr.* **137**(7), 1713–1717.

Kim, Y. I. (1999). Folate and carcinogenesis: Evidence, mechanisms, and implications. *J. Nutr. Biochem.* **10**(2), 66–88.

Kim, Y. I. (2000). Methylenetetrahydrofolate reductase polymorphisms, folate, and cancer risk: A paradigm of gene-nutrient interactions in carcinogenesis. *Nutr. Rev.* **58**(7), 205–209.

Kim, Y. I. (2004a). Folate and DNA methylation: A mechanistic link between folate deficiency and colorectal cancer? *Cancer Epidemiol. Biomarkers Prev.* **13**(4), 511–519.

Kim, Y. I. (2004b). Folate, colorectal carcinogenesis, and DNA methylation: Lessons from animal studies. *Environ. Mol. Mutagen.* **44**(1), 10–25.

Kim, Y. I. (2004c). Will mandatory folic acid fortification prevent or promote cancer? *Am. J. Clin. Nutr.* **80**(5), 1123–1128.

Kim, Y. I. (2005). Nutritional epigenetics: Impact of folate deficiency on DNA methylation and colon cancer susceptibility. *J. Nutr.* **135**(11), 2703–2709.

Kim, Y. I., Christman, J. K., Fleet, J. C., Cravo, M. L., Salomon, R. N., Smith, D., Ordovas, J., Selhub, J., and Mason, J. B. (1995). Moderate folate deficiency does not cause global hypomethylation of hepatic and colonic DNA or c-myc-specific hypomethylation of colonic DNA in rats. *Am. J. Clin. Nutr.* **61**(5), 1083–1090.

Kim, Y. I., Pogribny, I. P., Salomon, R. N., Choi, S. W., Smith, D. E., James, S. J., and Mason, J. B. (1996). Exon-specific DNA hypomethylation of the p53 gene of rat colon induced by dimethylhydrazine. Modulation by dietary folate. *Am. J. Pathol.* **149**(4), 1129–1137.

Kim, Y. I., Pogribny, I. P., Basnakian, A. G., Miller, J. W., Selhub, J., James, S. J., and Mason, J. B. (1997). Folate deficiency in rats induces DNA strand breaks and hypomethylation within the p53 tumor suppressor gene. *Am. J. Clin. Nutr.* **65**(1), 46–52.

Kim, Y. I., Fawaz, K., Knox, T., Lee, Y. M., Norton, R., Arora, S., Paiva, L., and Mason, J. B. (1998). Colonic mucosal concentrations of folate correlate well with blood measurements of folate status in persons with colorectal polyps. *Am. J. Clin. Nutr.* **68**(4), 866–872.

Kim, Y. I., Baik, H. W., Fawaz, K., Knox, T., Lee, Y. M., Norton, R., Libby, E., and Mason, J. B. (2001a). Effects of folate supplementation on two provisional molecular markers of colon cancer: A prospective, randomized trial. *Am. J. Gastroenterol.* **96**(1), 184–195.

Kim, Y. I., Fawaz, K., Knox, T., Lee, Y. M., Norton, R., Libby, E., and Mason, J. B. (2001b). Colonic mucosal concentrations of folate are accurately predicted by blood measurements of folate status among individuals ingesting physiologic quantities of folate. *Cancer Epidemiol. Biomarkers Prev.* **10**(6), 715–719.

Knock, E., Deng, L., Wu, Q., Lawrance, A. K., Wang, X. L., and Rozen, R. (2008). Strain differences in mice highlight the role of DNA damage in neoplasia induced by low dietary folate. *J. Nutr.* **138**(4), 653–658.

Kristal, A. R., and Potter, J. D. (2006). Not the time to abandon the food frequency questionnaire: Counterpoint. *Cancer Epidemiol. Biomarkers Prev.* **15**(10), 1759–1760.

Kucharski, R., Maleszka, J., Foret, S., and Maleszka, R. (2008). Nutritional control of reproductive status in honeybees via DNA methylation. *Science* **319**(5871), 1827–1830.

Le Leu, R. K., Young, G. P., and McIntosh, G. H. (2000a). Folate deficiency diminishes the occurrence of aberrant crypt foci in the rat colon but does not alter global DNA methylation status. *J. Gastroenterol. Hepatol.* **15**(10), 1158–1164.

Le Leu, R. K., Young, G. P., and McIntosh, G. H. (2000b). Folate deficiency reduces the development of colorectal cancer in rats. *Carcinogenesis* **21**(12), 2261–2265.

Le Marchand, L., Donlon, T., Hankin, J. H., Kolonel, L. N., Wilkens, L. R., and Seifried, A. (2002). B-vitamin intake, metabolic genes, and colorectal cancer risk (United States). *Cancer Causes Control* **13**(3), 239–248.

Libbus, B. L., Borman, L. S., Ventrone, C. H., and Branda, R. F. (1990). Nutritional folate-deficiency in Chinese hamster ovary cells. Chromosomal abnormalities associated with perturbations in nucleic acid precursors. *Cancer Genet. Cytogenet.* **46**(2), 231–242.

Likogianni, V., Janel, N., Ledru, A., Beaune, P., Paul, J. L., and Demuth, K. (2006). Thiol compounds metabolism in mice, rats and humans: Comparative study and potential explanation of rodents protection against vascular diseases. *Clin. Chim. Acta* **372**(1–2), 140–146.

Lillycrop, K. A., Phillips, E. S., Jackson, A. A., Hanson, M. A., and Burdge, G. C. (2005). Dietary protein restriction of pregnant rats induces and folic acid supplementation prevents epigenetic modification of hepatic gene expression in the offspring. *J. Nutr.* **135**(6), 1382–1386.

Lillycrop, K. A., Slater-Jefferies, J. L., Hanson, M. A., Godfrey, K. M., Jackson, A. A., and Burdge, G. C. (2007). Induction of altered epigenetic regulation of the hepatic glucocorticoid receptor in the offspring of rats fed a protein-restricted diet during pregnancy suggests that reduced DNA methyltransferase-1 expression is involved in impaired DNA methylation and changes in histone modifications. *Br. J. Nutr.* **97**(6), 1064–1073.

Lim, U., Flood, A., Choi, S. W., Albanes, D., Cross, A. J., Schatzkin, A., Sinha, R., Katki, H. A., Cash, B., Schoenfeld, P., *et al.* (2008). Genomic methylation of leukocyte DNA in relation to colorectal adenoma among asymptomatic women. *Gastroenterology* **134**(1), 47–55.

Linhart, H. G., Troen, A., Bell, G. W., Cantu, E., Chao, W. H., Moran, E., Steine, E., He, T., and Jaenisch, R. (2009). Folate deficiency induces genomic uracil misincorporation and hypomethylation but does not increase DNA point mutations. *Gastroenterology* **136**(1), 227–235, (e223).

Lucock, M. (2000). Folic acid: Nutritional biochemistry, molecular biology, and role in disease processes. *Mol. Genet. Metab.* **71**(1–2), 121–138.

Lucock, M. D., Daskalakis, I., Lumb, C. H., Schorah, C. J., and Levene, M. I. (1998). Impaired regeneration of monoglutamyl tetrahydrofolate leads to cellular folate depletion in mothers affected by a spina bifida pregnancy. *Mol. Genet. Metab.* **65**(1), 18–30.

Lucock, M., Daskalakis, I., Hinkins, M., and Yates, Z. (2001). An examination of polymorphic genes and folate metabolism in mothers affected by a spina bifida pregnancy. *Mol. Genet. Metab.* **73**(4), 322–332.

Lue, P. F., and Dizon, S. E. (1967). Studies in the mode of action of royal jelly in honeybee development. VII. The free amino acids in the haemolymph of developing larvae. *Can. J. Zool.* **45**(2), 205–214.

Ma, J., Stampfer, M. J., Giovannucci, E., Artigas, C., Hunter, D. J., Fuchs, C., Willett, W. C., Selhub, J., Hennekens, C. H., and Rozen, R. (1997). Methylenetetrahydrofolate reductase polymorphism, dietary interactions, and risk of colorectal cancer. *Cancer Res.* **57**(6), 1098–1102.

Mason, J. B., and Choi, S. W. (2000a). The mechanisms by which folate depletion enhances colorectal carcinogenesis: A unified scheme. *Nestle Nutr. Workshop Ser. Clin. Perform. Programme* **4**, 87–99, (discussion: 99–101).

Mason, J. B., and Choi, S. W. (2000b). Folate and carcinogenesis: Developing a unifying hypothesis. *Adv. Enzyme Regul.* **40**, 127–141.

McDowell, M. A., Lacher, D. A., Pfeiffer, C. M., Mulinare, J., Picciano, M. F., Rader, J. I., Yetley, E. A., Kennedy-Stephenson, J., and Johnson, C. L. (2008). Blood folate levels: The latest NHANES results. *NCHS Data Brief* **6,** 1–8.

McKay, J. A., Williams, E. A., and Mathers, J. C. (2004). Folate and DNA methylation during in utero development and aging. *Biochem. Soc. Trans.* **32**(Pt 6), 1006–1007.

Meenan, J., O'Hallinan, E., Scott, J., and Weir, D. G. (1997). Epithelial cell folate depletion occurs in neoplastic but not adjacent normal colon mucosa. *Gastroenterology* **112**(4), 1163–1168.

Mills, J. L., Scott, J. M., Kirke, P. N., McPartlin, J. M., Conley, M. R., Weir, D. G., Molloy, A. M., and Lee, Y. J. (1996). Homocysteine and neural tube defects. *J. Nutr.* **126**(3), 756S–760S.

Molloy, P. L. (2009). DNA hypomethylation in cancer. *In* "Cancer Epigenetics" (T. Tollefsbol, ed.). CRC Press, Boca Raton.

Molloy, A. M., Daly, S., Mills, J. L., Kirke, P. N., Whitehead, A. S., Ramsbottom, D., Conley, M. R., Weir, D. G., and Scott, J. M. (1997). Thermolabile variant of 5, 10-methylenetetrahydrofolate reductase associated with low red-cell folates: Implications for folate intake recommendations. *Lancet* **349**(9065), 1591–1593.

Morgan, H. D., Santos, F., Green, K., Dean, W., and Reik, W. (2005). Epigenetic reprogramming in mammals. *Hum. Mol. Genet.* **14**(Spec No 1), R47–R58.

Morris, M. S., Jacques, P. F., Rosenberg, I. H., and Selhub, J. (2007). Folate and vitamin B-12 status in relation to anemia, macrocytosis, and cognitive impairment in older Americans in the age of folic acid fortification. *Am. J. Clin. Nutr.* **85**(1), 193–200.

Niculescu, M. D., and Zeisel, S. H. (2002). Diet, methyl donors and DNA methylation: Interactions between dietary folate, methionine and choline. *J. Nutr.* **132**(8 Suppl.), 2333S–2335S.

Nijhout, H. F., Reed, M. C., Budu, P., and Ulrich, C. M. (2004). A mathematical model of the folate cycle: New insights into folate homeostasis. *J. Biol. Chem.* **279**(53), 55008–55016.

Nijhout, H. F., Reed, M. C., Anderson, D. F., Mattingly, J. C., James, S. J., and Ulrich, C. M. (2006). Long-range allosteric interactions between the folate and methionine cycles stabilize DNA methylation reaction rate. *Epigenetics* **1**(2), 81–87.

Nijhout, H. F., Reed, M. C., and Ulrich, C. M. (2008). Chapter 2 mathematical models of folate-mediated one-carbon metabolism. *Vitam. Horm.* **79**, 45–82.

Park, E. I., and Garrow, T. A. (1999). Interaction between dietary methionine and methyl donor intake on rat liver betaine-homocysteine methyltransferase gene expression and organization of the human gene. *J. Biol. Chem.* **274**(12), 7816–7824.

Parker, N., Turk, M. J., Westrick, E., Lewis, J. D., Low, P. S., and Leamon, C. P. (2005). Folate receptor expression in carcinomas and normal tissues determined by a quantitative radioligand binding assay. *Anal. Biochem.* **338**(2), 284–293.

Paz, M. F., Avila, S., Fraga, M. F., Pollan, M., Capella, G., Peinado, M. A., Sanchez-Cespedes, M., Herman, J. G., and Esteller, M. (2002). Germ-line variants in methyl-group metabolism genes and susceptibility to DNA methylation in normal tissues and human primary tumors. *Cancer Res.* **62**(15), 4519–4524.

Pfeiffer, C. M., Johnson, C. L., Jain, R. B., Yetley, E. A., Picciano, M. F., Rader, J. I., Fisher, K. D., Mulinare, J., and Osterloh, J. D. (2007). Trends in blood folate and vitamin B-12 concentrations in the United States, 1988–2004. *Am. J. Clin. Nutr.* **86**(3), 718–727.

Piedrahita, J. A., Oetama, B., Bennett, G. D., van Waes, J., Kamen, B. A., Richardson, J., Lacey, S. W., Anderson, R. G., and Finnell, R. H. (1999). Mice lacking the folic acid-binding protein Folbp1 are defective in early embryonic development. *Nat. Genet.* **23**(2), 228–232.

Pogribny, I. P., Basnakian, A. G., Miller, B. J., Lopatina, N. G., Poirier, L. A., and James, S. J. (1995). Breaks in genomic DNA and within the p53 gene are associated with hypomethylation in livers of folate/methyl-deficient rats. *Cancer Res.* **55**(9), 1894–1901.

Pogribny, I. P., James, S. J., Jernigan, S., and Pogribna, M. (2004). Genomic hypomethylation is specific for preneoplastic liver in folate/methyl deficient rats and does not occur in non-target tissues. *Mutat. Res.* **548**(1–2), 53–59.

Poschl, G., Stickel, F., Wang, X. D., and Seitz, H. K. (2004). Alcohol and cancer: Genetic and nutritional aspects. *Proc. Nutr. Soc.* **63**(1), 65–71.

Potten, C. S., and Allen, T. D. (1977). Ultrastructure of cell loss in intestinal mucosa. *J. Ultrastruct. Res.* **60**(2), 272–277.

Pufulete, M., Al-Ghnaniem, R., Leather, A. J., Appleby, P., Gout, S., Terry, C., Emery, P. W., and Sanders, T. A. (2003). Folate status, genomic DNA hypomethylation, and risk of colorectal adenoma and cancer: A case control study. *Gastroenterology* **124**(5), 1240–1248.

Pufulete, M., Al-Ghnaniem, R., Khushal, A., Appleby, P., Harris, N., Gout, S., Emery, P. W., and Sanders, T. A. (2005a). Effect of folic acid supplementation on genomic DNA methylation in patients with colorectal adenoma. *Gut* **54**(5), 648–653.

Pufulete, M., Al-Ghnaniem, R., Rennie, J. A., Appleby, P., Harris, N., Gout, S., Emery, P. W., and Sanders, T. A. (2005b). Influence of folate status on genomic DNA methylation in colonic mucosa of subjects without colorectal adenoma or cancer. *Br. J. Cancer* **92**(5), 838–842.

Quinlivan, E. P., Davis, S. R., Shelnutt, K. P., Henderson, G. N., Ghandour, H., Shane, B., Selhub, J., Bailey, L. B., Stacpoole, P. W., and Gregory, J. F., 3 rd (2005). Methylenetetrahydrofolate reductase 677C→T polymorphism and folate status affect one-carbon incorporation into human DNA deoxynucleosides. *J. Nutr.* **135**(3), 389–396.

Quinlivan, E. P., Hanson, A. D., and Gregory, J. F. (2006). The analysis of folate and its metabolic precursors in biological samples. *Anal. Biochem.* **348**(2), 163–184.

Rampersaud, G. C., Kauwell, G. P., Hutson, A. D., Cerda, J. J., and Bailey, L. B. (2000). Genomic DNA methylation decreases in response to moderate folate depletion in elderly women. *Am. J. Clin. Nutr.* **72**(4), 998–1003.

Reed, M. C., Nijhout, H. F., Neuhouser, M. L., Gregory, J. F., 3 rd, Shane, B., James, S. J., Boynton, A., and Ulrich, C. M. (2006). A mathematical model gives insights into nutritional and genetic aspects of folate-mediated one-carbon metabolism. *J. Nutr.* **136**(10), 2653–2661.

Richardson, B. (2003). Impact of aging on DNA methylation. *Ageing Res. Rev.* **2**(3), 245–261.

Richardson, R. E., Healy, M. J., and Nixon, P. F. (1979). Folates of rat tissue. Bioassay of tissue folylpolyglutamates and a relationship of liver folypolyglutamates to nutritional folate sufficiency. *Biochim. Biophys. Acta* **585**(1), 128–133.

Rogers, A. E. (1995). Methyl donors in the diet and responses to chemical carcinogens. *Am. J. Clin. Nutr.* **61**(3 Suppl.), 659S–665S.

Ross, J. F., Chaudhuri, P. K., and Ratnam, M. (1994). Differential regulation of folate receptor isoforms in normal and malignant tissues in vivo and in established cell lines. Physiologic and clinical implications. *Cancer* **73**(9), 2432–2443.

Ryan, B. M., and Weir, D. G. (2001). Relevance of folate metabolism in the pathogenesis of colorectal cancer. *J. Lab. Clin. Med.* **138**(3), 164–176.

Schofield, P. N., Joyce, J. A., Lam, W. K., Grandjean, V., Ferguson-Smith, A., Reik, W., and Maher, E. R. (2001). Genomic imprinting and cancer; new paradigms in the genetics of neoplasia. *Toxicol. Lett.* **120**(1–3), 151–160.

Schwahn, B. C., Laryea, M. D., Chen, Z., Melnyk, S., Pogribny, I., Garrow, T., James, S. J., and Rozen, R. (2004a). Betaine rescue of an animal model with methylenetetrahydrofolate reductase deficiency. *Biochem. J.* **382**(Pt 3), 831–840.

Schwahn, B. C., Wendel, U., Lussier-Cacan, S., Mar, M. H., Zeisel, S. H., Leclerc, D., Castro, C., Garrow, T. A., and Rozen, R. (2004b). Effects of betaine in a murine model of mild cystathionine-beta-synthase deficiency. *Metabolism* **53**(5), 594–599.

Selhub, J., and Miller, J. W. (1992). The pathogenesis of homocysteinemia: Interruption of the coordinate regulation by S-adenosylmethionine of the remethylation and transsulfuration of homocysteine. *Am. J. Clin. Nutr.* **55**(1), 131–138.

Shane, B. (1996). Folate chemistry and metabolism. *In* "Folate in Health and Disease" (L. B. Bailey, ed.), pp. 1–22. Marcel Dekker Inc., New York.

Sharp, L., and Little, J. (2004). Polymorphisms in genes involved in folate metabolism and colorectal neoplasia: A HuGE review. *Am. J. Epidemiol.* **159**(5), 423–443.

Shen, F., Wu, M., Ross, J. F., Miller, D., and Ratnam, M. (1995). Folate receptor type gamma is primarily a secretory protein due to lack of an efficient signal for glycosylphosphatidylinositol modification: Protein characterization and cell type specificity. *Biochemistry* **34**(16), 5660–5665.

Sie, K. K., Chen, J., Sohn, K. J., Croxford, R., Thompson, L. U., and Kim, Y. I. (2009). Folic acid supplementation provided in utero and during lactation reduces the number of terminal end buds of the developing mammary glands in the offspring. *Cancer Lett.* **280**(1), 72–77.

Slattery, M. L., Curtin, K., Sweeney, C., Levin, T. R., Potter, J., Wolff, R. K., Albertsen, H., and Samowitz, W. S. (2007). Diet and lifestyle factor associations with CpG island methylator phenotype and BRAF mutations in colon cancer. *Int. J. Cancer* **120**(3), 656–663.

Smith, A. D., Kim, Y. I., and Refsum, H. (2008). Is folic acid good for everyone? *Am. J. Clin. Nutr.* **87** (3), 517–533.

Sohn, K. J., Stempak, J. M., Reid, S., Shirwadkar, S., Mason, J. B., and Kim, Y. I. (2003). The effect of dietary folate on genomic and p53-specific DNA methylation in rat colon. *Carcinogenesis* **24**(1), 81–90.

Sohn, K. J., Jang, H., Campan, M., Weisenberger, D. J., Dickhout, J., Wang, Y. C., Cho, R. C., Yates, Z., Lucock, M., Chiang, E. P., *et al.* (2009). The methylenetetrahydrofolate reductase C677T mutation induces cell-specific changes in genomic DNA methylation and uracil mis-incorporation: A possible molecular basis for the site-specific cancer risk modification. *Int. J. Cancer* **124**(9), 1999–2005.

Song, J., Sohn, K. J., Medline, A., Ash, C., Gallinger, S., and Kim, Y. I. (2000). Chemopreventive effects of dietary folate on intestinal polyps in Apc+/−Msh2−/− mice. *Cancer Res.* **60**(12), 3191–3199.

Stempak, J. M., Sohn, K. J., Chiang, E. P., Shane, B., and Kim, Y. I. (2005). Cell and stage of transformation-specific effects of folate deficiency on methionine cycle intermediates and DNA methylation in an in vitro model. *Carcinogenesis* **26**(5), 981–990.

Stern, L. L., Mason, J. B., Selhub, J., and Choi, S. W. (2000). Genomic DNA hypomethylation, a characteristic of most cancers, is present in peripheral leukocytes of individuals who are homozygous for the C677T polymorphism in the methylenetetrahydrofolate reductase gene. *Cancer Epidemiol. Biomarkers Prev.* **9**(8), 849–853.

Stover, P. J. (2004). Physiology of folate and vitamin B12 in health and disease. *Nutr. Rev.* **62**(6 Pt 2), S3–S12, (discussion: S13).

Suh, J. R., Herbig, A. K., and Stover, P. J. (2001). New perspectives on folate catabolism. *Annu. Rev. Nutr.* **21**, 255–282.

Thuesen, B. H., Husemoen, L. L., Ovesen, L., Jorgensen, T., Fenger, M., and Linneberg, A. (2009). Lifestyle and genetic determinants of folate and vitamin B_{12} levels in a general adult population. *Br. J. Nutr.* **103**, 1195–1204.

Toffoli, G., Cernigoi, C., Russo, A., Gallo, A., Bagnoli, M., and Boiocchi, M. (1997). Overexpression of folate binding protein in ovarian cancers. *Int. J. Cancer* **74**(2), 193–198.

Trimble, K. C., Molloy, A. M., Scott, J. M., and Weir, D. G. (1993). The effect of ethanol on one-carbon metabolism: Increased methionine catabolism and lipotrope methyl-group wastage. *Hepatology* **18**(4), 984–989.

Trinh, B. N., Ong, C. N., Coetzee, G. A., Yu, M. C., and Laird, P. W. (2002). Thymidylate synthase: A novel genetic determinant of plasma homocysteine and folate levels. *Hum. Genet.* **111**(3), 299–302.

Ulrich, C. M. (2005). Nutrigenetics in cancer research—Folate metabolism and colorectal cancer. *J. Nutr.* **135**(11), 2698–2702.

van den Donk, M., van Engeland, M., Pellis, L., Witteman, B. J., Kok, F. J., Keijer, J., and Kampman, E. (2007). Dietary folate intake in combination with MTHFR C677T genotype and promoter methylation of tumor suppressor and DNA repair genes in sporadic colorectal adenomas. *Cancer Epidemiol. Biomarkers Prev.* **16**(2), 327–333.

van der Put, N. M., Thomas, C. M., et al. (1997). "Altered folate and vitamin B12 metabolism in families with spina bifida offspring." *Qjm* **90**(8), 505–510.

van Engeland, M., Weijenberg, M. P., Roemen, G. M., Brink, M., de Bruine, A. P., Goldbohm, R. A., van den Brandt, P. A., Baylin, S. B., de Goeij, A. F., and Herman, J. G. (2003). Effects of dietary folate and alcohol intake on promoter methylation in sporadic colorectal cancer: The Netherlands cohort study on diet and cancer. *Cancer Res.* **63**(12), 3133–3137.

Vanyushin, B. F., Mazin, A. L., Vasilyev, V. K., and Belozersky, A. N. (1973). The content of 5-methylcytosine in animal DNA: The species and tissue specificity. *Biochim. Biophys. Acta* **299**(3), 397–403.

Varela-Moreiras, G., and Selhub, J. (1992). Long-term folate deficiency alters folate content and distribution differentially in rat tissues. *J. Nutr.* **122**(4), 986–991.

Wagner, C. (1995). Biochemical role of folate in cellular metabolism. *In* "Folate in Health and Disease" (L. B. Bailey, ed.), pp. 23–42. Marcel Dekker Inc., New York.

Wagner, C., Briggs, W. T., and Cook, R. J. (1985). Inhibition of glycine N-methyltransferase activity by folate derivatives: Implications for regulation of methyl group metabolism. *Biochem. Biophys. Res. Commun.* **127**(3), 746–752.

Wainfan, E., and Poirier, L. A. (1992). Methyl groups in carcinogenesis: Effects on DNA methylation and gene expression. *Cancer Res.* **52**(7 Suppl.), 2071s–2077s.

Wainfan, E., Dizik, M., Stender, M., and Christman, J. K. (1989). Rapid appearance of hypomethylated DNA in livers of rats fed cancer-promoting, methyl-deficient diets. *Cancer Res.* **49**(15), 4094–4097.

Wald, D. S., Law, M., and Morris, J. K. (2002). Homocysteine and cardiovascular disease: Evidence on causality from a meta-analysis. *BMJ* **325**(7374), 1202.

Wang, X. L., Duarte, N., Cai, H., Adachi, T., Sim, A. S., Cranney, G., and Wilcken, D. E. (1999). Relationship between total plasma homocysteine, polymorphisms of homocysteine metabolism related enzymes, risk factors and coronary artery disease in the Australian hospital-based population. *Atherosclerosis* **146**(1), 133–140.

Wang, H. X., Wahlin, A., Basun, H., Fastbom, J., Winblad, B., and Fratiglioni, L. (2001). Vitamin B (12) and folate in relation to the development of Alzheimer's disease. *Neurology* **56**(9), 1188–1194.

Wasson, G. R., McGlynn, A. P., McNulty, H., O'Reilly, S. L., McKelvey-Martin, V. J., McKerr, G., Strain, J. J., Scott, J., and Downes, C. S. (2006). Global DNA and p53 region-specific hypomethylation in human colonic cells is induced by folate depletion and reversed by folate supplementation. *J. Nutr.* **136**(11), 2748–2753.

Waterland, R. A., and Jirtle, R. L. (2004). Early nutrition, epigenetic changes at transposons and imprinted genes, and enhanced susceptibility to adult chronic diseases. *Nutrition* **20**(1), 63–68.

Waterland, R. A., and Michels, K. B. (2007). Epigenetic epidemiology of the developmental origins hypothesis. *Annu. Rev. Nutr.* **27**, 363–388.

Weir, D. G., and Scott, J. M. (1998). Colonic mucosal folate concentrations and their association with colorectal cancer. *Am. J. Clin. Nutr.* **68**(4), 763–764.

Whitelaw, E., and Martin, D. I. (2001). Retrotransposons as epigenetic mediators of phenotypic variation in mammals. *Nat. Genet.* **27**(4), 361–365.

Wilcken, D. E., and Wilcken, B. (2001). Historical overview and recent perspectives. *In* "Homocysteine in health and disease" (R. Carmel and D. W. Jacobsen, eds.), pp. 1–8. Cambridge University Press, Cambridge.

Wilson, M. J., Shivapurkar, N., and Poirier, L. A. (1984). Hypomethylation of hepatic nuclear DNA in rats fed with a carcinogenic methyl-deficient diet. *Biochem. J.* **218**(3), 987–990.

Wilson, A. S., Power, B. E., and Molloy, P. L. (2007). DNA hypomethylation and human diseases. *Biochim. Biophys. Acta* **1775**(1), 138–162.

Wolff, G. L., Kodell, R. L., Moore, S. R., and Cooney, C. A. (1998). Maternal epigenetics and methyl supplements affect agouti gene expression in A^{vy}/a mice. *FASEB J.* **12**(11), 949–957.

Yi, P., Melnyk, S., Pogribna, M., Pogribny, I. P., Hine, R. J., and James, S. J. (2000). Increase in plasma homocysteine associated with parallel increases in plasma S-adenosylhomocysteine and lymphocyte DNA hypomethylation. *J. Biol. Chem.* **275**(38), 29318–29323.

Zapisek, W. F., Cronin, G. M., Lyn-Cook, B. D., and Poirier, L. A. (1992). The onset of oncogene hypomethylation in the livers of rats fed methyl-deficient, amino acid-defined diets. *Carcinogenesis* **13**(10), 1869–1872.

Epigenetic Biomarkers

5

Epigenetic Alterations as Cancer Diagnostic, Prognostic, and Predictive Biomarkers

Dajun Deng, Zhaojun Liu, and Yantao Du

Key Laboratory of Carcinogenesis and Translational Research (Ministry of Education), Peking University School of Oncology, Beijing Cancer Hospital and Institute, Fu-Cheng-Lu, Haidian District, Beijing, 100142, PR China

ABSTRACT

Alterations of DNA methylation and transcription of microRNAs (miRNAs) are very stable phenomena in tissues and body fluids and suitable for sensitive detection. These advantages enable us to translate some important discoveries on epigenetic oncology into biomarkers for control of cancer. A few promising epigenetic biomarkers are emerging. Clinical trials using methylated CpG islands

0065-2660/10 $35.00
DOI: 10.1016/S0065-2660(10)71005-1

of *p16*, *Septin9*, and *MGMT* as biomarkers are carried out for predication of cancer development, diagnosis, and chemosensitivity. Circulating miRNAs are promising biomarkers, too. Breakthroughs in the past decade imply that epigenetic biomarkers may be useful in reducing the burden of cancer.

I. INTRODUCTION

The ultimate goal of science is to benefit human society. Researches on molecular biology and cancer in the past 40 years have greatly improved our understanding on life science, especially on molecular carcinogenesis. Although it is said that we have lost the war against cancer, we still get something from the war. Based on accumulation of knowledge on carcinogenesis and tumor biology, some breakthroughs on prevention, diagnosis, and therapy of cancer have been obtained.

Epigenetic mechanisms play crucial roles in cell differentiation, embryo development, and host adaptations to environmental factors. Epigenetic alterations also play undisputed roles in initiation, promotion, and progression of cancers. DNA methylation is one of the most stable epigenetic modifications in mammalian cells and presents stably in various kinds of stored tissue samples, including formalin-fixed and paraffin-embedded tissue blocks. Moreover, alterations of DNA methylation can be sensitively detected even if they are present only in a very limited number of cells in the tested tissues. Using DNA methylation as a predictor to represent transcription availability of a target gene has showed advantages over other kinds of assays such as quantitative RT-PCR, Western blot, and immunohistochemistry (IHC), which are used to analyze expression status of genes in the majorities of cells in the tissue samples. These advantages enable us to obtain DNA methylation signatures of a few critical cells such as cancer stem cells related with malignant transformation of precancerous lesions, metastasis and recurrence of cancer, chemosensitivity, and patients' overall survival. Although studies on relationships between epigenetics and cancer have been initiated for only about a decade, some important epigenetic biomarkers including methylated *H19/Igf2*, *hMLH1*, *MGMT*, *p16*, and *Septin9* are emerging for early diagnosis, prognosis, and therapy response of cancers and their related diseases (Fig. 5.1). Circulating microRNA (miRNA) may be another kind of promising epigenetic biomarker because of its high stability in tissues and body fluids and availability of convenient detection assays. In this review, we summarized current status of development and investigations of potential epigenetic biomarkers and related challenges.

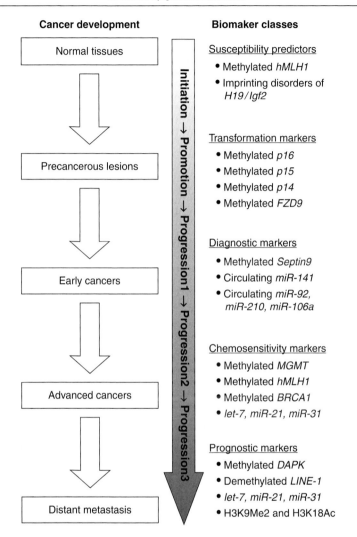

Figure 5.1. Illustration of multiple stages of cancer development and epigenetic biomarker classes used in the corresponding stages. The clinical and pathologic changes during carcinogenesis are displayed in the rectangles. The long gradient dark arrow indicates that carcinogenesis is a long-term continuous process. The representative candidates of each class of epigenetic biomarker are listed along the process. Different class biomarkers overlap with each other.

II. DNA METHYLATION

There are numerous aberrant DNA methylations in cancer tissues, compared with methylation status of these genes in their corresponding normal-appearing tissues. These methylation changes may result from diverse reasons. Some of them are reflections of morphological changes in tumor tissues such as loss of glands, immigration of inflammation cells, and metaplasia of epithelium. Some of them reflect the adaptation of cells in tumor tissues to niche hypoxic and malnutritional conditions such as angiogenesis, anaerobic glycolysis, and cachexia. These changes are the consequences of or in association with carcinogenesis, thus called as "passenger" alterations. Most alterations of DNA methylation in tumor tissues are "passenger" events. There is another kind of alteration of DNA methylation called as "driver" event causing or accelerating carcinogenesis. The "driver" events mainly contribute to dedifferentiation, active proliferation, and antiapoptosis of cancer cells. It is difficult to distinguish a driver alteration from a passenger one in many cases. A driver change may become an optimal biomarker. Besides driver events, a passenger alteration could also be used as cancer biomarker if its sensitivity and specificity are good enough.

A. Predictors for development of cancers

Development of cancer involves multiple processes including initiation, promotion, and progression. Human cancers in adults mainly develop from precancerous lesions such as epithelial dysplasia/intraepithelial neoplasia, intestinal metaplasia, Barrett's esophagus, myelodysplasia syndrome (MDS). Risk of cancer for epithelial dysplasia is about 100 times of that for normal epithelium. However, most precancerous lesions will persist or regress and only a small proportion of them will progress to malignant diseases. Because it is impossible to identify the malignant potential of these lesions on histopathological grounds alone, development of biomarker predicting the potential of progression of these lesions is eagerly awaited. In addition, although epigenetic inactivation of *hMLH1* and imprinting disorders of *H19/Igf2* are sporadic or heritable events that lead to sporadic or familial cancers, methylation tests of these genes have been offered to high-risk subjects now.

1. Methylated *p16* (*Cdkn2a*, *Ink4a*, *Mts-1*)

Tumor suppressor gene (TSG) *p16* locates in the 9p21 region. This region also includes other TSG genes such as *Arf*, *p15*, and *MTAP* and two long noncoding RNAs (ncRNAs) *p16AS* and *p15AS*. P16 protein regulates a critical cell-cycle regulatory pathway P16-CDK4-RB, which plays important roles in controlling of stem cell self-renewal, cell life span, and cell senescence. Loss of this region is the most frequent deletion event in cancer cell lines and cancer tissues (Beroukhim *et al.*, 2010; Bignell *et al.*, 2010; Merlo *et al.*, 1995). It is reported that loss of heterozygous of the *p16* locus increases risk of cancer in oral precancerous lesion (Mao *et al.*, 1996). However, genetic changes are not the only mechanism leading to *p16* inactivation in cancers. Methylation of the CpG island around transcription starting site (TSS) blocks transcription of *p16* completely (Merlo *et al.*, 1995). Frequency of inactivation of *p16* by DNA methylation is even higher than that by genetic changes in many cancers. In gastric carcinomas, for example, the frequency of *p16* inactivation by homozygous deletions ranges from 0% to 9%; by point mutations, from 0% to 2%; by methylation, from 32% to 42% (Günther *et al.*, 1998; Lee *et al.*, 1997; Luo *et al.*, 2006; Shim *et al.*, 2000; Takaoka *et al.*, 1997; Toyota *et al.*, 1999b; Wu *et al.*, 1998). In fact, *p16* is one of the most frequently inactivated genes by DNA methylation (Esteller, 2002).

In addition to cancer tissues, *p16* methylation can also be detected in precancerous lesions and inflammatory lesions in the stomach, colon, lung, liver, oral rinses, etc. (Dong *et al.*, 2009; Jicai *et al.*, 2006; Kukitsu *et al.*, 2008; Licchesi *et al.*, 2008; López *et al.*, 2003). In experimental animals, *p16* methylation can be observed with methylation-specific PCR (MSP) at cancerous and precancerous lesions in the lung, liver, and glandular stomach of rats induced by chemical carcinogens (Bai *et al.*, 2003; Belinsky *et al.*, 1998, 2002; Pogribny and James, 2002). In a large population based study, it is concluded that *p16* methylation is very prevalent in gastritis tissues and correlates with *Helicobacter pylori* infection significantly (Dong *et al.*, 2009). These phenomena indicate that *p16* methylation is an early frequent event during carcinogenesis. Recently, we find that methylated CpG sites are mainly distributed at the exon-1 within *p16* CpG islands in gastritis lesions with a methylation-enrichment PCR and clone sequencing. Location of methylated CpG sites extends to the *p16* promoter region in gastric carcinoma tissues subsequently (Lu *et al.* unpublished).

Results of several nested case–control studies indicate that *p16* methylation correlates with malignant transformation of epithelial precancerous lesions. In a population based nested case–control study in China, Sun *et al.* (2004) reported that *p16* methylation was detectable by MSP in 24% (5/21) of low-grade gastric dysplasia lesions that progressed to gastric carcinomas within 5 years. However, *p16* methylation was not detected in any of 21 control lesions that remain as dysplasia at the same grade during the follow-up ($P < 0.048$, two-sides). The sensitivity 24% is low, but the specificity is up to 100% in that study. In a nested case–control study in the United States, Belinsky *et al.* (2006) investigated relationship between methylation status of 14 genes in proximal sputum samples by MSP and risk of lung cancer, and found that *p16* methylation was the only significant gene to increase lung cancer risk [40% (39/98) of cases vs. 27% (25/92) of controls, adjusted odds ratio 1.9 (95% of CI, 1.0–3.7)].

In another nested case–control study in the United States, Schulmann *et al*. (2005) analyzed correlation of methylation level of eight genes by quantitative MSP and progression of Barrett's esophagus among 53 patients, and found that methylation of *p16*, *HPP1*, and *RUNX3* in esophagus biopsies correlated with esophagus cancer risk significantly. The predictive value of *p16* methylation on progression of Barrett's esophagus is also reported by others (Wang *et al*., 2009b), and further validated in a multicenter nested case–control study among 195 patients with progressive or nonprogressive Barrett's esophagus (Jin *et al*., 2009).

Oral epithelial dysplasia (OED) is one kind of precancerous lesion of oral cancer. Oral cancer risk is significantly high among elder patients. To validate the application value of *p16* methylation as a biomarker to predict malignant potential of the lesions ultimately, a prospective cohort study should be carried out. Hall *et al*. (2008) reported that *p16* methylation positive rate by a methylation-enrichment pyrosequencing was significantly higher in OED progressed to oral squamous cell carcinoma than that without progression in British (8 of 14 patients vs. 2 of 24, $P < 0.004$; age information not provided). Furthermore, in a prospective cohort study in China, Cao *et al*. (2009) found that the frequency of malignant progression of mild/moderate OED with *p16* methylation by MSP was significantly higher than that without *p16* methylation within the follow-up time ranged from 3 months to 124 months [44% (14/32) patients vs. 17% (8/46): adjusted odd ratio $= 3.7$; $P = 0.01$]. Among patients older than 60 years or with moderate OED, oral cancer risk of the OED lesions with *p16* methylation was further increased (adjusted odd ratio is 12.0 and 15.6, respectively) in their study. The sensitivity of *p16* methylation to predict malignant progression of OED is up to 77% at 73% specificity for patients over 60 years old, and 86% at 72% specificity for patients with moderate OED. Shorter cancer-free survival time was also observed among patients with OED containing methylated *p16*. Taken together, the association between *p16* methylation and malignant potential of OED is confirmed in two prospective cohorts. It is recommended that these patients with OEDs be offered a detection of *p16* methylation (Cao *et al*., 2009). It is necessary to study if similar associations could be obtained prospectively among patients with epithelial dysplasia in the lung, esophagus, and other organs. A clinical diagnosis kit for detection of *p16* methylation in human tissues is under development in a Chinese biotech company.

A specific attention should be paid to HPV-infection-related cancers. P16 protein is weakly expressed in most normal tissues generally. It is well known that HPV-E7 inhibits function of RB protein by formation of E7-RB complex. Khleif *et al*. (1996) reported that expression of the viral E7 resulted in P16 overexpression through inhibition of RB. Interestingly, both *p16* methylation and P16 overexpression can be observed during the development of HPV-related cervical cancers (Nehls *et al.*, 2008; Queiroz *et al.*, 2006; Xu *et al.*, 2007). Using *in situ* MSP and IHC assays, it is found that *p16* methylation can be observed only in neoplastic cells of the cervix and associated with loss of P16 expression (Nuovo *et al.*, 1999). Moreover, *p16* methylation occurs heterogeneously within early cervical tumor cells without HPV-E7 transcripts. In advanced cervical cancers, the majorities of cells have *p16* methylation, lack P16, and no longer express HPV-E7. These data suggest that HPV-E7 expression may induce P16 overexpression at initiation stage and that *p16* inactivation may favor the growth of transformed cells during evolution of advanced cancers from early lesions. Because P16 overexpression promotes cell senescence, it needs to investigate whether *p16* is methylated consequently in the cells with previous P16 over-expression, or whether they are independent responses of different cells to exposure of environmental factors. In a prospective cohort among 113 Chinese females with mild cervical dysplasia (CIN1), correlation between baseline *p16* methylation and progression of CIN1 is not observed during 6 years' follow-up (Hu *et al.*, 2009), though HPV infection correlates with progression of these CIN1 lesions well in this cohort (Qiao *et al.*, unpublished). However, it is reported that P16 overexpression correlates with progression of CIN1 to CIN3 significantly (Negri *et al.*, 2004). These data suggest that molecular changes directly induced by HPV infection may be much stronger factors than *p16* methylation in determining progression of cervical dysplasia.

Furthermore, *p16* methylation correlates with persistent HBV infection consistently (Jicai *et al.*, 2006; Matsuda *et al.*, 1999). Deletion of 13q containing the *Rb* locus is a frequent event in HBV-related hepatocarcinomas (Nishida *et al.*, 1994; Zhang *et al.*, 1994). It is not clear whether deletion of the *Rb* locus results in P16 overexpression during the HBV-related hepatocarcinogenesis. It is reported that expression of anion exchanger 1 (AE1) and its interaction with P16 sequestrates P16 in the cytoplasm in gastric and colonic adenocarcinomas (Shen *et al.*, 2007). Downregulation of AE1 restores function of P16 to inhibit cell proliferation. P16 is often overexpressed in islets and constrains islet prolif-eration and regeneration in an age-dependent manner (Krishnamurthy *et al.*, 2006). We also find overexpression of P16 proteins in the nucleus of epithelial cells in normal minor salivary gland tissue (Fig. 5.2A and B). However, the overexpressed P16 is observed only in the cytoplasm of oral cancer cells (Fig. 5.2C). It suggests that overexpression of P16 in the nucleus is one kind of normal physiological phenomena in some cell types, whether connected to cell senescence or not. But inactivation of *p16* by methylation or sequestration of P16 in the cytoplasm is adaption of target cells to environmental factors and may stably promote the affected cells to escape from senescence and lead to immortalization.

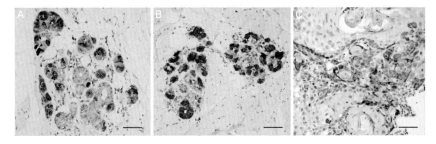

Figure 5.2. Immunohistochemical staining of P16 protein in oral minor salivary gland tissues adjacent to oral squamous cell carcinomas. (A) P16 overexpression in the nucleus in oral minor salivary gland sample from patient A aged 70 years; (B) P16 overexpression in the nucleus in oral minor salivary gland sample from patient B aged 38 years; (C) P16 overexpression in the cytoplasm in part of oral squamous cancer cells from the patient B; black bar, 50 μm of length.

2. Methylated *hMLH1*

hMLH1 is a mismatch repair gene located at the 3p21.3 region. Heterozygous *hMLH1* mutations result in hereditary nonpolyposis colorectal cancer (HNPCC; Papadopoulos *et al.*, 1994). Methylation of CpG islands within the *hMLH1* promoter is found in most sporadic primary colorectal cancers (CRCs) with high microsatellite instability (MSI-positive) and often associated with loss of hMLH1 protein expression (Herman *et al.*, 1998). Similar phenomena are also observed in the MSI-positive endometrial cancers and malignant insulinomas (Mei *et al.*, 2009; Simpkins *et al.*, 1999).

Region-A of CpG island within the *hMLH1* promoter is partially methylated in a number of cancer cell lines and normal tissues with hMLH1 expression (Fig. 5.3A; Deng *et al.*, 1999; Menigatti *et al.*, 2009). However, methylation at Region-C within the core promoter region near TSS is observed only in the *hMLH1*-silenced cell lines. The Region-C methylation and its correlation with gene expression are also detected in CRC tissues *in vivo* (Deng *et al.*, 2002). In normal colorectal mucosa, the Region-A methylation is an age-related event, whereas the Region-C methylation is not detectable (Menigatti *et al.*, 2009). In CRC tissues, however, the Region-C methylation is detectable and relates to age and sex, too (Poynter *et al.*, 2008). Unfortunately, it is not well recognized that only methylation of Region-C relates to the epigenetic inactivation of *hMLH1*. Primers used in the majorities of previous published works are not designed to detect the Region-C methylation (Capel *et al.*, 2007). Thus, information on *hMLH1* methylation in those works cannot connect with the epigenetic inactivation of *hMLH1* completely.

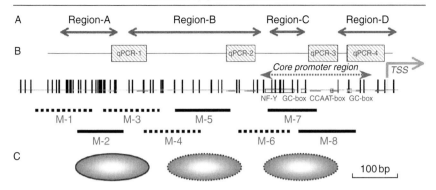

Figure 5.3. Nucleosome positioning of CpG island within the *hMLH1* promoter. Distribution of CpG sites (short bars) and location of the core promoter region are marked. (A) Four regions proposed by Deng *et al.* (1999); the core promoter region covers eight key CpG sites within Region-C; (B) nucleosome signals in *hMLH1*-methylated RKO cells reported by Lin *et al.* (2007b); (C) nucleosome signals in *hMLH1*-active MGC803 and *hMLH1*-inactive RKO cells by Bai *et al.* (2008). TSS, transcription start site; M1~M8, amplicons of DNA fragments used for characterization of nucleosome-free regions (dashed lines) and core nucleosome regions (solid lines); solid ellipsoid, nucleosome existing in both MGC803 and RKO cells; dashed ellipsoid, nucleosomes existing in RKO cells only.

Recently, the exact nucleosome occupancy within the CpG island of *hMLH1* is characterized (Fig. 5.3B and C; Bai *et al.*, 2008, Lin *et al.*, 2007b). DNA fragment within the proximal nucleosomes overlaps with the core promoter region including Region-C and Region-D, in which nucleosome signals are detectable by quantitative PCR only in the *hMLH1* fully methylated RKO cell line, but not in the *hMLH1*-active MGC803 cell line. These results suggest that DNA methylation may be nucleosome dependent and that methylation of the core promoter region should account for epigenetic repression of *hMLH1* transcription.

Moreover, Deng *et al.* (2003a) reported that only fully methylated-*hMLH1* alleles were detectable by denatured-HPLC among cancer cells in a xenograft from a fresh primary gastric carcinoma in which both methylated and unmethylated-*hMLH1* alleles were detectable. That *hMLH1* is fully methylated in all cancer cells in the xenografts implies that inactivation of *hMLH1* by methylation should favor the growth of cancer cells *in vivo* and be an initiation event in carcinogenesis.

Most importantly, constitutional epimutation without sequence change is involved in abolishing expression of *hMLH1*. Unlike above biallelic methylation silence of *hMLH1* in gastric cancer cells, methylation of one allele of *hMLH1* in somatic cells throughout the body is observed in 24 individuals who have a predisposition for the development of HNPCC or endometrial cancer (Hitchins and Ward, 2009; Hitchins *et al.*, 2007; Suter *et al.*, 2004). Transmission of the epimutation from two mothers to their sons is observed (but is erased in their spermatozoa). Although *hMLH1* epimutation is likely reversible between generations, it is advised that family members of the *hMLH1* epimutation carriers, especially the first degree relatives of probands, are counseled and offered a molecular testing to determine their carrier status.

3. Imprinting disorders of the *H19/Igf2* locus and other genes

The phenomenon, one allele active and another allele silenced in tissues, is called gene imprinting. That both paternal and maternal alleles of imprinted genes are expressed or not expressed in a tissue is called loss of imprinting (LOI) or gain of imprinting (GOI). LOI and GOI can be induced by genetic and epigenetic alterations, which lead to a number of pediatric diseases, such as *H19/Igf2*-related Silver–Russell syndrome (SRS) and Beckwith–Wiedemann syndrome (BWS), *SNRPN/LIS1*-related Prader–Willi syndrome (PWS), and Angelman syndrome (AS; Gurrieri and Accadia, 2009; Weksberg *et al.*, 2010).

Both *H19* and *Igf2* locate in the 11p15.5 region. In normal adult human tissues, maternal *H19* allele is actively transcribed, whereas the paternal allele is silenced by methylation. In contrast, paternal allele of *Igf2* is active; the maternal allele is inactivated by methylation (Rachmilewitz *et al.*, 1992; Rainier *et al.*, 1993). Generally, long ncRNAs will target to sequences on their antisense strands such as *Xsix* ncRNA to *Xist*, *p15AS* to the *p15* promoter, and *Luc7L mutant* ncRNA to the *HbA2* promoter (Lee and Lu, 1999; Shibata *et al.*, 2008; Tufarelli *et al.*, 2003; Yu *et al.*, 2008b). Maternal *H19* encodes a 2.3-kb untranslating ncRNA that is antisense of *Igf2* (Pfeifer *et al.*, 1996). Whether the paternal *Igf2* sense strand or *H19* antisense strand is the direct target of the maternal *H19*, ncRNA is unknown, though a number of potential downstream targets of the *H19* ncRNA are identified (Matouk *et al.*, 2007). *Igf2* encodes a 67-amino acid peptide mitogen. The neighboring genes *H19* and *Igf2* share the same enhancer. The paternal-specific methylation region between *H19* and *Igf2* allele is likely required for the imprinting of both genes because a deletion within the methylation region results in LOI of both of them (Thorvaldsen *et al.*, 1998). LOI and GOI of *H19/Igf2* cause SRS, a growth retardation disorder, or BWS, an overgrowth syndrome. Wilms tumor is a common feature of BWS. In patients with BWS and Wilms tumors, both paternal and maternal alleles are methylated, resulting in biallelic activation of *Igf2* and biallelic silencing of *H19* (Riccio *et al.*, 2009). Upregulation of *H19* can also be observed in cancers of the bladder and liver (Ariel *et al.*, 1997; Takai *et al.*, 2001; Wu *et al.*, 2008). BWS occurs sporadically in about 85% of BWS cases, but familial transmission occurs in 15% of the cases. Thus, in the cases of prenatal testing, especially for families with the history of child with severe manifestations of BWS, cytogenetic, and epigenetic analysis of amniocytes can be undertaken (Weksberg et al., 2010). This strategy is also suitable for families with the history of PWS and AS.

4. Other potential predictive methylation biomarkers

MDS is a clonal hematological disorder disease that will transform into acute myeloid leukemia (AML) at frequency of 10%–35%. p15 Methylation can be detected in 75% of cases of AML evolved from MDS and 30%–50% of cases with high-risk MDS (Aggerholm et al., 2006; Aoki et al., 2000; Chen and Wu, 2002; Tien et al., 2001; Vidal et al., 2007). It has been reported that p15 methylation in MDS is correlated with blast bone marrow involvement and increases with disease evolution toward AML (Quesnel et al., 1998; Tien et al., 2001). Recently, it is found that FZD9 methylation is another independent predictor of prognosis of MDS/AML (Jiang et al., 2009). Attractively, DNA methylation inhibitor 5-aza-2'-deoxycytidine (decitabine) is an effective chemical used for treatment of MDS (Musolino et al., 2010; Shen et al., 2010; Steensma et al., 2010). Analysis of methylation status of p15 and FZD9 may be useful for prediction of the prognosis and chemosensitivity of MDS.

As mentioned above, methylation of HPP1 and RUNX3 correlates with progression of Barrett's esophagus in nested case–control study. Moriyama et al. (2007) reported that p14 methylation was detected in 100% (5/5) of colorectal dysplasia lesions from patients with ulcerative colitis and that dysplasia was detected in two of eight patients with p14 methylation, but in none of 28 patients with no p14 methylation during follow-up surveillance ($P = 0.044$). More possible predictors need to be screened out for different precancerous lesions.

B. Early diagnosis of cancers

The challenge remains in early detection or screen for cancer. There are many feasible physical approaches to screen the majority of cancers at early stages. However, most physical examinations are expensive, and even invasive. In population, because the examination costs outweigh the healthy benefits, most physical examinations cannot be widely used to screen cancers. They are only offered to a small proportion of patients. Thus, more efficient, convenient, noninvasive, and economic molecular tests are eagerly awaited to benefit the main proportion of people. Some aberrant DNA methylation may become such biomarkers.

DNA methylation signatures are tissue and cell type specific. Tissue samples should be used in DNA methylation tests for detection of cancers generally. For example, DNA methylation markers in tissue biopsies can be used for diagnosis of cancer at early or precancerous stages. Methylated DNA markers in blood, milk, saliva, gastric juice, urine, feces, and sputum, etc. containing the exfoliated cancerous/precancerous cells and noncancerous host cells would be very useful for early detection of cancers. Although DNA methylation in the surgical resected tissues is not a good diagnostic marker, it may be an optimal predictor of prognosis, metastasis, and chemosensitivity of cancers. Moreover, circulating *Septin9* methylation may become a biomarker for screening CRC. A few potential methylation candidates are emerging for detection of cancers at early stages and prediction of prognosis of cancers now.

1. Methylated *Septin9*

As an evolutionarily conserved group of GTP-binding proteins involved in the cell-cycle control, polarity determination, cytokinesis, and other essential cellular functions (Hall and Russell, 2004), Septins are expressed extensively in eukaryotic cells except plant cells (Hall *et al.*, 2005). Thirteen *Septin* genes have been identified within the human genome. The previous researches have proved that there are mutiple-splicing transcript variants for most *Septin* genes. For example, 37 splice patterns have been characterized among transcripts of human *Septin9*, according to the summary on the GeneCards web site (Fig. 5.4). Furthermore, Septin can interact with other Septin(s), as well as with components of the cytoskeleton such as Actin and Tubulin. The genomic, transcriptional, and isoform complexity has hindered progress in our understanding of these genes (Russell and Hall, 2005).

Septin9 at 17q25.3 is expressed ubiquitously, with the isoform showing tissue-specific expression. Septin9 is overexpressed in diverse human tumors. Some of spliced forms of transcripts of *Septin9* are associated with cancer (Hall *et al.*, 2005; McDade *et al.*, 2007; Scott *et al.*, 2005). For example, *Septin9-v4* (v-4 or isoform-d; NM_001113495.1) is the predominant transcript in normal cells while *Septin9-v4** is the predominant one in tumors (McDade *et al.*, 2007). Both *Septin9-v4* and *Septin9-v4** have their own 5'UTRs and are translated with different efficiencies, although they encode the identical proteins. The human *Septin9* locus has also identified as a common site for allelic imbalance in sporadic ovarian and breast cancer (Kalikin *et al.*, 2000; Russell *et al.*, 2000). However, no direct evidence is obtained to indicate Septin9 functioning as an oncogene.

Interestingly, based on the data of methylation-specific arbitrarily primed PCR and methylated CpG island amplification methods, it has proved that methylation of *Septin9* CpG island around the exon-4 in CRC tissues is

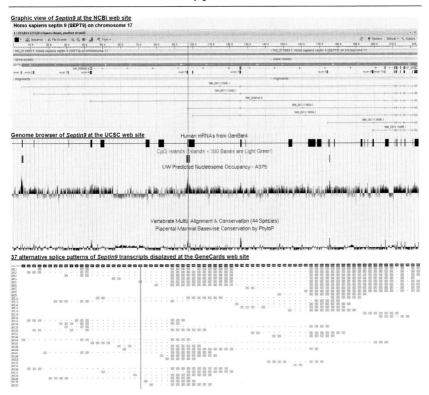

Figure 5.4. Distribution of 18 exons, various transcript isoforms, three CpG islands, predicted nucleosomes, and conservation of the *Septin9* gene within the 17q25.3 locus. The yellow (or grey) vertical line marks location of 8 CpGs tested with the HeavyMethyl methylation assay. This image is mainly composed of the graphic views of the human *Septin9* gene at three web sites (http://www.ncbi.nlm.nih.gov, http://genome.ucsc.edu, and http://www.genecards.org) (For interpretation of the references to color in this figure legend, the reader is referred to the Web version of this chapter.).

significantly higher than that in the corresponding normal colon tissues (Lofton-Day *et al.*, 2008). Moreover, *Septin9* methylation is detected in 69% of plasma samples from CRC patients ($N = 133$), but not detected in 86% of the control samples from healthy individuals ($N = 179$) by quantitative PCR. Using a HeavyMethyl *Septin9* DNA methylation assay, Grützmann *et al.* (2008) identified 48% of CRCs at 93% specificity in the training study (252 patients and 102 controls) and 58% of CRCs at 90% specificity in the testing study (126 patients and 183 controls). In addition, the methylated *Septin9* positive rates for plasmas from patients with the other cancers (11/96) and noncancerous conditions (41/315) were low. This result indicates that circulating methylated *Septin9* might be a CRC-specific marker. Recently, an optimized HeavyMethyl *Septin9* DNA methylation assay called ^{m}*SEPT9* was developed for plasma at Epigenetics Inc. The assay identifies 72% of CRCs at 93% specificity in the training study (97 cases and 172 controls) and 68% of CRCs at 89% specificity in the testing study (90 cases and 155 controls; deVos *et al.*, 2009). In that study, the sensitivity at 91% specificity under the conditional qualitative algorithm is correlated with the CRC stages: 46% for stage-I ($T_{1,2}N_0M_0$, 19/41), 77% for stage-II ($T_{3,4}N_0M_0$, 60/ 78) and stage-III ($T_{1,2,3,4}N_{1,2}M_0$, 47/61), and 100% for stage-IV ($T_{1,2,3,4}N_{0,1,2}M_1$, 7/7). These data indicate that detectable plasma methylated *Septin9* is likely associated with the CRC size more than the lymph node involvement. Moreover, nearly half of CRCs at the stage-I could be identified with the noninvasive test. It might be useful for screening CRCs at early stages. From what have been showed above, we may draw the conclusion that plasma methylated *Septin9* would be a specific biomarker for CRCs.

It was reported that the rate of polyp detection (>1 cm) was $\sim 20\%$ with the plasma methylated *Septin9* test (Grützmann *et al.*, 2008). Tänzer *et al.* (2010) also reported that plasma methylated *Septin9* positive rate is 9% of controls (3/33), 29% of patients with colorectal precancerous lesions (27/94), and 73% of CRC patients (24/33). Whether plasma methylated *Septin9* could be used to predict malignant transformation of colorectal precancerous lesions is worth to be studied further with nested case–control and cohort follow-up studies. The sensitivity and specificity of a methylated *Septin9* test in tissue biopsies may be greatly improved in the cases of precancerous lesions.

Generally, DNA methylation around TSS represses gene transcription epigenetically. There are three CpG islands, which locate near TSS of different transcripts of *Septin9* (Fig. 5.4). Methylation status of the region [chr17:75369567– 75369657 in the 2009 assembly; negative strand, including 8 CpG sites] within CpG island around the exon-5 of *Septin9* is analyzed in the HeavyMethyl *Septin9* DNA methylation assay (deVos *et al.*, 2009). This region is mapped on 5′UTR of the transcript variant 2 [v-2 or isoform-b; NM_001113493.1]. Whether the methylation of these eight CpGs represses transcription of the transcript is unknown. It is necessary to investigate the biological functions of Septin9 encoded by this transcript. Furthermore, it is also worth to confirm whether the methylated *Septin9* in plasma comes from CRC cells.

2. Other possible DNA methylation markers

Many candidates of DNA methylation markers, such as *hMLH1*, *p16*, *MGMT*, *IGF1*, and *ID4*, for diagnosis and prognosis of cancers are emerging, as reviewed by Kim *et al.* (2010) and McCabe *et al.* (2009). However, only few of them are validated in case–control and/or prospective cohort studies. Gonzalgo *et al.* (2003) found that postbiopsy urinary *GSTP1* methylation was detected in 58% (7/12) of patients with evidence of prostate adenocarcinoma at 76% specificity (14/21). Chen *et al.* (2005) reported that fecal methylated *vementin* was detectable in 46% (43/94) of CRC patients at 90% specificity (169/198). Tänzer *et al.* (2010) reported that 45% (22/49) of polyp was identified at 82% specificity (18/22) by plasma *ALX4* methylation test. There are numerous reports on association of promoter methylation in urine DNA and bladder cancer detection (Chan *et al.*, 2002; Friedrich *et al.*, 2004; Hoque *et al.*, 2006; Yu *et al.*, 2007b).

It is reported that more methylated *TMEFF2* and *NFGR* was detected in CRC tissue samples than in the corresponding normal samples (Lofton-Day *et al.*, 2008). In addition, plasma methylated *TMEFF2* and *NFGR* levels were also significantly higher in CRC patients than in healthy controls. However, little complementarity is observed between *Septin9* and *TMEFF2* and *NGR* methylation in that study. Similar phenomenon is also observed between *Septin9* and *ALX4* methylation for identification of colorectal polyp (Tänzer *et al.*, 2010). Thus, whether combinations of different methylation markers improve the performance of methylation tests should be investigated comprehensively.

According to a subset of cancers that exhibit widespread promoter methylation, the concept of CpG island methylator phenotype (CIMP) is created to describe the phenomenon (Issa, 2004; Toyota *et al.*, 1999a,b). CIMP phenotype is associated with MSI and *BRAF* mutations in colon cancer (Weisenberger *et al.*, 2006). Five methylation markers (*CACNA1G, IGF2, NEUROG1, RUNX3,* and *SOCS1*) are used to distinguish CIMP-high from CIMP-low CRCs. In a comprehensive statistical analysis, it is demonstrated that CIMP-high CRCs is independently associated with older age, proximal location, poor differentiation, MSI-high, *BRAF* mutation, and inversely with long interspersed nuclear element 1 (*LINE*-1) hypomethylation and β-catenin activation (Nosho *et al.*, 2008). It is not known whether CIMP is a universal phenomenon for diverse cancers or a cancer type-specific phenotype. Frequency of some methylation markers is much higher than that of CIMP-high in advanced CRCs. For example, the methylated *Septin9* positive rate in advanced CRC tissues is 77% for stage-III and 100% for stage-IV CRCs (deVos *et al.*, 2009). But the rate of CIMP-high is 11% (26/229) for stage-III and 15% (16/106) for stage-IV CRCs (Nosho *et al.*, 2008). Apparently, not all of the cancer-related methylation alterations are CIMP dependent. It is worth to study whether combination of CIMP-high and CIMP-independent methylation markers will improve the performance of methylation markers in early detection of cancer and prediction of prognosis of cancers.

C. Prediction of prognosis of cancers

Advanced cancer is still the main cause of human death because of refractory to chemotherapy/radiotherapy, recurrence, metastasis of tumors. Identification of reliable prognostic factors for cancer recurrence, metastasis, and patients' survival could have significant clinical importance of cancer patients' clinical managements. For example, cancer patients with high-risk of recurrence/metastasis would be appropriate candidates for novel adjuvant or chemoprevention strategies, whereas patients with low-risk could be avoided overtreatments.

The prognostic biomarkers primarily include histologic/cytologic atypia, cell proliferation, hormones expression, and molecular biomarkers, including epigenetic ones. Aberrant DNA methylation of CpG islands within 5-prime of genes occurs almost in every type of cancer and easy to measure. Potential of gene-specific DNA methylation as a predictor of important clinical features has been explored in number of studies now. It is likely that small sets of methylated genes could readily be harnessed as clinically useful biomarkers for some cancers (Figueroa *et al.*, 2010). Genes involved in cell-cycle regulation, tumor cell invasion, DNA repair, chromatin remodeling, cell signaling, transcription, and apoptosis could be aberrantly methylated and silenced in nearly all tumor types. These provide tumor cells with a growth advantage, increase their genetic instability (allowing them to acquire further advantageous genetic changes), and allow them to metastasize (Robertson, 2005). Most of these genes have been validated among different cancer patient groups as the prognostic markers such as death-associated protein kinase gene (*DAPK*), *LINE*-1, *p16*, and *APC*.

1. Methylated *DAPK*

DAPK located at chromosome 9q21.33 encodes a calcium/calmodulin-regulated serine/threonine protein kinase. It is composed of several functional domains, including a kinase domain, an ankyrin repeat domain, and a death domain (Bialik and Kimchi, 2006; Raveh *et al.*, 2000). Identified as a mediator of interferon-c-mediated apoptosis initially (Deiss *et al.*, 1995), DAPK takes part in many kinds of apoptosis-inducing pathways (Cohen *et al.*, 1999; Jang *et al.*, 2002; Raveh *et al.*, 2001) and the control of autophagy (Inbal *et al.*, 2002). It may play an important role during the initiation of neoplastic transformation (Michie *et al.*, 2010).

DAPK methylation has been reported in many tumor types. It is associated with the tumor recurrence, metastasis, and patients' survival. *DAPK* methylation has been frequently reported in non-small-cell lung cancer (NSCLC). Tang *et al.* (2000) reported that methylation of the *DAPK* promoter was detected by MSP in 44% (59/135) of patients with pathologic stage-I NSCLCs who had undergone curative surgery, and significantly associated with poorer overall and disease-specific survival. This result is confirmed by several independent studies listed below. Lu *et al.* (2004) performed a six-biomarker analysis in 94 patients with stage-I NSCLCs and a minimum follow-up period of 5 years. Results of multivariable analysis in their study indicate that *DAPK* methylation (by MSP), detected in 47% of NSCLCs, is an independent predictor of poorer disease-specific survival. Buckingham *et al.* (2010) also reported that *DAPK* methylation (by pyrosequencing) was associated with shorter time to recurrence in patients with stage-I and-II NSCLCs. Similar results are also reported by Kim *et al.* (2001) and Harden *et al.* (2003), although statistical significance is not obtained in their studies. In contrast, Brock *et al.* (2008) did not find the *DAPK* methylation difference (by nested MSP) between the stage-I NSCLC patients with recurrence ($N = 50$) and patients without recurrence ($N = 104$) prospectively. But methylated *p16*, *H-cadherin*, *RASSF1A*, and *APC* correlate with recurrence of these patients. Methodological difference in these studies might account for different results.

Besides the studies in NSCLC, the association between *DAPK* methylation and the outcomes of the cancer patients is also seen in many other types of cancers such as bladder cancer (Jarmalaite *et al.*, 2008; Tada *et al.*, 2002), advanced cervical cancer (Iliopoulos *et al.*, 2009), response to treatment and overall survival in B-cell lymphoma (Amara *et al.*, 2008), and others (Dansranjavin *et al.*, 2006; Kato *et al.*, 2008). Further prospective cohorts are necessary to validate if *DAPK* methylation is a good prognostic biomarker for these cancers.

2. Hypomethylated *LINE*-1

LINE-1 represents a family of nonlong terminal repeat retroposons that are interspersed throughout genomic DNA. *LINE*-1 elements present at over 500,000 copies in the human genome and comprise about 20% of the human genome. Most of them are truncated in the $5'$ region or mutated, making transposition impotent (Sassaman *et al.*, 1997). Only 3000–4000 copies are full-length and 30–100 are active retrotransposonses (Brouha *et al.*, 2003; Deininger *et al.*, 2003).

Usually *LINE*-1 elements are heavily methylated in normal human tissues. *LINE*-1 hypomethylation is a common characteristic of human cancers and accounts for a substantial proportion of the genomic hypomethylation observed in this disease. Hypomethylation in the promoter region of potent full-length *LINE*-1 elements causes transcriptional activation of *LINE*-1, resulting in transposition of the retroelement and chromosomal alteration such as deletion, amplification, and translocation. Therefore, the status of *LINE*-1 methylation may be a key factor linking global hypomethylation with genomic instability (Kazazian and Goodier, 2002; Symer *et al*., 2002).

Although the prevailing view is that global DNA hypomethylation changes occur extremely early in all human cancers, recent studies have showed that global DNA hypomethylation can occur very late and play roles in cancer progression. It is well recognized that genomic instability relates to the outcome of cancer patients (Choma *et al*., 2001; Kronenwett *et al*., 2006; Walther *et al*., 2008). The close relationship between *LINE*-1 methylation and genomic instability also suggests that it might be a possible prognostic marker in cancer.

Global hypomethylation and subsequent chromosomal instability are well-characterized features of advanced chronic myeloid leukemia (CML). Roman-Gomez *et al*. (2005) investigated the methylation status of the *LINE*-1 promoter in CML samples, including chronic phase ($N = 140$) and the blast crisis ($N = 47$) and found that *LINE*-1 hypomethylation activated sense/antisense transcription. In addition, *LINE*-1 hypomethylation was more frequent in blast crisis CML than in chronic phase CML (74.5% vs. 38%, $P < 0.0001$) and was a poor prognostic factor in terms of progression-free survival. They suggested that hypomethylation of various repetitive sequences in the genome might be a hallmark of CML progression. Then they studied the DNA methylation changes of the most important DNA repetitive elements, such as *LINE*-1, *Alu*, Satellite-alpha, and Satellite-2, during the progression of CML from chronic phase to blast crisis and achieved the same results (Roman-Gomez *et al*., 2008). Similar phenomena were also detected in ovarian and prostate cancers (Pattamadilok *et al*., 2008; Yegnasubramanian *et al*., 2008).

Moreover, results from two investigations using large-scale CRC and NSCLC patients should be more powerful. Using 643 CRC patients in two independent prospective cohorts, Ogino *et al*. (2008) found that *LINE*-1 hypomethylation was significantly associated with the increase in CRC-specific mortality and overall mortality. The association in their study was consistent across two independent cohorts and strata of clinical and molecular characteristic, such as sex, age, tumor location, stage, CIMP, microsatellite instability, and so on. In another study using 379 NSCLC patients, methylation levels of *LINE*-1, *APC*, *CDH13*, and *RASSF1* were quantitatively assessed in a large series of unselected NSCLC and matching normal lung tissues (Saito *et al*., 2010). In 364 NSCLC cases, *LINE*-1 and *APC* methylation were promising prognostic factors in both univariate and multivariate analyses. Of these 364 cases, 128 are stage-IA NSCLCs. Only *LINE*-1 methylation is an independent poor prognostic factor for the stage-IA NSCLCs. Accurate prediction of the likely outcome of stage-IA NSCLC patients using *LINE*-1 methylation as a biomarker may be very useful for their postoperative managements, including the adjuvant chemotherapy decisions and the frequency of follow-up examination.

3. Other candidates

Besides *DAPK* and *LINE*-1, there are a number of other methylation biomarkers which associate with cancer recurrence, metastasis, survival time of the patients (Table 5.1). Methylated *p16* and *APC* are additional examples of prognostic biomarkers in a panel of cancers. As mentioned above, *p16* methylation is a prognostic factor of NSCLCs (Brock *et al.*, 2008; Ota *et al.*, 2006). Methylated *p16* also relates to poor prognosis of esophageal adenocarcinomas and metastasis of gastric carcinomas (Brock *et al.*, 2003; Luo *et al.*, 2006). Circulating methylated *p16* is detected in 46% (12/26) of patients with the methylated-*p16* positive gastric carcinomas and disappeared after surgical dissection (Liu *et al.*, 2005). Methylated *APC* is a prognostic factor of NSCLCs too (Brock *et al.*, 2008; Saito *et al.*, 2010). High plasma levels of methylated *APC* DNA were statistically significantly associated with reduced survival of patients with esophageal adeno-carcinomas (Kawakami *et al.*, 2000).

    Alu elements are the most abundant gene in the human genome (about 1,000,000 copies per haploid) and comprise about 10% of the genome mass. Like *LINE*-1, hypomethylation of *Alu* elements is another indicator of lower genome stability, which is necessary for gene recombination and chromosome transloca-tion (Daskalos *et al.*, 2009). Retrotranscriptase encoded by *LINE*-1 helps retro-transcription and transposition of *Alu* elements into the genome (Dewannieux *et al.*, 2003). Total methylation content of *Alu* elements and *LINE*-1 sequences is highly correlated with global DNA methylation content (Ehrlich, 2002). How-ever, Xiang *et al.* (2010) did not find association of total *Alu* methylation level and clinical-pathological characteristics of gastric cancers ($N = 48$), although *Alu* hypomethylation was observed during gastric carcinogenesis.

    The association between the level of DNA methylation and poor prog-nosis is a recurrent theme in tumor biology. Multigene methylation profiling is also used to predict prognosis of cancers. Methylation signatures affect overall survival, tumor recurrence, and progression (Brock *et al.*, 2008; Hernandez-Vargas *et al.*, 2010; Rosenbaum *et al.*, 2005; Safar *et al.*, 2005). Combinations of methylation status of different gene promoters may result in stronger prognostic value than individual methylation markers, and improve sensitivity.

D. Prediction of sensitivity of cancers to therapies

The methylation silence of some particular genes is a potential predictor of the response of cancers to various treatments. Genes related to chemosensitivity are always involved in cell cycle, apoptosis, the metabolism of carcinogens, carcino-genesis, and DNA repair, all of which are involved in the development of cancer (Esteller, 2007; Herman and Baylin, 2003). The methylation-associated

Table 5.1. List of Methylated Genes Related to Prognosis of Cancer

Gene	Cancer type	Associated characteristics (case number)	Reference [PubMed ID]
APC	Nonsmall-cell lung cancer	Poor survival ($N = 99$)	Usadel et al. (2002) [11809682]
	Esophageal cancer	Poor survival ($N = 126$)	Kawakami et al. (2000) [11078757]
	Nonsmall-cell lung cancer	Poor survival ($N = 91$)	Brabender et al. (2001) [11429699]
ASC/TMS1	Nonsmall-cell lung cancer	Invasion ($N = 152$)	Machida et al. (2006) [16778195]
	Glioma	Poor survival ($N = 23$)	Stone et al. (2004) [15466382]
CD44	Prostate cancer	Metastasis ($N = 74$)	Kito et al. (2001) [11582589]
CD44 and PTGS2	Prostate cancer	Recurrence ($N = 60$)	Woodson et al. (2006) [17998819]
CDH1	Nonsmall-cell lung cancer	Longer survival ($N = 88$)	de Maat et al. (2007) [17960794]
	Gastric cancer	Recurrence ($N = 38$)	House et al. (2003) [14675710]
COX2	Gastric cancer	Improved survival ($N = 177$)	de Maat et al. (2007) [17971584]
DAPK	Nonsmall-cell lung cancer	Poor survival ($N = 135$)	Tang et al. (2000) [10995806]
	Nonsmall-cell lung cancer	Poor survival ($N = 185$)	Kim et al. (2001) [11313923]
	Nonsmall-cell lung cancer	Poor survival ($N = 90$)	Harden et al. (2003) [12684406]
	Nonsmall-cell lung cancer	Poor survival ($N = 94$)	Lu et al. (2004) [15542809]
	Nonsmall-cell lung cancer	No effect ($N = 154$)	Brock et al. (2008) [18337602]
	Nonsmall-cell lung cancer	Poor survival ($N = 132$)	Buckingham et al. (2010) [19795445]
DAPK and APC	Nonsmall-cell lung cancer	Poor prognosis ($N = 90$)	Harden et al. (2003) [12684406]
ER	Acute myeloid leukemia	Improved survival ($N = 268$)	Li et al. (1999) [10353741]
ESR1	Prostate cancer	Progression ($N = 41$)	Li et al. (2000) [10676656]
FHIT	Nonsmall-cell lung cancer	Poor survival ($N = 124$)	Maruyama et al. (2004) [15042681]
	Nonsmall-cell lung cancer	Vascular invasion ($N = 120$)	Tomizawa et al. (2004) [15541815]
	Gastric cancer	Node metastasis ($N = 75$)	Oshimo et al. (2004) [15386345]
	Melanoma	Advanced stage ($N = 122$)	Tanemura et al. (2009) [19223509]
GSTP1	Prostate cancer	Recurrence ($N = 74$)	Rosenbaum et al. (2005) [16322291]
	Prostate cancer	Advanced stage ($N = 118$)	Jerónimo et al. (2004) [15623627]
	Nonsmall-cell lung cancer	Poor prognosis ($N = 379$)	Saito et al. (2010) [20371677]

Gene	Cancer type	Outcome	Reference
HLTF	Colorectal cancer	Recurrence, poor prognosis (N = 106)	Herbst et al. (2009) [19282772]
hMLH1	Breast cancer	Metastasis and poor survival (N = 78)	Karray-Chouayekh et al. (2009) [19644562]
ID-4	Colorectal cancer	Poor survival (N = 92)	Umetani et al. (2004) [15569977]
IGFBP3	Nonsmall-cell lung cancer	Poor prognosis (N = 83)	Chang et al. (2002) [12473575]
LAMA3/B3/C2	Prostate cancer	Poor survival (N = 101)	Sathyanarayana et al. (2003) [14695140]
LINE-1 demethylation	Prostate cancer	Metastasis (N = 76)	Yegnasubramanian et al. (2008) [18974140]
p16	Colorectal cancer	Lymphatic invasion (N = 50)	Goto et al. (2009) [19311161]
	Gastric cancer	Metastasis (N = 82)	Luo et al. (2006) [16534497]
	Lung cancer	Metastasis (N = 29)	Seike et al. (2000) [11106248]
MGMT	Nonsmall-cell lung cancer	Poor survival (N = 90)	Brabender et al. (2003) [12538473]
	Glioma	Poor survival (N = 66)	Liu et al. (2010a) [20154338]
	Glioma	Better prognosis (N = 67)	Felsberg et al. (2009) [19861461]
PTGS2	Prostate cancer	Poor prognosis (N = 53)	Bastian et al. (2005) [15930345]
PTGS2	Prostate cancer	Recurrence (N = 164)	Yegnasubramanian et al. (2004) [15026333]
RARβ	Bladder cancer	Invasion (N = 58)	Jarmalaite et al. (2008) [18824877]
RASSF1	Nonsmall-cell lung cancer	Recurrence (N = 132)	Buckingham et al. (2010) [19795445]
	Bladder cancer	Poor survival (N = 96)	Yates et al. (2007) [17404085]
	Neuroblastoma	Poor survival (N = 61)	Yang et al. (2004) [15623630]
	Ovarian cancer	Invasion (N = 76)	Makarla et al. (2005) [16061849]
RUNX3	Bladder cancer	Invasion, poor survival (N = 149)	Kim et al. (2008) [18639281]
SLC18A2	Prostate cancer	Poor survival (N = 738)	Sørensen et al. (2009) [19228741]
TIMP3	Bladder cancer	Metastasis (N = 175)	Hoque et al. (2008) [18082200]
Vimentin	Colorectal cancer	Liver metastasis (N = 48)	Shirahata et al. (2009) [19331162]

silencing of the MGMT gene for the DNA-repair protein O^6-alkylguanine DNA alkyltransferase in gliomas is used as a predictor of chemosensitivity to alkylating agents in clinical trials.

MGMT locates in the chromosome 10q26 region. MGMT removes mutagenic, cytotoxic adducts from O^6-guanine in DNA, the preferred point of attack of alkylating agents. Expression level of MGMT in normal tissues is very low under physiological condition, but upregulated after exposed to alkylating agents or radiation (Grombacher _et al._, 1996; Vielhauer _et al._, 2001). MGMT overexpression protects normal tissues from the toxic effects of alkylating carcinogens including agents used in chemotherapy. Transcriptional silence of MGMT by the promoter methylation of CpG island is detectable in diverse cancer types such as gliomas, lymphomas, breast cancers, prostate tumors, and retinoblastomas (Esteller _et al._, 1999; Watts _et al._, 1997).

Because of roles of MGMT in the DNA repair, many attentions have been focused on the relationship between MGMT methylation and prediction of chemosensitivity. It is found that MGMT methylation is an independent predictor of a favorable response of gliomas to guanine methylation agents carmustine [BCNU, _N_,_N′_-bis(2-chloroethyl)-_N_-nitroso-urea] and temozolomide (4-methyl-5-oxo-2,3,4,6,8-pentazabicyclo[4.3.0]nona-2,7,9-triene-9-carboxamide). Esteller _et al._ (2000) analyzed methylation patterns in the CpG island of MGMT in 47 diagnosed grade-III/IV gliomas and found that 63% (12/19) of patients with methylated MGMT in tumors had a partial or complete response to BCNU, as compared with 4% (1/28) of patients without methylated MGMT ($P < 0.001$). They also found that MGMT methylation was associated with an increase in overall survival and the time to progression of disease. Subsequently, in a randomized clinical trial of 206 glioma patients, MGMT methylation was detected in 45% of tumor samples and a survival benefit was observed in patients treated with temozolomide and radiotherapy (Hegi _et al._, 2005). That survival benefit is mainly observed in patients treated with nitrosourea alkylating agents indicates that MGMT methylation predicts benefit from alkylating agent chemotherapy rather than simply being a prognostic marker (Gerson, 2004).

Carmustine and temozolomide are the first-line chemicals that have been used in the treatment of malignant gliomas regardless of methylation status of the MGMT promoter for a long time. The discovery of significance of MGMT methylation in predicting chemosensitivity suggests that patients with unfavorable MGMT methylation status might be selected for other regimens or combinations of carmustine/temozolomide with MGMT inhibitors. Besides its effect on response to the chemotherapy, MGMT methylation can also be used to predict the response to radiotherapy and prognostic in the absence of adjuvant alkylating chemotherapy for glioblastoma (Rivera _et al._, 2010). Association between serum MGMT methylation and glioma sensitivity to BCNU was reported previously (Balaña _et al._, 2003). Recently, Liu _et al._ (2010a) detected

MGMT methylation in glioma tumor tissue, serum, and cerebrospinal fluid (CSF) quantitatively, and found that MGMT methylation in serum and CSF was all accompanied with the methylation in the corresponding tumor tissues with 100% specificity. Detection circulating MGMT methylation may be a promising non-/mini-invasive assay, which is special usefulness in clinical managements of patients from whom tumor tissue is unaccessible.

The potential of the methylation status of other genes for predicting the response to chemotherapy has also been seen between cisplatin and methylation of DNA mismatch repair gene hMLH1 (Strathdee et al., 1999), between methotrexate and RFC methylation (Ferreri et al., 2004), between irinotecan and WRN methylation (Agrelo et al., 2006; Fiegl et al., 2006), between tamoxifen and PGR/HSD17B4/CDH13/MYOD1/BRCA1 methylation (Fiegl et al., 2006). We summarize briefly information of partial published methylation markers related to chemosensitivity (Table 5.2).

III. MicroRNAs

miRNAs are an abundant class of small single-strand noncoding regulatory RNAs. It is estimated that the human genome may contain up to 1000 miRNA genes, which account for 2%–5% of human genes (Mirnezami et al., 2009). Up to date, 721 human miRNAs are listed in miRBase (Release 14, http://www.mirbase.org; Griffiths-Jones, et al., 2008). miRNAs can serve as master regulators. A single miRNA can regulate hundreds of different target genes at transcription or translation level. Due to their broad targeting ability, miRNAs may play critical roles in regulating diverse biological processes, including cell proliferation, differentiation, and apoptosis (reviewed by Du and Pertsemlidis, 2010). A substantial number of miRNA genes are located in genomic regions that are frequently amplified, deleted, or rearranged in cancer, providing evidence of a role for miRNAs in cancer pathogenesis (Calin et al., 2002; Nairz et al., 2006; Zhang et al., 2006). Alteration of miRNA expression and its correlation with the development and progression of cancers is an emerging field (Iorio et al., 2008; Lu et al., 2005; Sarkar et al., 2010).

It is well recognized that miRNAs could play roles in the formation and maintenance of cancer stem/initiating cells, in acquisition of epithelial–mesenchymal transition ability, and in response of cancer cells to chemotherapy or radiotherapy. Knockdown or reexpression of specific miRNAs by synthetic antisense oligonucleotides or pre-miRNAs could induce drug sensitivity, leading to increased inhibition of cancer cell growth, invasion, and metastasis. MiRNA is very stable in tissues and body fluids (Chen et al., 2008; Mitchell et al., 2008; Szafranska et al., 2008). In addition, expression of a number of miRNAs could be silenced by DNA methylation (Ando et al., 2009; Bandres et al., 2009; Lujambio

Table 5.2. List of Methylated Genes Related to Chemosensitivity of Cancers

Gene	Cancer type	Associated characteristics (case number)	Reference (PubMed ID)
ABCB1 and *GSTP1*	Breast cancer	Increased sensitivity to doxorubicin ($N=238$)	Dejeux et al. (2010) [20338046]
ASC/TMS1	Gastric cancer	Increased resistance to 5-fluorouriacil ($N=81$)	Kato et al. (2008) [17943730]
BRCA1	Breast cancer	Increased sensitivity to cisplatin ($N=28$)	Silver et al. (2010) [20100965]
BRCA1 and *BRCA2*	Ovarian cancer	Increased sensitivity to cisplatin ($N=115$)	Swisher et al. (2009) [19602291]
CDK10	Breast cancer	Increased resistance to tamoxifen ($N=87$)	Iorns et al. (2008) [18242510]
CHFR	Endometrial cancer	Increased sensitivity to taxanes ($N=50$)	Yanokura et al. (2007) [17143476]
	Gastric cancer	Increased sensitivity to microtubule inhibitors ($N=61$)	Satoh et al. (2003) [14695171]
hMLH1	Ovarian cancer	Increased resistance to cisplatin ($N=36$)	Watanabe et al. (2007) [17595760]
	Ovarian cancer	Increased resistance to carboplatin/taxoid ($N=138$)	Gifford et al. (2004) [15240532]
LINE1 demethylation	Chronic myeloid leukemia	Increased sensitivity to decitabine ($N=16$)	Oki et al. (2008) [18055864]
	Chronic myeloid leukemia	Increased sensitivity to decitabine ($N=35$)	Issa et al. (2005) [15883410]
MGMT	Glioma	Increased sensitivity to carmustine ($N=47$)	Esteller et al. (2000) [11070098]
	Glioma	Increased sensitivity to temozolomide ($N=206$)	Hegi et al. (2005) [15758010]
	Glioma	Increased sensitivity to temozolomide ($N=38$)	Hegi et al. (2004) [15041700]
RASSF1A and *HIC1*	Germ cell tumor	Increased resistance to cisplatin ($N=70$)	Koul et al. (2004) [15149548]
WRN	Werner syndrome	Increased resistance to irinotecan ($N=630$)	Agrelo et al. (2006) [16723399]

et al., 2008). Thus, downregulation of a target miRNA by methylation within a specific cell type even at a few of cells in tested tissues is also detectable with methylation assays, as mentioned above. These advantages suggest that miRNA may become another kind of useful cancer biomarkers for early detection of cancers and for prediction of prognosis and responses of cancer to chemotherapy.

A. Diagnosis of cancers

Because most miRNAs in mammalian embryonic stem cells are induced during cellular maturation in tissue-specific expression patterns, they may play key roles in the maintenance of cell lineage characteristics (Darr and Benvenisty, 2006; Houbaviy *et al.*, 2003; Suh *et al.*, 2004). The remaining lineage-specific miRNAs in cancer cells may be useful for identification of their lineage origins. Both upregulation of oncogenic miRNAs and downregulation of oncosuppressor miRNAs can be observed in cancer tissues.

Lu *et al.* (2005) report that 129 of 217 miRNAs are expressed to a lower extent in all tumor samples compared with the normal tissue and that expression profiles of only a few hundred miRNAs correctly classify the tumors of different origins. Their miRNA classifier clusters the cancers from an endothelial lineage such as colon, liver, pancreas, and stomach together, and cancers of hematopoietic origin as separated groups. In contrast, profiling of 15,000 mRNA genes fails to group the cancers accurately. Since then on, there are a number of reports on cancers and miRNA expression. For example, a seven-miRNA signature (*miR-187, miR-221, miR-222, miR-146b, miR-155, miR-224,* and *miR-197*) is differentially overexpressed in thyroid tumors and validated in the fine-needle aspiration samples, showing 95% accuracy of thyroid cancer detection (Nikiforova *et al.*, 2008).

RNAs including miRNAs are degraded by intracellular exosomes, a multiprotein complex with activity of $3' \rightarrow 5'$ exoribonuclease in mammalian cells. There is another kind of exosome that is 40–100 nm membrane vesicle secreted by a wide range of mammalian cell types. This kind of exosome contains proteins and functional RNA, such as mRNA and miRNA, which can be shuttled from one cell to another through the exocytosis → endocytosis pathway, affecting the recipient cell's protein production (Valadi *et al.*, 2007). Tumor cells actively release exosomes into the peripheral circulation through a ceramide-dependent secretory machinery (Kosaka *et al.*, 2010). It is found that total amount of both exosomal proteins and miRNAs in plasma samples from patients with lung adenocarcinomas are significantly higher than those from healthy controls, and that miRNA signatures in circulating exosomes and tumor tissues correlate with each other very well (Rabinowits *et al.*, 2009). The similar phenomenon on exosomal miRNAs is also observed among patients with ovarian cancers and benign diseases (Taylor and Gercel-Taylor, 2008).

　　　　The fact that miRNAs are maintained in a protected state in serum and plasma makes them particularly attractive as biomarkers, thus allowing the detection of miRNA expression patterns directly from serum. Unlike detection of exosomal miRNA assays in which circulating exosomes should be isolated, detection of serum and plasma miRNAs is very convenient, because serum and plasma samples themselves can be used directly as the RNA substrates for synthesis of cDNA templates in quantitative RT-PCR without isolation processes of miRNAs (Chen *et al.*, 2008; Lodes *et al.*, 2009; Mitchell *et al.*, 2008).

　　　　Mitchell *et al.* (2008) report that serum levels of *miR-141* could detect individuals with prostate cancer with 60% (15/25) sensitivity at 100% (25/25) specificity. It is found that plasma *miR-92* marker yields a receiver operating characteristic (ROC) curve area of 88.5% for identification of CRCs (Ng *et al.*, 2009). The sensitivity of this test is up to 89% (80/90) at 70% specificity (63/90, including 50 healthy controls, 20 patients with inflammatory bowel diseases, and 20 patients with gastric carcinomas). Significant increase of circulating *miR-210* is also observed among patients with pancreatic carcinomas and breast cancers (Ho *et al.*, 2010; Wang *et al.*, 2009a). Tsujiura *et al.* (2010) report that the ROC curve values are 0.72 for the circulating *miR-106b* assay and 0.88 for the *miR-106a/let-7a* ratio assay for diagnosis of gastric carcinomas. Taken together, more and more data on circulating miRNAs and diagnosis of cancers are emerging. The high stability of miRNAs in plasma and serum and convenient detection assays make circulating miRNAs attractive biomarkers in screen for cancers. Further works should be done to investigate whether they are suitable for detection of cancers at early stages and prediction of malignant potential of precancerous lesions. The application value of miRNAs in other body fluids is also worth to be explored.

B. Prediction of prognosis of cancers

There are a number of association studies on miRNAs and prognosis of cancers. We use three miRNAs as examples, oncosuppressor *let-7*, oncogenic *miR-21*, and cancer type-dependent *miR-31* below. Other potential prognosis-related miRNAs are listed in Table 5.3.

1. *let-7* miRNA

let-7 miRNAs are encoded by 11 *let-7* genes (*let-7a-1/2/3*, *let-7b/c/d/e/g/i*, *let-7f-1/2*), which are transcribed in tissue-specific patterns. Takamizawa *et al.* (2004) first report that low level of *let-7a* expression in lung cancer tissue is significantly associated with shorter survival of patients ($N = 143$). Yanaihara *et al.* (2006) also find that low *let-7a-2* and high *miR-155* expression independently correlate

Table 5.3. Alterations of miRNA Expression Related to Prognosis of Cancers

miRNA	Cancer type	Associated characteristics (case number)	Reference [PubMed ID]
let-7	Lung cancers	Downregulation, poor survival ($N = 143$)	Takamizawa et al. (2004) [15172979]
	Lung cancer	Downregulation, poor survival ($N = 104$)	Yanaihara et al. (2006) [16530703]
	Squamous cell lung cancer	Downregulation, poor survival ($N = 121$)	Landi et al. (2010) [20068076]
	Ovarian carcinoma	Downregulation, poor survival ($N = 214$)	Lu et al. (2007) [17974952]
	Ovarian carcinoma	Downregulation, poor survival ($N = 72 + 53$)	Yang et al. (2008) [19074899]
	Ovarian carcinoma	Downregulation, poor prognosis ($N = 28$)	Nam et al. (2008) [18451233]
	Head and neck cancers	Downregulation, poor survival ($N = 104$)	Childs et al. (2009) [19179615]
	Gastric carcinoma	Downregulation, poor survival ($N = 353$)	Ueda et al. (2010) [20022810]
miR-15b	Melanoma	Upregulation, poor survival and recurrence ($N = 128$)	Satzger et al. (2010) [19830692]
miR-18a	Colorectal cancer	Upregulation, poor prognosis ($N = 69$)	Motoyama et al. (2009) [19287964]
	Ovarian carcinoma	Upregulation, poor prognosis($N = 28$)	Nam et al. (2008) [18451233]
miR-21	Nonsmall-cell lung cancer	Upregulation, poor prognosis ($N = 48$)	Markou et al. (2008) [18719201]
	Nonsmall-cell lung cancer	Upregulation, poor prognosis ($N = 47$)	Gao et al. (2010) [20363096]
	Head and neck carcinoma	Upregulation, poor survival ($N = 169$)	Avissar et al. (2009a) [19901002]
	Head and neck carcinoma	Upregulation, poor survival ($N = 113$)	Avissar et al. (2009b) [19351747]
	Squamous cell esophagus cancer	Upregulation, poor prognosis ($N = 170$)	Mathé et al. (2009) [19789312]
	Colorectal cancer	Upregulation, poor prognosis ($N = 113$)	Schetter et al. (2008) [18230780]
	Breast cancer	Upregulation, poor prognosis ($N = 113$)	Yan et al. (2008) [18812439]
	Pancreas carcinoma	Upregulation, liver metastasis ($N = 56$)	Roldo et al. (2006) [16966691]
	Diffuse large B-cell lymphoma	Upregulation, good survival ($N = 103$)	Lawrie et al. (2008) [18318758]
miR-29c* (miR-29c)	Malignant pleural mesothelioma	Upregulation, longer survival ($N = 142$)	Pass et al. (2010) [20160038]
	Chronic lymphocytic leukemia	Downregulation, poor prognosis ($N = 110$)	Stamatopoulos et al. (2009) [19144983]
miR-31	Colorectal cancer	Upregulation, in advanced cancers ($N = 12$)	Bandrés et al. (2006) [16854228]
	Colorectal cancer	Downregulation, poor differentiation ($N = 35$)	Slaby et al. (2007) [18196926]
	Breast cancer	Downregulation, more metastasis ($N = 56$)	Valastyan et al. (2009) [19524507]

(Continues)

Table 5.3. (*Continued*)

miRNA	Cancer type	Associated characteristics (case number)	Reference [PubMed ID]
miR-96	Prostate carcinoma	Upregulation, cancer recurrence ($N = 79$)	Schaefer et al. (2010) [19676045]
miR-146b	Squamous cell lung cancer	Upregulation, poor survival ($N = 71$)	Raponi et al. (2009) [19584273]
miR-155	Lung cancer	Downregulation, poor survival ($N = 104$)	Yanaihara et al. (2006) [16530703]
	Diffuse large B-cell lymphoma	Upregulation, longer relapse-free survival ($N = 103$)	Lawrie et al. (2008) [18318758]
miR-181a	Nonsmall-cell lung cancer	Upregulation, good survival ($N = 47$)	Gao et al. (2010) [20363096]
miR-196a-2	Pancreatic cancer	Upregulation, poor survival ($N = 107$)	Bloomston et al. (2007) [17473300]
miR-200	Ovarian carcinoma	Downregulation, poor survival ($N = 55$)	Hu et al. (2009) [19501389]
	Ovarian carcinoma	Upregulation, poor prognosis ($N = 28$)	Nam et al. (2008) [18451233]
	Colorectal cancer	Upregulation, poor prognosis ($N = 24$)	Xi et al. (2006) [18079988]
miR-205	Head and neck cancers	Downregulation, poor survival ($N = 104$)	Childs et al. (2009) [19179615]
miR-210	Breast cancer	Upregulation, poor survival ($N = 219$)	Camps et al. (2008) [18316553]
miR-214 and miR-433	Gastric cancer	Upregulation, poor survival ($N = 353$)	Ueda et al. (2010) [20022810]

with poor survival for patients with lung adenocarcinomas ($N = 104$). Again, Yu et al. (2008a) identify a five-miRNA signature (*let-7a, miR-221, miR-137, miR-372,* and *miR-182*) that can predict survival in patients with NSCLCs ($N = 174$). In addition, Landi et al. (2010) also set up a five-miRNA signature (*let-7e, miR-25, miR-34c-5p, miR-191,* and *miR-34a*) that can predict survival in patients with lung squamous cell carcinomas ($N = 121$), but not in patients with lung adenocarcinomas ($N = 165$).

Functionally, it is found that *let-7* miRNA is markedly reduced in breast tumor-initiating cells (BT-IC) and increased with differentiation (Yu et al., 2007a). In the same work, they also find that upregulation of *let-7* in BT-IC cells with *let-7a*-lentivirus reduces proliferation, mammosphere formation, and the proportion of undifferentiated cells *in vitro* and tumor formation and metastasis in NOD/SCID mice. Antagonizing *let-7a* by antisense oligonucleotides enhances *in vitro* self-renewal of non-BT-IC cells. In another work, it is also found that the RNA-binding proteins LIN28 and LIN28b, which block *let-7* precursors from being processed to mature miRNAs, are overexpressed in multiple primary tumor types and derepress *let-7* targets and that upregulation of these proteins facilitate transformation *in vitro* (Viswanathan et al., 2009). Thus, low *let-7a* miRNA in cancer tissues may be a potential indicator of poor prognosis of lung cancer.

2. miR-21

miR-21 is an oncogenic miRNA that attracts lots of attention from cancer biologists. Chan et al. (2005) initially showed that *miR-21* was highly expressed in human glioma cells. They demonstrated that *miR-21* in glioma cells acted as an antiapoptotic factor by blocking expression of key apoptosis-enabling genes, such as *caspases-3* and *caspases-7*. Meng et al. (2007) indicated that *miR-21* was overexpressed in hepatocarcinomas and involved in the PTEN signaling pathway. Si et al. (2007) reported that *miR-21* was overexpressed in breast cancer. This indirectly caused downregulation of *Bcl-2* and decreased apoptosis of cancer cells. They also reported that *miR-21* downregulated expression of the TSG tropomyosin-1 through targeting the 3′UTR of its mRNA (Zhu et al., 2007).

miR-21 may be an independent prognosis factor in many cancers. Yan et al. (2008) found that the upregulation of *miR-21* correlated with the survival of 113 patients with breast cancers. It is also reported that *miR-21* is an independent prognosis factor for patients with other cancer types such as NSCLC (Gao et al., 2010; Markou et al., 2008), diffuse large B-cell lymphoma (Lawrie et al., 2008), head and neck squamous cell carcinoma (Avissar et al.,

2009a,b), squamous cell esophagus cancer (Mathé *et al.*, 2009), CRC (Schetter *et al.*, 2008). All of these results show that *miR-21* may be a potential prognosis factor for human cancer.

3. miR-31

miR-31 locates at 9p21.3 (chr9:21502114–21502184) near the *p16-Arf-p15* locus, a genomic region deleted mostly in diverse cancer types. Valastyan *et al.* (2009) reported that *miR-31* expression correlated with metastasis of breast cancers negatively. They found that overexpression of *miR-31* in otherwise-aggressive breast tumor cells suppressed metastasis and stable inhibition of *miR-31* expression *in vivo* promotes otherwise-nonaggressive breast cancer cells to metastasize. These phenotypes are specifically attributable to *miR-31*-mediated inhibition of several steps of metastasis, including local invasion, extravasation, or initial survival at a distant site and metastatic colonization. In contrast, Liu *et al.* (2010d) reported that *miR-31* functioned as an oncogenic miRNA in mouse with engineered knockdown of *miR-31* and human lung cancer cells by repressing *LATS2* and *PPP2R2A*. Sarver *et al.* (2009) also found that *miR-31* was upregulated in CRC tissues, especially in tumors that were deficient for mismatch repair. Furthermore, the plasma *miR-31* level is increased in patients with oral squamous cell carcinomas (Liu *et al.*, 2010b). Another study shows that *miR-31* is underexpressed in serous ovarian carcinomas and that forced *miR-31* overexpression inhibits proliferation and induces apoptosis only in cancer cell lines with a dysfunctional p53 pathway, but not in p53 wild-type cells (Creighton *et al.*, 2010). These contradictive data suggest that role of *miR-31* in carcinogenesis may be cancer type dependent. Both overexpression and underexpression of *miR-31* in cancer tissues would be informative for prognosis of cancers.

C. Prediction of responses of cancer to chemotherapy

Evidences show that cancer stem/initiating cells may be more resistant to chemical agents than differentiated cells. There are different miRNA signatures within different cell lineages, including stem cells. In addition, it is supposed that cancer cells resistant to chemotherapy may also have special expression alteration of miRNA(s). If the chemosensitivity-related miRNA signature is detectable in tumor tissue or body fluid samples directly, it will very useful for personalized treatment of cancers.

It is verified that inhibition of *miR-21* expression in malignant cholangiocytes increases the sensitivity to gemcitabine (Meng *et al.*, 2006). Upregulation of *miR-21* expression is an independent prognostic marker not only for poor survival but also for poor responsiveness to adjuvant chemotherapy (Schetter *et al.*, 2008). Yang *et al.* (2008) found that *let-7i* in epithelial ovarian cancers was downregulated in chemotherapy-resistant patients. Malumbres *et al.* (2009) reported that downregulation of *miR-222* in cholangio carcinoma was a better progression-free survival factor for patients ($N = 106$) with diffuse large B-cell lymphoma treated with R-CHOP chemotherapy. Zenz *et al.* (2009) confirmed that the low expression of *miR-34* could reduce resistance to fludarabine-refractory in 60 patients with chronic lymphocytic leukemia. Besides, Svoboda *et al.* (2008) also indicated that *miR-125b* and *miR-137* were upregulated in rectal cancer and resulted in worse response to capecitabine chemoradiotherapy ($N = 66$). Because of the short history of researches on miRNA and cancers, only a few reports are available right now. More evidences on miRNAs and chemosensitivity would be published in the coming years.

IV. OTHER EPIGENETIC ALTERATIONS

Histone modification patterns are called "histone codes," playing important roles in regulation of transcription of genes transiently and epigenetically. Although histone modifications are closely connected with transcription activity of genes, only part of them such as H3K9 di- or trimethylation (H3K9Me2 or H3K9Me3) are directly associated with stably epigenetic inactivation of genes. Global histone modifications, such as methylation or acetylation (Ac) of lysine and arginine residues, can be analyzed with IHC and Western blot assays. Given gene-specific and genome-wide histone modifications can be detected with chromatin immunoprecipitation (ChIP) based assays, such as PCR, microarray, and deep sequencing. Up to date, only a few global histone modifications by IHC are associated with prognosis of cancers.

Seligson *et al.* (2005) have reported that histone modifications such as H3K18Ac, H3K4Me2, H4K12Ac, and H4R3Me2 correlate with recurrence of prostate cancer among 183 patients. Recently, Ellinger *et al.* (2010) report that H3K4Me1, H3K9Me2, H3K9Me3, H3Ac, and H4Ac by IHC are significantly reduced in prostate cancer tissues compared to nonmalignant prostate tissues. In addition, H3Ac and H3K9Me2 levels allow discrimination of prostate cancers from noncancerous tissues sensitively ($>78\%$) and specifically ($>91\%$). Again, it is reported that low H3K4Me2, H3K9Me2, and H3K18Ac levels are significant independent predictors for poor survival of 315 patients with pancreatic adenocarcinomas (Manuyakorn *et al.*, 2010). It is also found that the levels of H3K18Ac, H4R3Me2, and H3K27Me3 correlate positively with differentiation of esophageal squamous cell carcinomas and low H3K27Me3 is an independent and good survival factor (Tzao *et al.*, 2009). Although the biological basis of these correlations is unclear, it is suggested that higher levels of these modifications might be related to the increased proliferative capacity of dedifferentiated tumors. These histone codes within certain parts of chromatin of cells in a tissue may change dynamically during the development, aging, and adaption to environmental factors. The complexity and instability of most histone modifications and lack of full understanding of these codes hamper their applications as cancer biomarkers at present.

There are also some reports that overexpression of epigenetic-related enzymes DNA methyltransferase (DNMT), histone deacetylase (HDAC), and downregulation of histone acetytlransferease (HAT) are associated with cancers. Saito *et al.* (2003) reported that DNMT1 protein level was increased in 43% (23/53) of hapatocellular carcinomas and associated with poor tumor differentiation and portal vein involvement and overall survival. Patients with DNMT1 overexpression in lung squamous carcinomas have a trend of poor prognosis than those without such overexpression (Lin *et al.*, 2007a). Upregulation of *DNMT3b* transcription is also observed in sporadic breast carcinomas and hapatocellular carcinomas and associated with poor overall survival (Girault *et al.*, 2003; Oh *et al.*, 2007).

Krusche *et al.* (2005) have reported that HDAC-1 expression correlates significantly with estrogen and progesterone receptor expression and is an independent prognostic marker of better disease free survival of 200 breast cancer patients. In a retrospective analysis, it is found that overexpression of HDAC1, HDAC2, and HDAC3 is significantly associated with nodal spread and is an independent prognostic marker for gastric carcinomas (Weichert *et al.*, 2008a). Similar result is also obtained among patients with prostate cancers (Weichert *et al.*, 2008b). In addition, Pfister *et al.* (2008) found that expression of HAT hMOF and its product H4K16Ac was markedly reduced in both breast cancer and medulloblastomas. For medulloblastomas, downregulation of hMOF expression is associated with lower survival rate of 102 patients, identifying hMOF as an independent prognostic marker for clinical outcome.

Chromatin conformation, nucleosome occupancy, and chromosome territory are also important epigenetic regulatory mechanisms of gene expression and cell differentiation. It is unknown whether they could be used as cancer biomarkers in the future.

V. PROSPECTIVE AND CHALLENGES

An optimal cancer biomarker should represent the majorities of cancers in an organ among populations. Both α-fetal protein (AFP) and prostate-specific antigen (PSA) are such kinds of markers. Although active researches on epigenetics and cancers have been for 15 years, practical epigenetic biomarkers such as *p16* and *Septin9* methylation for prediction of cancer development, diagnosis, and prognosis of cancers are emerging. DNA methylation markers represent long-term inactivation of genes and are very stable in diverse tissues samples such as paraffin-embedded and body fluids. They can be sensitively detected, even in very limited number of cells. miRNAs are also very stable in tissue and blood samples. Circulating miRNA levels can be quantified very conveniently. These advantages make DNA methylation and miRNA optimal biomarkers for clinical management of cancers.

A number of potential DNA methylation and miRNAs markers have being screened out based on their associations with tumor biological character-istics and patients' survival (Hernandez-Vargas et al., 2010; Lofton-Day et al., 2008; Shames et al., 2006; Tan et al., 2009). Huge association raw data are continuously flowing out from different kinds of high throughput platforms. It is difficult to select right genes from those data for further validation. The strategy of combination of array data and published information is often used in candidate selection in most recently published works. However, only few of them have been validated in prospective cohort studies. There are too many data to be validated! How to benefit cancer patients from these data becomes a challenge for both biologists and oncologists. With accumulating more and more association data, it is necessary to evaluate these cancer-associated markers systemically and make a suggested list of selected marker candidates for global oncologists and biotechnique companies for further validations in multicentral case–control and follow-up cohort studies.

Unlike using surgical tissues to study predictors of prognosis of cancers, it is very difficult to validate a panel of molecular markers in precancer-ous lesions because the amount of genomic DNA from the biopsies is very limited. An optimal way to sample the paraffin-embedded biopsies for extraction of DNA is to collect the cut-off-part that is not suitable for prepara-tion of slides for regular histopathologic examinations. A few methylation markers may be analyzed using such material from biopsy samples with current assays. More sensitive detection assays or practical techniques to detect methylation status of multiple CpG islands simultaneously need to be developed for establishment of a methylation pattern for diagnosis of malignant progression potential of different kinds of precancerous lesions sensitively and specifically.

Selection of right CpG sites within CpG islands is also a crucial issue during analysis of DNA methylation. As illustrated in the cases of *hMLH1* methylation, not methylation at any CpG site relates to transcriptional repres-sion of target genes. The same situation may be also suitable to other genes (Deng et al., 2003b; Ushijima, 2005). Unfortunately, the crucial CpG sites correlated with transcription are not investigated for most CpG islands around the target gene promoters. The detail extension process or pattern of methylation sites within CpG islands is largely unknown *in vivo*. These essential data should be very useful for designing proper methylation assays during development of optimal DNA methylation markers.

As to miRNAs, combination of data of miRNA expression in tumor tissues with histopathological diagnosis should be an optimal way for prediction of prognosis and therapy response of cancers. The significance of circulating miRNAs on prediction of prognosis of cancers also promotes us to consider whether some kinds of the circulating miRNAs can be used to detect potential of malignant transformation of precancerous lesions or to screen early cancers in population. The example of *miR-31* illustrates the complexity of miRNA functions and clinical applications. More investigations are necessary for understanding of the feasibility of clinical usages of miRNA expression in cancer tissues and body fluids.

Molecular classifiers, including epigenetic ones, are awaited not only for early diagnosis of cancer, but also for personalized therapy of cancers. Knowledge on the human genome and its regulation networks including epigenetic mechanisms accelerate efficiency of discovery of biomarkers and novel drugs targeted to crucial genes in cancer. We are opening the curtain on the performance to reduce the burden of cancer greatly.

References

Aggerholm, A., Holm, M. S., Guldberg, P., Olesen, L. H., and Hokland, P. (2006). Promoter hypermethylation of p15INK4B, HIC1, CDH1, and ER is frequent in myelodysplastic syndrome and predicts poor prognosis in early-stage patients. *Eur. J. Haematol.* **76,** 23–32.

Agrelo, R., Cheng, W. H., Setien, F., Ropero, S., Espada, J., Fraga, M. F., Herranz, M., Paz, M. F., Sanchez-Cespedes, M., Artiga, M. J., *et al.* (2006). Epigenetic inactivation of the premature aging Werner syndrome gene in human cancer. *Proc. Natl. Acad. Sci. USA* **103,** 8822–8827.

Amara, K., Trimeche, M., Ziadi, S., Laatiri, A., Hachana, M., and Korbi, S. (2008). Prognostic significance of aberrant promoter hypermethylation of CpG islands in patients with diffuse large B-cell lymphomas. *Ann. Oncol.* **19,** 1774–1786.

Ando, T., Yoshida, T., Enomoto, S., Asada, K., Tatematsu, M., Ichinose, M., Sugiyama, T., and Ushijima, T. (2009). DNA methylation of microRNA genes in gastric mucosae of gastric cancer patients: Its possible involvement in the formation of epigenetic field defect. *Int. J. Cancer* **124,** 2367–2374.

Aoki, E., Uchida, T., Ohashi, H., Nagai, H., Murase, T., Ichikawa, A., Yamao, K., Hotta, T., Kinoshita, T., Saito, H., *et al.* (2000). Methylation status of the p15INK4B gene in hematopoietic progenitors and peripheral blood cells in myelodysplastic syndromes. *Leukemia* **14,** 586–593.

Ariel, I., Ayesh, S., Perlman, E. J., Pizov, G., Tanos, V., Schneider, T., Erdmann, V. A., Podeh, D., Komitowski, D., Quasem, A. S., *et al.* (1997). The product of the imprinted H19 gene is an oncofetal RNA. *Mol. Pathol.* **50,** 34–44.

Avissar, M., McClean, M. D., Kelsey, K. T., and Marsit, C. J. (2009a). MicroRNA expression in head and neck cancer associates with alcohol consumption and survival. *Carcinogenesis* **30,** 2059–2063.

Avissar, M., Christensen, B. C., Kelsey, K. T., and Marsit, C. J. (2009b). MicroRNA expression ratio is predictive of head and neck squamous cell carcinoma. *Clin. Cancer Res.* **15,** 2850–2855.

Bai, H., Gu, L. K., Zhou, J., and Deng, D. J. (2003). p16 hypermethylation during gastric carcinogenesis of Wistar rats by N-methyl-N'-nitro-N-nitrosoguanidine. *Mutat. Res.* **535,** 73–78.

Bai, H., Zhou, J., and Deng, D. J. (2008). Nucleosome positions and differential methylation status of various regions within the MLH1 CpG island. *Chin. J. Cancer Res.* **20,** 237–242.

Balaña, C., Ramirez, J. L., Taron, M., Roussos, Y., Ariza, A., Ballester, R., Sarries, C., Mendez, P., Sanchez, J. J., and Rosell, R. (2003). O6-methyl-guanine-DNA methyltransferase methylation in serum and tumor DNA predicts response to 1, 3-bis(2-chloroethyl)-1-nitrosourea but not to temozolamide plus cisplatin in glioblastoma multiforme. *Clin. Cancer Res.* **9,** 1461–1468.

Bandrés, E., Cubedo, E., Agirre, X., Malumbres, R., Zárate, R., Ramirez, N., Abajo, A., Navarro, A., Moreno, I., Monzó, M., *et al.* (2006). Identification by Real-time PCR of 13 mature microRNAs differentially expressed in colorectal cancer and non-tumoral tissues. *Mol. Cancer* **5,** 29.

Bandres, E., Agirre, X., Bitarte, N., Ramirez, N., Zarate, R., Roman-Gomez, J., Prosper, F., and Garcia-Foncillas, J. (2009). Epigenetic regulation of microRNA expression in colorectal cancer. *Int. J. Cancer* **125,** 2737–2743.

Bastian, P. J., Ellinger, J., Wellmann, A., Wernert, N., Heukamp, L. C., Müller, S. C., and von Ruecker, A. (2005). Diagnostic and prognostic information in prostate cancer with the help of a small set of hypermethylated gene loci. *Clin. Cancer Res.* **11,** 4097–4106.

Belinsky, S. A., Nikula, K. J., Palmisano, W. A., Michels, R., Saccomanno, G., Gabrielson, E., Baylin, S. B., and Herman, J. G. (1998). Aberrant methylation of p16(INK4a) is an early event in lung cancer and a potential biomarker for early diagnosis. *Proc. Natl. Acad. Sci. USA* **95,** 11891–11896.

Belinsky, S. A., Snow, S. S., Nikula, K. J., Finch, G. L., Tellez, C. S., and Palmisano, W. A. (2002). Aberrant CpG island methylation of the p16(INK4a) and estrogen receptor genes in rat lung tumors induced by particulate carcinogens. *Carcinogenesis* **23,** 335–339.

Belinsky, S. A., Liechty, K. C., Gentry, F. D., Wolf, H. J., Rogers, J., Vu, K., Haney, J., Kennedy, T. C., Hirsch, F. R., Miller, Y., *et al.* (2006). Promoter hypermethylation of multiple genes in sputum precedes lung cancer incidence in a high-risk cohort. *Cancer Res.* **66,** 3338–3344.

Beroukhim, R., Mermel, C. H., Porter, D., Wei, G., Raychaudhuri, S., Donovan, J., Barretina, J., Boehm, J. S., Dobson, J., Urashima, M., *et al.* (2010). The landscape of somatic copy-number alteration across human cancers. *Nature* **463,** 899–905.

Bialik, S., and Kimchi, A. (2006). The death-associated protein kinases: Structure, function, and beyond. *Annu. Rev. Biochem.* **75,** 189–210.

Bignell, G. R., Greenman, C. D., Davies, H., Butler, A. P., Edkins, S., Andrews, J. M., Buck, G., Chen, L., Beare, D., Latimer, C., *et al.* (2010). Signatures of mutation and selection in the cancer genome. *Nature* **463,** 893–898.

Bloomston, M., Frankel, W. L., Petrocca, F., Volinia, S., Alder, H., Hagan, J. P., Liu, C. G., Bhatt, D., Taccioli, C., and Croce, C. M. (2007). MicroRNA expression patterns to differentiate pancreatic adenocarcinoma from normal pancreas and chronic pancreatitis. *JAMA* **297,** 1901–1908.

Brabender, J., Usadel, H., Danenberg, K. D., Metzger, R., Schneider, P. M., Lord, R. V., Wickramasinghe, K., Lum, C. E., Park, J., Salonga, D., *et al.* (2001). Adenomatous polyposis coli gene promoter hypermethylation in non-small cell lung cancer is associated with survival. *Oncogene* **20,** 3528–3532.

Brabender, J., Usadel, H., Metzger, R., Schneider, P. M., Park, J., Salonga, D., Tsao-Wei, D. D., Groshen, S., Lord, R. V., Takebe, N., *et al.* (2003). Quantitative O(6)-methylguanine DNA methyltransferase methylation analysis in curatively resected non-small cell lung cancer: associations with clinical outcome. *Clin. Cancer Res.* **9,** 223–227.

Brock, M. V., Gou, M., Akiyama, Y., Muller, A., Wu, T. T., Montgomery, E., Deasel, M., Germonpre, P., Rubinson, L., Heitmiller, R. F., *et al.* (2003). Prognostic importance of promoter hypermethylation of multiple genes in esophageal adenocarcinoma. *Clin. Cancer Res.* **9,** 2912–2919.

Brock, M. V., Hooker, C. M., Ota-Machida, E., Han, Y., Guo, M., Ames, S., Glöckner, S., Piantadosi, S., Gabrielson, E., Pridham, G., *et al.* (2008). DNA methylation markers and early recurrence in stage I lung cancer. *N. Engl. J. Med.* **358**, 1118–1128.

Brouha, B., Schustak, J., Badge, R. M., Lutz-Prigge, S., Farley, A. H., Moran, J. V., and Kazazian, H. H., Jr. (2003). Hot L1s account for the bulk of retrotransposition in the human population. *Proc. Natl. Acad. Sci. USA* **100**, 5280–5285.

Buckingham, L., Penfield, F. L., Kim, A., Liptay, M., Barger, C., Basu, S., Fidler, M., Walters, K., Bonomi, P., and Coon, J. (2010). PTEN, RASSF1 and DAPK site-specific hypermethylation and outcome in surgically treated stage I and II nonsmall cell lung cancer patients. *Int. J. Cancer* **126**, 1630–1639.

Calin, G. A., Dumitru, C. D., Shimizu, M., Bichi, R., Zupo, S., Noch, E., Aldler, H., Rattan, S., Keating, M., Rai, K., *et al.* (2002). Frequent deletions and down-regulation of micro-RNA genes miR15 and miR16 at 13q14 in chronic lymphocytic leukemia. *Proc. Natl. Acad. Sci. USA* **99**, 15524–15529.

Camps, C., Buffa, F. M., Colella, S., Moore, J., Sotiriou, C., Sheldon, H., Harris, A. L., Gleadle, J. M., and Ragoussis, J. (2008). hsa-miR-210 Is induced by hypoxia and is an independent prognostic factor in breast cancer. *Clin. Cancer Res.* **14**, 1340–1348.

Cao, J., Zhou, J., Gao, Y., Gu, L., Meng, H., Liu, H., and Deng, D. (2009). Methylation of p16 CpG island associated with malignant progression of oral epithelial dysplasia: A prospective cohort study. *Clin. Cancer Res.* **15**, 5178–5183.

Capel, E., Fléjou, J. F., and Hamelin, R. (2007). Assessment of MLH1 promoter methylation in relation to gene expression requires specific analysis. *Oncogene* **26**, 7596–7600.

Chan, M. W., Chan, L. W., Tang, N. L., Tong, J. H., Lo, K. W., Lee, T. L., Cheung, H. Y., Wong, W. S., Chan, P. S., Lai, F. M., *et al.* (2002). Hypermethylation of multiple genes in tumor tissues and voided urine in urinary bladder cancer patients. *Clin. Cancer Res.* **8**, 464–470.

Chan, J. A., Krichevsky, A. M., and Kosik, K. S. (2005). MicroRNA-21 is an antiapoptotic factor in human glioblastoma cells. *Cancer Res.* **65**, 6029–6033.

Chang, Y. S., Wang, L., Liu, D., Mao, L., Hong, W. K., Khuri, F. R., and Lee, H. Y. (2002). Correlation between insulin-like growth factor-binding protein-3 promoter methylation and prognosis of patients with stage I non-small cell lung cancer. *Clin. Cancer Res.* **8**, 3669–3675.

Chen, H., and Wu, S. (2002). Hypermethylation of the p15(INK4B) gene in acute leukemia and myelodysplastic syndromes. *Chin. Med. J. (Engl.)* **115**, 987–990.

Chen, W. D., Han, Z. J., Skoletsky, J., Olson, J., Sah, J., Myeroff, L., Platzer, P., Lu, S., Dawson, D., Willis, J., *et al.* (2005). Detection in fecal DNA of colon cancer-specific methylation of the nonexpressed vimentin gene. *J. Natl. Cancer Inst.* **97**, 1124–1132.

Chen, X., Ba, Y., Ma, L., Cai, X., Yin, Y., Wang, K., Guo, J., Zhang, Y., Chen, J., Guo, X., *et al.* (2008). Characterization of microRNAs in serum: A novel class of biomarkers for diagnosis of cancer and other diseases. *Cell Res.* **18**, 997–1006.

Childs, G., Fazzari, M., Kung, G., Kawachi, N., Brandwein-Gensler, M., McLemore, M., Chen, Q., Burk, R. D., Smith, R. V., Prystowsky, M. B., *et al.* (2009). Low-level expression of microRNAs let-7d and miR-205 are prognostic markers of head and neck squamous cell carcinoma. *Am. J. Pathol.* **174**, 736–745.

Choma, D., Daures, J. P., Quantin, X., and Pujol, J. L. (2001). Aneuploidy and prognosis of non-small-cell lung cancer: A meta-analysis of published data. *Br. J. Cancer* **85**, 14–22.

Cohen, O., Inbal, B., Kissil, J. L., Raveh, T., Berissi, H., Spivak-Kroizaman, T., Feinstein, E., and Kimchi, A. (1999). DAP-kinase participates in TNF-alpha- and Fas-induced apoptosis and its function requires the death domain. *J. Cell Biol.* **146**, 141–148.

Creighton, C. J., Fountain, M. D., Yu, Z., Nagaraja, A. K., Zhu, H., Khan, M., Olokpa, E., Zariff, A., Gunaratne, P. H., Matzuk, M. M., *et al.* (2010). Molecular profiling uncovers a p53-associated role for microRNA-31 in inhibiting the proliferation of serous ovarian carcinomas and other cancers. *Cancer Res.* **70**, 1906–1915.

Dansranjavin, T., Mobius, C., Tannapfel, A., Bartels, M., Wittekind, C., Hauss, J., and Witzigmann, H. (2006). E-cadherin and DAP kinase in pancreatic adenocarcinoma and corresponding lymph node metastases. *Oncol. Rep.* **15,** 1125–1131.

Darr, H., and Benvenisty, N. (2006). Factors involved in selfrenewal and pluripotency of embryonic stem cells. *Handb. Exp. Pharmacol.* **174,** 1–19.

Daskalos, A., Nikolaidis, G., Xinarianos, G., Savvari, P., Cassidy, A., Zakopoulou, R., Kotsinas, A., Gorgoulis, V., Field, J. K., and Liloglou, T. (2009). Hypomethylation of retrotransposable elements correlates with genomic instability in non-small cell lung cancer. *Int. J. Cancer* **124,** 81–87.

Deininger, P. L., Moran, J. V., Batzer, M. A., and Kazazian, H. H., Jr. (2003). Mobile elements and mammalian genome evolution. *Curr. Opin. Genet. Dev.* **13,** 651–658.

Deiss, L. P., Feinstein, E., Berissi, H., Cohen, O., and Kimchi, A. (1995). Identification of a novel serine/threonine kinase and a novel 15-kD protein as potential mediators of the gamma interferon-induced cell death. *Genes Dev.* **9,** 15–30.

Dejeux, E., Rønneberg, J. A., Solvang, H., Bukholm, I., Geisler, S., Aas, T., Gut, I. G., Børresen-Dale, A. L., Lønning, P. E., Kristensen, V. N., *et al.* (2010). DNA methylation profiling in doxorubicin treated primary locally advanced breast tumours identifies novel genes associated with survival and treatment response. *Mol. Cancer* **9,** 68.

Deng, G., Chen, A., Hong, J., Chae, H. S., and Kim, Y. S. (1999). Methylation of CpG in a small region of the hMLH1 promoter invariably correlates with the absence of gene expression. *Cancer Res.* **59,** 2029–2033.

Deng, G., Peng, E., Gum, J., Terdiman, J., Sleisenger, M., and Kim, Y. S. (2002). Methylation of hMLH1 promoter correlates with the gene silencing with a region-specific manner in colorectal cancer. *Br. J. Cancer* **86,** 574–579.

Deng, D. J., Zhou, J., Zhu, B. D., Ji, J. F., Harper, J. C., and Powell, S. M. (2003a). Silencing-specific methylation and single nucleotide polymorphism of hMLH1 promoter in gastric carcinomas. *World J. Gastroenterol.* **9,** 26–29.

Deng, D., El-Rifai, W., Ji, J., Zhu, B., Trampont, P., Li, J., Smith, M. F., and Powel, S. M. (2003b). Hypermethylation of metallothionein-3 CpG island in gastric carcinoma. *Carcinogenesis* **24,** 25–29.

deVos, T., Tetzner, R., Model, F., Weiss, G., Schuster, M., Distler, J., Steiger, K. V., Grützmann, R., Pilarsky, C., Habermann, J. K., *et al.* (2009). Circulating methylated SEPT9 DNA in plasma is a biomarker for colorectal cancer. *Clin. Chem.* **55,** 1337–1346.

Dewannieux, M., Esnault, C., and Heidmann, T. (2003). LINE-mediated retrotransposition of marked Alu sequences. *Nat. Genet.* **35,** 41–48.

Dong, C. X., Deng, D. J., Pan, K. F., Zhang, L., Zhang, Y., Zhou, J., and You, W. C. (2009). Promoter methylation of p16 associated with *Helicobacter pylori* infection in precancerous gastric lesions: A population-based study. *Int. J. Cancer* **124,** 434–439.

Du, L., and Pertsemlidis, A. (2010). microRNAs and lung cancer: Tumors and 22-mers. *Cancer Metastasis Rev.* **29,** 109–122.

Ehrlich, M. (2002). DNA methylation in cancer: Too much, but also too little. *Oncogene* **21,** 5400–5413.

Ellinger, J., Kahl, P., von der Gathen, J., Rogenhofer, S., Heukamp, L. C., Gütgemann, I., Walter, B., Hofstädter, F., Büttner, R., Müller, S. C., *et al.* (2010). Global levels of histone modifications predict prostate cancer recurrence. *Prostate* **70,** 61–69.

Esteller, M. (2002). CpG island hypermethylation and tumor suppressor genes: A booming present, a brighter future. *Oncogene* **21,** 5427–5440.

Esteller, M. (2007). Cancer epigenomics: DNA methylomes and histone-modification maps. *Nat. Rev. Genet.* **8,** 286–298.

Esteller, M., Hamilton, S. R., Burger, P. C., Baylin, S. B., and Herman, J. G. (1999). Inactivation of the DNA repair gene O6-methylguanine-DNA methyltransferase by promoter hypermethylation is a common event in primary human neoplasia. *Cancer Res.* **59,** 793–797.

Esteller, M., Garcia-Foncillas, J., Andion, E., Goodman, S. N., Hidalgo, O. F., Vanaclocha, V., Baylin, S. B., and Herman, J. G. (2000). Inactivation of the DNA-repair gene MGMT and the clinical response of gliomas to alkylating agents. *N. Engl. J. Med.* **343,** 1350–1354.

Felsberg, J., Rapp, M., Loeser, S., Fimmers, R., Stummer, W., Goeppert, M., Steiger, H. J., Friedensdorf, B., Reifenberger, G., and Sabel, M. C. (2009). Prognostic significance of molecular markers and extent of resection in primary glioblastoma patients. *Clin. Cancer Res.* **15,** 6683–6693.

Ferreri, A. J., Dell'Oro, S., Capello, D., Ponzoni, M., Iuzzolino, P., Rossi, D., Pasini, F., Ambrosetti, A., Orvieto, E., Ferrarese, F., *et al.* (2004). Aberrant methylation in the promoter region of the reduced folate carrier gene is a potential mechanism of resistance to methotrexate in primary central nervous system lymphomas. *Br. J. Haematol.* **126,** 657–664.

Fiegl, H., Millinger, S., Goebel, G., Muller-Holzner, E., Marth, C., Laird, P. W., and Widschwendter, M. (2006). Breast cancer DNA methylation profiles in cancer cells and tumor stroma: Association with HER-2/neu status in primary breast cancer. *Cancer Res.* **66,** 29–33.

Figueroa, M. E., Lugthart, S., Li, Y., Erpelinck-Verschueren, C., Deng, X., Christos, P. J., Schifano, E., Booth, J., van Putten, W., Skrabanek, L., *et al.* (2010). DNA methylation signatures identify biologically distinct subtypes in acute myeloid leukemia. *Cancer Cell* **17,** 13–27.

Friedrich, M. G., Weisenberger, D. J., Cheng, J. C., Chandrasoma, S., Siegmund, K. D., Gonzalgo, M. L., Toma, M. I., Huland, H., Yoo, C., Tsai, Y. C., *et al.* (2004). Detection of methylated apoptosis-associated genes in urine sediments of bladder cancer patients. *Clin. Cancer Res.* **10,** 7457–7465.

Gao, W., Yu, Y., Cao, H., Shen, H., Li, X., Pan, S., and Shu, Y. (2010). Deregulated expression of miR-21, miR-143 and miR-181a in non small cell lung cancer is related to clinicopathologic characteristics or patient prognosis. *Biomed. Pharmacother.* doi:10.1016/j.biopha.2010.01.018.

Gerson, S. L. (2004). MGMT: Its role in cancer aetiology and cancer therapeutics. *Nat. Rev. Cancer* **4,** 296–307.

Gifford, G., Paul, J., Vasey, P. A., Kaye, S. B., and Brown, R. (2004). The acquisition of hMLH1 methylation in plasma DNA after chemotherapy predicts poor survival for ovarian cancer patients. *Clin. Cancer Res.* **10,** 4420–4426.

Girault, I., Tozlu, S., Lidereau, R., and Bièche, I. (2003). Expression analysis of DNA methyltransferases 1, 3A, and 3B in sporadic breast carcinomas. *Clin. Cancer Res.* **9,** 4415–4422.

Gonzalgo, M. L., Pavlovich, C. P., Lee, S. M., and Nelson, W. G. (2003). Prostate cancer detection by GSTP1 methylation analysis of postbiopsy urine specimens. *Clin. Cancer Res.* **9,** 2673–2677.

Goto, T., Mizukami, H., Shirahata, A., Sakata, M., Saito, M., Ishibashi, K., Kigawa, G., Nemoto, H., Sanada, Y., and Hibi, K. (2009). Aberrant methylation of the p16 gene is frequently detected in advanced colorectal cancer. *Anticancer Res.* **29,** 275–277.

Griffiths-Jones, S., Saini, H. K., van Dongen, S., and Enright, A. J. (2008). miRBase: Tools for microRNA genomics. *Nucleic Acids Res.* **36,** D154–D158, (Database issue).

Grombacher, T., Mitra, S., and Kaina, B. (1996). Induction of the alkyltransferase (MGMT) gene by DNA damaging agents and the glucocorticoid dexamethasone and comparison with the response of base excision repair genes. *Carcinogenesis* **17,** 2329–2336.

Grützmann, R., Molnar, B., Pilarsky, C., Habermann, J. K., Schlag, P. M., Saeger, H. D., Miehlke, S., Stolz, T., Model, F., Roblick, U. J., *et al.* (2008). Sensitive detection of colorectal cancer in peripheral blood by Septin 9 DNA methylation assay. *PLoS ONE* **3,** e3759.

Günther, T., Schneider-Stock, R., Pross, M., Manger, T., Malfertheiner, P., Lippert, H., and Roessner, A. (1998). Alterations of the p16/MTS1-tumor suppressor gene in gastric cancer. *Pathol. Res. Pract.* **194,** 809–813.

Gurrieri, F., and Accadia, M. (2009). Genetic imprinting: The paradigm of Prader–Willi and Angelman syndromes. *Endocr. Dev.* **14,** 20–28.

Hall, P. A., and Russell, S. E. (2004). The pathobiology of the septin gene family. *J. Pathol.* **204,** 489–505.

Hall, P. A., Jung, K., Hillan, K. J., and Russell, S. E. (2005). Expression profiling the human septin gene family. *J. Pathol.* **206,** 269–278.

Hall, G. L., Shaw, R. J., Field, E. A., Rogers, S. N., Sutton, D. N., Woolgar, J. A., Lowe, D., Liloglou, T., Field, J. K., and Risk, J. M. (2008). p16 Promoter methylation is a potential predictor of malignant transformation in oral epithelial dysplasia. *Cancer Epidemiol. Biomarkers Prev.* **17,** 2174–2179.

Harden, S. V., Tokumaru, Y., Westra, W. H., Goodman, S., Ahrendt, S. A., Yang, S. C., and Sidransky, D. (2003). Gene promoter hypermethylation in tumors and lymph nodes of stage I lung cancer patients. *Clin. Cancer Res.* **9,** 1370–1375.

Hegi, M. E., Diserens, A. C., Godard, S., Dietrich, P. Y., Regli, L., Ostermann, S., Otten, P., Van Melle, G., de Tribolet, N., and Stupp, R. (2004). Clinical trial substantiates the predictive value of O-6-methylguanine-DNA methyltransferase promoter methylation in glioblastoma patients treated with temozolomide. *Clin. Cancer Res.* **10,** 1871–1874.

Hegi, M. E., Diserens, A. C., Gorlia, T., Hamou, M. F., de Tribolet, N., Weller, M., Kros, J. M., Hainfellner, J. A., Mason, W., Mariani, L., *et al.* (2005). MGMT gene silencing and benefit from temozolomide in glioblastoma. *N Engl J. Med.* **352,** 997–1003.

Herbst, A., Wallner, M., Rahmig, K., Stieber, P., Crispin, A., Lamerz, R., and Kolligs, F. T. (2009). Methylation of helicase-like transcription factor in serum of patients with colorectal cancer is an independent predictor of disease recurrence. *Eur. J. Gastroenterol. Hepatol.* **21,** 565–569.

Herman, J. G., and Baylin, S. B. (2003). Gene silencing in cancer in association with promoter hypermethylation. *N Engl J. Med.* **349,** 2042–2054.

Herman, J. G., Umar, A., Polyak, K., Graff, J. R., Ahuja, N., Issa, J.-P. J., Markowitz, S., Willson, J. K. V., Hamilton, S. R., Kinzler, K. W., *et al.* (1998). Incidence and functional consequences of hMLH1 promoter hypermethylation in colorectal carcinoma. *Proc. Natl. Acad. Sci. USA* **95,** 6870–6875.

Hernandez-Vargas, H., Lambert, M.-P., Le Calvez-Kelm, F., Gouysse, G., McKay-Chopin, S., Tavtigian, S. V., Scoazec, J.-Y., and Hercep, Z. (2010). Hepatocellular carcinoma displays distinct DNA methylation signatures with potential as clinical predictors. *PLoS ONE* **5,** e9749.

Hitchins, M. P., and Ward, R. L. (2009). Constitutional (germline) MLH1 epimutation as an aetiological mechanism for hereditary non-polyposis colorectal cancer. *J. Med. Genet.* **46,** 793–802.

Hitchins, M. P., Wong, J. J. L., Suthers, G., Suter, C. M., Martin, D. I. K., Hawkins, N. J., and Ward, R. L. (2007). Inheritance of a cancer-associated MLH1 germ-line epimutation. *N. Engl. J. Med.* **356,** 697–705.

Ho, A. S., Huang, X., Cao, H., Christman-Skieller, C., Bennewith, K., Le, Q. T., and Koong, A. C. (2010). Circulating miR-210 as a novel hypoxia marker in pancreatic cancer. *Transl. Oncol.* **3,** 109–113.

Hoque, M. O., Begum, S., Topaloglu, O., Chatterjee, A., Rosenbaum, E., Van Criekinge, W., Westra, W. H., Schoenberg, M., Zahurak, M., Goodman, S. N., *et al.* (2006). Quantitation of promoter methylation of multiple genes in urine DNA and bladder cancer detection. *J. Natl Cancer Inst.* **98,** 996–1004.

Hoque, M. O., Begum, S., Brait, M., Jeronimo, C., Zahurak, M., Ostrow, K. L., Rosenbaum, E., Trock, B., Westra, W. H., Schoenberg, M., *et al.* (2008). Tissue inhibitor of metalloproteinases-3 promoter methylation is an independent prognostic factor for bladder cancer. *J. Urol.* **179,** 743–747.

Houbaviy, H. B., Murray, M. F., and Sharp, P. A. (2003). Embryonic stem cell-specific MicroRNAs. *Dev. Cell* **5,** 351–358.

House, M. G., Guo, M., Efron, D. T., Lillemoe, K. D., Cameron, J. L., Syphard, J. E., Hooker, C. M., Abraham, S. C., Montgomery, E. A., Herman, J. G., *et al.* (2003). Tumor suppressor gene hypermethylation as a predictor of gastric stromal tumor behavior. *J. Gastrointest. Surg.* **7,** 1004–1014.

Hu, S., Wang, X., Zhou, J., Chen, F., Zhang, X., Chen, W., Pan, Q., Zhao, F., Deng, D., and Qiao, Y. (2009). Natural History of low-grade cervical intraepithelial neoplasia and its relationship with p16 hypermethylation. *Chin. Cancer* **18,** 420–423.

Iliopoulos, D., Oikonomou, P., Messinis, I., and Tsezou, A. (2009). Correlation of promoter hypermethylation in hTERT, DAPK and MGMT genes with cervical oncogenesis progression. *Oncol. Rep.* **22,** 199–204.

Inbal, B., Bialik, S., Sabanay, I., Shani, G., and Kimchi, A. (2002). DAP kinase and DRP-1 mediate membrane blebbing and the formation of autophagic vesicles during programmed cell death. *J. Cell Biol.* **157,** 455–468.

Iorio, M. V., Casalini, P., Tagliabue, E., Ménard, S., and Croce, C. M. (2008). MicroRNA profiling as a tool to understand prognosis, therapy response and resistance in breast cancer. *Eur. J. Cancer* **44,** 2753–2759.

Iorns, E., Turner, N. C., Elliott, R., Syed, N., Garrone, O., Gasco, M., Tutt, A. N., Crook, T., Lord, C. J., and Ashworth, A. (2008). Identification of CDK10 as an important determinant of resistance to endocrine therapy for breast cancer. *Cancer Cell* **13,** 91–104.

Issa, J. P. (2004). CpG island methylator phenotype in cancer. *Nat. Rev. Cancer* **4,** 988–993.

Issa, J. P., Gharibyan, V., Cortes, J., Jelinek, J., Morris, G., Verstovsek, S., Talpaz, M., Garcia-Manero, G., and Kantarjian, H. M. (2005). Phase II study of low-dose decitabine in patients with chronic myelogenous leukemia resistant to imatinib mesylate. *J. Clin. Oncol.* **23,** 3948–3956.

Jang, C. W., Chen, C. H., Chen, C. C., Chen, J. Y., Su, Y. H., and Chen, R. H. (2002). TGF-beta induces apoptosis through Smad-mediated expression of DAP-kinase. *Nat. Cell Biol.* **4,** 51–58.

Jarmalaite, S., Jankevicius, F., Kurgonaite, K., Suziedelis, K., Mutanen, P., and Husgafvel-Pursiainen, K. (2008). Promoter hypermethylation in tumour suppressor genes shows association with stage, grade and invasiveness of bladder cancer. *Oncology* **75,** 145–151.

Jerónimo, C., Henrique, R., Hoque, M. O., Mambo, E., Ribeiro, F. R., Varzim, G., Oliveira, J., Teixeira, M. R., Lopes, C., and Sidransky, D. (2004). A quantitative promoter methylation profile of prostate cancer. *Clin. Cancer Res.* **10,** 8472–8478.

Jiang, Y., Dunbar, A., Gondek, L. P., Mohan, S., Rataul, M., O'Keefe, C., Sekeres, M., Saunthararajah, Y., and Maciejewski, J. P. (2009). Aberrant DNA methylation is a dominant mechanism in MDS progression to AML. *Blood* **113,** 1315–1325.

Jicai, Z., Zongtao, Y., Jun, L., Haiping, L., Jianmin, W., and Lihua, H. (2006). Persistent infection of hepatitis B virus is involved in high rate of p16 methylation in hepatocellular carcinoma. *Mol. Carcinog.* **45,** 530–536.

Jin, Z., Cheng, Y., Gu, W., Zheng, Y., Sato, F., Mori, Y., Olaru, A. V., Paun, B. C., Yang, J., Kan, T., *et al.* (2009). A multicenter, double-blinded validation study of methylation biomarkers for progression prediction in Barrett's esophagus. *Cancer Res.* **69,** 4112–4115.

Kalikin, L. M., Sims, H. L., and Petty, E. M. (2000). Genomic and expression analyses of alternatively spliced transcripts of the MLL septin-like fusion gene (MSF) that map to a 17q25 region of loss in breast and ovarian tumors. *Genomics* **63,** 165–172.

Karray-Chouayekh, S., Trifa, F., Khabir, A., Boujelbane, N., Sellami-Boudawara, T., Daoud, J., Frikha, M., Gargouri, A., and Mokdad-Gargouri, R. (2009). Clinical significance of epigenetic inactivation of hMLH1 and BRCA1 in Tunisian patients with invasive breast carcinoma. *J. Biomed. Biotechnol.* 2009, 369129.

Kato, K., Iida, S., Uetake, H., Takagi, Y., Yamashita, T., Inokuchi, M., Yamada, H., Kojima, K., and Sugihara, K. (2008). Methylated TMS1 and DAPK genes predict prognosis and response to chemotherapy in gastric cancer. *Int. J. Cancer* **122**, 603–608.

Kawakami, K., Brabender, J., Lord, R. V., Groshen, S., Greenwald, B. D., Krasna, M. J., Yin, J., Fleisher, A. S., Abraham, J. M., Beer, D. G., *et al.* (2000). Hypermethylated APC DNA in plasma and prognosis of patients with esophageal adenocarcinoma. *J. Natl. Cancer Inst.* **92**, 1805–1811.

Kazazian, H. H., Jr., and Goodier, J. L. (2002). LINE drive. retrotransposition and genome instability. *Cell* **110**, 277–280.

Khleif, S. N., DeGregori, J., Yee, C. L., Otterson, G. A., Kaye, F. J., Nevins, J. R., and Howley, P. M. (1996). Inhibition of cyclin D-CDK4/CDK6 activity is associated with an E2F-mediated induction of cyclin kinase inhibitor activity. *Proc. Natl. Acad. Sci. USA* **93**, 4350–4354.

Kim, D. H., Nelson, H. H., Wiencke, J. K., Christiani, D. C., Wain, J. C., Mark, E. J., and Kelsey, K. T. (2001). Promoter methylation of DAP-kinase: Association with advanced stage in non-small cell lung cancer. *Oncogene* **20**, 1765–1770.

Kim, D. S., Kim, M. J., Lee, J. Y., Kim, Y. Z., Kim, E. J., and Park, J. Y. (2007). Aberrant methylation of E-cadherin and H-cadherin genes in nonsmall cell lung cancer and its relation to clinicopathologic features. *Cancer* **110**, 2785–2792.

Kim, E. J., Kim, Y. J., Jeong, P., Ha, Y. S., Bae, S. C., and Kim, W. J. (2008). Methylation of the RUNX3 promoter as a potential prognostic marker for bladder tumor. *J. Urol.* **180**, 1141–1145.

Kim, M. S., Lee, J., and Sidransky, D. (2010). DNA methylation markers in colorectal cancer. *Cancer Metastasis Rev.* **29**, 181–206.

Kito, H., Suzuki, H., Ichikawa, T., Sekita, N., Kamiya, N., Akakura, K., Igarashi, T., Nakayama, T., Watanabe, M., Harigaya, K., *et al.* (2001). Hypermethylation of the CD44 gene is associated with progression and metastasis of human prostate cancer. *Prostate* **49**, 110–115.

Kosaka, N., Iguchi, H., Yoshioka, Y., Takeshita, F., Matsuki, Y., and Ochiya, T. (2010). Secretory mechanisms and intercellular transfer of microRNAs in living cells. *J. Biol. Chem.* doi:10.1074/jbc.M110.107821.

Koul, S., McKiernan, J. M., Narayan, G., Houldsworth, J., Bacik, J., Dobrzynski, D. L., Assaad, A. M., Mansukhani, M., Reuter, V. E., Bosl, G. J., *et al.* (2004). Role of promoter hypermethylation in Cisplatin treatment response of male germ cell tumors. *Mol. Cancer* **3**, 16.

Krishnamurthy, J., Ramsey, M. R., Ligon, K. L., Torrice, C., Koh, A., Bonner-Weir, S., and Sharpless, N. E. (2006). p16INK4a induces an age-dependent decline in islet regenerative potential. *Nature* **443**, 453–457.

Kronenwett, U., Ploner, A., Zetterberg, A., Bergh, J., Hall, P., Auer, G., and Pawitan, Y. (2006). Genomic instability and prognosis in breast carcinomas. *Cancer Epidemiol. Biomarkers Prev.* **15**, 1630–1635.

Krusche, C. A., Wülfing, P., Kersting, C., Vloet, A., Böcker, W., Kiesel, L., Beier, H. M., and Alfer, J. (2005). Histone deacetylase-1 and -3 protein expression in human breast cancer: A tissue microarray analysis. *Breast Cancer Res. Treat.* **90**, 15–23.

Kukitsu, T., Takayama, T., Miyanishi, K., Nobuoka, A., Katsuki, S., Sato, Y., Takimoto, R., Matsunaga, T., Kato, J., Sonoda, T., *et al.* (2008). Aberrant crypt foci as precursors of the dysplasia-carcinoma sequence in patients with ulcerative colitis. *Clin. Cancer Res.* **14**, 48–54.

Landi, M. T., Zhao, Y., Rotunno, M., Koshiol, J., Liu, H., Bergen, A. W., Rubagotti, M., Goldstein, A. M., Linnoila, I., Marincola, F. M., *et al.* (2010). MicroRNA expression differentiates histology and predicts survival of lung cancer. *Clin. Cancer Res.* **16**, 430–441.

Lawrie, C. H., Soneji, S., Marafioti, T., Cooper, C. D., Palazzo, S., Paterson, J. C., Cattan, H., Enver, T., Mager, R., Boultwood, J., *et al.* (2008). MicroRNA expression distinguishes between germinal center B cell-like and activated B cell-like subtypes of diffuse large B cell lymphoma. *Int. J. Cancer* **121**, 1156–1161.

Lee, J. T., and Lu, N. (1999). Targeted mutagenesis of Tsix leads to nonrandom X inactivation. *Cell* **99,** 47–57.

Lee, Y. Y., Kang, S. H., Seo, J. Y., Jung, C. W., Lee, K. U., Choe, K. J., Kim, B. K., Kim, N. K., Koeffler, H. P., and Bang, Y. J. (1997). Alterations of p16INK4A and p15INK4B genes in gastric carcinomas. *Cancer* **80,** 1889–1896.

Li, Q., Kopecky, K. J., Mohan, A., Willman, C. L., Appelbaum, F. R., Weick, J. K., and Issa, J. P. (1999). Estrogen receptor methylation is associated with improved survival in adult acute myeloid leukemia. *Clin. Cancer Res.* **5,** 1077–1084.

Li, L. C., Chui, R., Nakajima, K., Oh, B. R., Au, H. C., and Dahiya, R. (2000). Frequent methylation of estrogen receptor in prostate cancer: correlation with tumor progression. *Cancer Res.* **60,** 702–706.

Licchesi, J. D., Westra, W. H., Hooker, C. M., and Herman, J. G. (2008). Promoter hypermethylation of hallmark cancer genes in atypical adenomatous hyperplasia of the lung. *Clin. Cancer Res.* **14,** 2570–2578.

Lin, R. K., Hsu, H. S., Chang, J. W., Chen, C. Y., Chen, J. T., and Wang, Y. C. (2007a). Alteration of DNA methyltransferases contributes to 5'CpG methylation and poor prognosis in lung cancer. *Lung Cancer* **55,** 205–213.

Lin, J. C., Jeong, S., Liang, G., Takai, D., Fatemi, M., Tsai, Y. C., Egger, G., Gal-Yam, E. N., and Jones, P. A. (2007b). Role of nucleosomal occupancy in the epigenetic silencing of the MLH1 CpG island. *Cancer Cell* **12,** 432–444.

Liu, Y., Zhang, L., Ren, H., Zhang, G., Qin, F., Kong, G., Deng, G., and Ji, J. (2005). Promoter hypermethylation of the p16 gene in pre- and post-gastrectomy plasma of patients with gastric carcinoma. *J. Peking Univ. Health Sci.* **37,** 257–260.

Liu, B. L., Cheng, J. X., Zhang, W., Zhang, X., Wang, R., Lin, H., Huo, J. L., and Cheng, H. (2010a). Quantitative detection of multiple gene promoter hypermethylation in tumor tissue, serum, and cerebrospinal fluid predicts prognosis of malignant gliomas. *Neuro Oncol.* doi:10.1093/neuonc/nop064.

Liu, C. J., Kao, S. Y., Tu, H. F., Tsai, M. M., Chang, K. W., and Lin, S. C. (2010b). Increase of microRNA miR-31 level in plasma could be a potential marker of oral cancer. *Oral Dis.* doi:10.1111/j.16012009.01646.x.

Liu, C. J., Tsai, M. M., Hung, P. S., Kao, S. Y., Liu, T. Y., Wu, K. J., Chiou, S. H., Lin, S. C., and Chang, K. W. (2010c). miR-31 ablates expression of the HIF regulatory factor FIH to activate the HIF pathway in head and neck carcinoma. *Cancer Res.* **70,** 1635–1644.

Liu, X., Sempere, L. F., Ouyang, H., Memoli, V. A., Andrew, A. S., Luo, Y., Demidenko, E., Korc, M., Shi, W., Preis, M., et al. (2010d). MicroRNA-31 functions as an oncogenic microRNA in mouse and human lung cancer cells by repressing specific tumor suppressors. *J. Clin. Invest.* **120,** 1298–1309.

Lodes, M. J., Caraballo, M., Suciu, D., Munro, S., Kumar, A., and Anderson, B. (2009). Detection of cancer with serum miRNAs on an oligonucleotide microarray. *PLoS ONE* **4,** e6229.

Lofton-Day, C., Model, F., DeVos, T., Tetzner, R., Distler, J., Schuster, M., Song, X., Lesche, R., Liebenberg, V., Ebert, M., et al. (2008). DNA methylation biomarkers for blood-based colorectal cancer screening. *Clin. Chem.* **54**(2), 414–423.

López, M., Aguirre, J. M., Cuevas, N., Anzola, M., Videgain, J., Aguirregaviria, J., and Martínez de Pancorbo, M. (2003). Gene promoter hypermethylation in oral rinses of leukoplakia patients— A diagnostic and/or prognostic tool? *Eur. J. Cancer* **39,** 2306–2309.

Lu, C., Soria, J. C., Tang, X., Xu, X. C., Wang, L., Mao, L., Lotan, R., Kemp, B., Bekele, B. N., Feng, L., et al. (2004). Prognostic factors in resected stage I non-small-cell lung cancer: A multivariate analysis of six molecular markers. *J. Clin. Oncol.* **22,** 4575–4583.

Lu, J., Getz, G., Miska, E. A., Alvarez-Saavedra, E., Lamb, J., Peck, D., Sweet-Cordero, A., Ebert, B. L., Mak, R. H., Ferrando, A. A., et al. (2005). MicroRNA expression profiles classify human cancers. *Nature* **435**, 834–838.

Lu, L., Katsaros, D., de la Longrais, I. A., Sochirca, O., and Yu, H. (2007). Hypermethylation of let-7a-3 in epithelial ovarian cancer is associated with low insulin-like growth factor-II expression and favorable prognosis. *Cancer Res.* **67**, 10117–10122.

Lujambio, A., Calin, G. A., Villanueva, A., Ropero, S., Sánchez-Céspedes, M., Blanco, D., Montuenga, L. M., Rossi, S., Nicoloso, M. S., Faller, W. J., et al. (2008). A microRNA DNA methylation signature for human cancer metastasis. *Proc. Natl. Acad. Sci. USA* **105**, 13556–13561.

Luo, D., Zhang, B., Lv, L., Xiang, S., Liu, Y., Ji, J., and Deng, D. (2006). Methylation of CpG islands of p16 associated with progression of primary gastric carcinomas. *Lab. Invest.* **86**, 591–598.

Machida, E. O., Brock, M. V., Hooker, C. M., Nakayama, J., Ishida, A., Amano, J., Picchi, M. A., Belinsky, S. A., Herman, J. G., Taniguchi, S., et al. (2006). Hypermethylation of ASC/TMS1 is a sputum marker for late-stage lung cancer. *Cancer Res.* **66**, 6210–6218.

Makarla, P. B., Saboorian, M. H., Ashfaq, R., Toyooka, K. O., Toyooka, S., Minna, J. D., Gazdar, A. F., and Schorge, J. O. (2005). Promoter hypermethylation profile of ovarian epithelial neoplasms. *Clin. Cancer Res.* **11**, 5365–6369.

Malumbres, R., Sarosiek, K. A., Cubedo, E., Ruiz, J. W., Jiang, X., Gascoyne, R. D., Tibshirani, R., and Lossos, I. S. (2009). Differentiation stage-specific expression of microRNAs in B lymphocytes and diffuse large B-cell lymphomas. *Blood* **113**, 3754–3764.

Manuyakorn, A., Paulus, R., Farrell, J., Dawson, N. A., Tze, S., Cheung-Lau, G., Hines, O. J., Reber, H., Seligson, D. B., and Horvath, S. (2010). Cellular histone modification patterns predict prognosis and treatment response in resectable pancreatic adenocarcinoma: Results from RTOG 9704. *J. Clin. Oncol.* **28**, 1358–1365.

Mao, L., Lee, J. S., Fan, Y. H., Ro, J. Y., Batsakis, J. G., Lippman, S., Hittelman, W., and Hong, W. K. (1996). Frequent microsatellite alterations at chromosomes 9p21 and 3p14 in oral premalignant lesions and their value in cancer risk assessment. *Nat. Med.* **2**, 682–685.

Markou, A., Tsaroucha, E. G., Kaklamanis, L., Fotinou, M., Georgoulias, V., and Lianidou, E. S. (2008). Prognostic value of mature microRNA-21 and microRNA-205 overexpression in non-small cell lung cancer by quantitative real-time RT-PCR. *Clin. Chem.* **54**, 1696–1704.

Maruyama, R., Sugio, K., Yoshino, I., Maehara, Y., and Gazdar, A. F. (2004). Hypermethylation of FHIT as a prognostic marker in nonsmall cell lung carcinoma. *Cancer* **100**, 1472–1477.

Mathé, E. A., Nguyen, G. H., Bowman, E. D., Zhao, Y., Budhu, A., Schetter, A. J., Braun, R., Reimers, M., Kumamoto, K., Hughes, D., et al. (2009). MicroRNA expression in squamous cell carcinoma and adenocarcinoma of the esophagus: Associations with survival. *Clin. Cancer Res.* **15**, 6192–6200.

Matouk, I. J., DeGroot, N., Mezan, S., Ayesh, S., Abu-lail, R., Hochberg, A., and Galun, E. (2007). The H19 non-coding RNA is essential for human tumor growth. *PLoS ONE* **2**, e845.

Matsuda, Y., Ichida, T., Matsuzawa, J., Sugimura, K., and Asakura, H. (1999). p16 (INK4) is inactivated by extensive CpG methylation in human hepatocellular carcinoma. *Gastroenterology* **116**, 394–400.

McCabe, M. T., Brandes, J. C., and Vertino, P. M. (2009). Cancer DNA methylation: Molecular mechanisms and clinical implications. *Clin. Cancer Res.* **15**, 3927–3937.

McDade, S. S., Hall, P. A., and Russell, S. E. (2007). Translational control of SEPT9 isoforms is perturbed in disease. *Hum. Mol. Genet.* **16**, 742–752.

Mei, M., Deng, D., Liu, T. H., Sang, X. T., Lu, X., Xiang, H. D., Zhou, J., Wu, H., Yang, Y., Chen, J., et al. (2009). Clinical implications of microsatellite instability and MLH1 gene inactivation in sporadic insulinomas. *J. Clin. Endocrinol. Metab.* **94**, 3448–3457.

Meng, F., Henson, R., Lang, M., Wehbe, H., Maheshwari, S., Mendell, J. T., Jiang, J., Schmittgen, T. D., and Patel, T. (2006). Involvement of human micro-RNA in growth and response to chemotherapy in human cholangiocarcinoma cell lines. *Gastroenterology* 130, 2113–2129.

Meng, F., Henson, R., Wehbe-Janek, H., Ghoshal, K., Jacob, S. T., and Patel, T. (2007). MicroRNA-21 regulates expression of the PTEN tumor suppressor gene in human hepatocellular cancer. *Gastroenterology* 133, 647–658.

Menigatti, M., Truninger, K., Gebbers, J. O., Marbet, U., Marra, G., and Schär, P. (2009). Normal colorectal mucosa exhibits sex- and segment-specific susceptibility to DNA methylation at the hMLH1 and MGMT promoters. *Oncogene* 28, 899–909.

Merlo, A., Herman, J. G., Mao, L., Lee, D. J., Gabrielson, E., Burger, P. C., Baylin, S. B., and Sidransky, D. (1995). 5-prime CpG island methylation is associated with transcriptional silencing of the tumour suppressor p16/CDKN2/MTS1 in human cancers. *Nat. Med.* 1, 686–692.

Michie, A. M., McCaig, A. M., Nakagawa, R., and Vukovic, M. (2010). Death-associated protein kinase (DAPK) and signal transduction: Regulation in cancer. *FEBS J.* 277, 74–80.

Mirnezami, A. H., Pickard, K., Zhang, L., Primrose, J. N., and Packham, G. (2009). MicroRNAs: Key players in carcinogenesis and novel therapeutic targets. *Eur. J. Surg. Oncol.* 35, 339–347.

Mitchell, P. S., Parkin, R. K., Kroh, E. M., Fritz, B. R., Wyman, S. K., Pogosova-Agadjanyan, E. L., Peterson, A., Noteboom, J., O'Briant, K. C., Allen, A., *et al.* (2008). Circulating microRNAs as stable bloodbased markers for cancer detection. *Proc. Natl. Acad. Sci. USA* 105, 10513–10518.

Moriyama, T., Matsumoto, T., Nakamura, S., Jo, Y., Mibu, R., Yao, T., and Iida, M. (2007). Hypermethylation of p14 (ARF) may be predictive of colitic cancer in patients with ulcerative colitis. *Dis. Colon Rectum* 50, 1384–1392.

Motoyama, K., Inoue, H., Takatsuno, Y., Tanaka, F., Mimori, K., Uetake, H., Sugihara, K., and Mori, M. (2009). Over- and under-expressed microRNAs in human colorectal cancer. *Int. J. Oncol.* 34, 1069–1075.

Musolino, C., Sant'antonio, E., Penna, G., Alonci, A., Russo, S., Granata, A., and Allegra, A. (2010). Epigenetic therapy in myelodysplastic syndromes. *Eur. J. Haematol.* doi:10.1111/j.16002010.01433.x.

Nairz, K., Rottig, C., Rintelen, F., Zdobnov, E., Moser, M., and Hafen, E. (2006). Overgrowth caused by misexpression of a microRNA with dispensable wild-type function. *Dev. Biol.* 291, 314–324.

Nam, E. J., Yoon, H., Kim, S. W., Kim, H., Kim, Y. T., Kim, J. H., Kim, J. W., and Kim, S. (2008). MicroRNA expression profiles in serous ovarian carcinoma. *Clin. Cancer Res.* 14, 2690–2695.

Negri, G., Vittadello, F., Romano, F., Kasal, A., Rivasi, F., Girlando, S., Mian, C., and Egarter-Vigl, E. (2004). p16INK4a expression and progression risk of low-grade intraepithelial neoplasia of the cervix uteri. *Virchows Arch.* 445, 616–620.

Nehls, K., Vinokurova, S., Schmidt, D., Kommoss, F., Reuschenbach, M., Kisseljov, F., Einenkel, J., von Knebel Doeberitz, M., and Wentzensen, N. (2008). p16 methylation does not affect protein expression in cervical carcinogenesis. *Eur. J. Cancer* 44, 2496–2505.

Ng, E. K., Chong, W. W., Jin, H., Lam, E. K., Shin, V. Y., Yu, J., Poon, T. C., Ng, S. S., and Sung, J. J. (2009). Differential expression of microRNAs in plasma of patients with colorectal cancer: A potential marker for colorectal cancer screening. *Gut* 58, 1375–1381.

Nikiforova, M. N., Tseng, G. C., Steward, D., Diorio, D., and Nikiforov, Y. E. (2008). MicroRNA expression profiling of thyroid tumors: Biological significance and diagnostic utility. *J. Clin. Endocrinol. Metab.* 93, 1600–1608.

Nishida, N., Fukuda, Y., Komeda, T., Kita, R., Sando, T., Furukawa, M., Amenomori, M., Shibagaki, I., Nakao, K., Ikenaga, M., *et al.* (1994). Amplification and overexpression of the cyclin D1 gene in aggressive human hepatocellular carcinoma. *Cancer Res.* 54, 3107–3110.

Nosho, K., Irahara, N., Shima, K., Kure, S., Kirkner, G. J., Schernhammer, E. S., Hazra, A., Hunter, D. J., Quackenbush, J., Spiegelman, D., *et al.* (2008). Comprehensive biostatistical analysis of CpG island methylator phenotype in colorectal cancer using a large population-based sample. *PLoS ONE* 3, e3698.

Nuovo, G. J., Plaia, T. W., Belinsky, S. A., Baylin, S. B., and Herman, J. G. (1999). In situ detection of the hypermethylation-induced inactivation of the p16 gene as an early event in oncogenesis. *Proc. Natl. Acad. Sci. USA* **96**, 12754–12759.

Ogino, S., Nosho, K., Kirkner, G. J., Kawasaki, T., Chan, A. T., Schernhammer, E. S., Giovannucci, E. L., and Fuchs, C. S. (2008). A cohort study of tumoral LINE-1 hypomethylation and prognosis in colon cancer. *J. Natl. Cancer Inst.* **100**, 1734–1738.

Oh, B. K., Kim, H., Park, H. J., Shim, Y. H., Choi, J., Park, C., and Park, Y. N. (2007). DNA methyltransferase expression and DNA methylation in human hepatocellular carcinoma and their clinicopathological correlation. *Int. J. Mol. Med.* **20**, 65–73.

Oki, Y., Jelinek, J., Shen, L., Kantarjian, H. M., and Issa, J. P. (2008). Induction of hypomethylation and molecular response after decitabine therapy in patients with chronic myelomonocytic leukemia. *Blood* **111**, 2382–2384.

Oshimo, Y., Kuraoka, K., Nakayama, H., Kitadai, Y., Yoshida, K., Chayama, K., and Yasui, W. (2004). Epigenetic inactivation of SOCS-1 by CpG island hypermethylation in human gastric carcinoma. *Int. J. Cancer* **112**, 1003–1009.

Ota, N., Kawakami, K., Okuda, T., Takehara, A., Hiranuma, C., Oyama, K., Ota, Y., Oda, M., and Watanabe, G. (2006). Prognostic significance of p16(INK4a) hypermethylation in non-small cell lung cancer is evident by quantitative DNA methylation analysis. *Anticancer Res.* **26**, 3729–3732.

Papadopoulos, N., Nicolaides, N. C., Wei, Y.-F., Ruben, S. M., Carter, K. C., Rosen, C. A., Haseltine, W. A., Fleischmann, R. D., Fraser, C. M., Adams, M. D., *et al.* (1994). Mutation of a mutL homolog in hereditary colon cancer. *Science* **263**, 1625–1629.

Pass, H. I., Goparaju, C., Ivanov, S., Donington, J., Carbone, M., Hoshen, M., Cohen, D., Chajut, A., Rosenwald, S., Dan, H., *et al.* (2010). hsa-miR-29c* is linked to the prognosis of malignant pleural mesothelioma. *Cancer Res.* **70**, 1916–1924.

Pattamadilok, J., Huapai, N., Rattanatanyong, P., Vasurattana, A., Triratanachat, S., Tresukosol, D., and Mutirangura, A. (2008). LINE-1 hypomethylation level as a potential prognostic factor for epithelial ovarian cancer. *Int. J. Gynecol. Cancer* **18**, 711–717.

Pfeifer, K., Leighton, P. A., and Tilghman, S. M. (1996). The structural H19 gene is required for transgene imprinting. *Proc. Natl. Acad. Sci. USA* **93**, 13876–13883.

Pfister, S., Rea, S., Taipale, M., Mendrzyk, F., Straub, B., Ittrich, C., Thuerigen, O., Sinn, H. P., Akhtar, A., and Lichter, P. (2008). The histone acetyltransferase hMOF is frequently down-regulated in primary breast carcinoma and medulloblastoma and constitutes a biomarker for clinical outcome in medulloblastoma. *Int. J. Cancer* **122**, 1207–1213.

Pogribny, I. P., and James, S. J. (2002). De novo methylation of the p16INK4A gene in early preneoplastic liver and tumors induced by folate/methyl deficiency in rats. *Cancer Lett.* **187**, 69–75.

Poynter, J. N., Siegmund, K. D., Weisenberger, D. J., Long, T. I., Thibodeau, S. N., Lindor, N., Young, J., Jenkins, M. A., Hopper, J. L., and Baron, J. A. (2008). Molecular characterization of MSI-H colorectal cancer by MLHI promoter methylation, immunohistochemistry, and mismatch repair germline mutation screening. *Cancer Epidemiol. Biomarkers Prev.* **17**, 3208–3215.

Queiroz, C., Silva, T. C., Alves, V. A., Villa, L. L., Costa, M. C., Travassos, A. G., Filho, J. B., Studart, E., Cheto, T., and de Freitas, L. A. (2006). P16(INK4a) expression as a potential prognostic marker in cervical pre-neoplastic and neoplastic lesions. *Pathol. Res. Pract.* **202**, 77–83.

Quesnel, B., Guillerm, G., Vereecque, R., Wattel, E., Preudhomme, C., Bauters, F., Vanrumbeke, M., and Fenaux, P. (1998). Methylation of the p15(INK4b) gene in myelodysplastic syndromes is frequent and acquired during disease progression. *Blood* **91**, 2985–2990.

Rabinowits, G., Gerçel-Taylor, C., Day, J. M., Taylor, D. D., and Kloecker, G. H. (2009). Exosomal microRNA: A diagnostic marker for lung cancer. *Clin. Lung Cancer* **10**, 42–46.

Rachmilewitz, J., Goshen, R., Ariel, I., Schneider, T., de Groot, N., and Hochberg, A. (1992). Parental imprinting of the human H19 gene. *FEBS Lett.* **309,** 25–28.

Rainier, S., Johnson, L. A., Dobry, C. J., Ping, A. J., Grundy, P. E., and Feinberg, A. P. (1993). Relaxation of imprinted genes in human cancer. *Nature* **362,** 747–749.

Raponi, M., Dossey, L., Jatkoe, T., Wu, X., Chen, G., Fan, H., and Beer, D. G. (2009). MicroRNA classifiers for predicting prognosis of squamous cell lung cancer. *Cancer Res.* **69,** 5776–5783.

Raveh, T., Berissi, H., Eisenstein, M., Spivak, T., and Kimchi, A. (2000). A functional genetic screen identifies regions at the C-terminal tail and death-domain of death-associated protein kinase that are critical for its proapoptotic activity. *Proc. Natl. Acad. Sci. USA* **97,** 1572–1577.

Raveh, T., Droguett, G., Horwitz, M. S., DePinho, R. A., and Kimchi, A. (2001). DAP kinase activates a p19ARF/p53-mediated apoptotic checkpoint to suppress oncogenic transformation. *Nat. Cell Biol.* **3,** 1–7.

Riccio, A., Sparago, A., Verde, G., De Crescenzo, A., Citro, V., Cubellis, M. V., Ferrero, G. B., Silengo, M. C., Russo, S., Larizza, L., *et al.* (2009). Inherited and Sporadic Epimutations at the IGF2-H19 locus in Beckwith–Wiedemann syndrome and Wilms' tumor. *Endocr. Dev.* **14,** 1–9.

Rivera, A. L., Pelloski, C. E., Gilbert, M. R., Colman, H., De La Cruz, C., Sulman, E. P., Bekele, B. N., and Aldape, K. D. (2010). MGMT promoter methylation is predictive of response to radiotherapy and prognostic in the absence of adjuvant alkylating chemotherapy for glioblastoma. *Neuro Oncol.* **12,** 116–121.

Robertson, K. D. (2005). DNA methylation and human disease. *Nat. Rev. Genet.* **6,** 597–610.

Roldo, C., Missiaglia, E., Hagan, J. P., Falconi, M., Capelli, P., Bersani, S., Calin, G. A., Volinia, S., Liu, C. G., Scarpa, A., *et al.* (2006). MicroRNA expression abnormalities in pancreatic endocrine and acinar tumors are associated with distinctive pathologic features and clinical behavior. *J. Clin. Oncol.* **24,** 4677–4684.

Roman-Gomez, J., Jimenez-Velasco, A., Agirre, X., Cervantes, F., Sanchez, J., Garate, L., Barrios, M., Castillejo, J. A., Navarro, G., Colomer, D., *et al.* (2005). Promoter hypomethylation of the LINE-1 retrotransposable elements activates sense/antisense transcription and marks the progression of chronic myeloid leukemia. *Oncogene* **24,** 7213–7223.

Roman-Gomez, J., Jimenez-Velasco, A., Agirre, X., Castillejo, J. A., Navarro, G., San Jose-Eneriz, E., Garate, L., Cordeu, L., Cervantes, F., Prosper, F., *et al.* (2008). Repetitive DNA hypomethylation in the advanced phase of chronic myeloid leukemia. *Leuk. Res.* **32,** 487–490.

Rosenbaum, E., Hoque, M. O., Cohen, Y., Zahurak, M., Eisenberger, M. A., Epstein, J. I., Partin, A. W., and Sidransky, D. (2005). Promoter hypermethylation as an independent prognostic factor for relapse in patients with prostate cancer following radical prostatectomy. *Clin. Cancer Res.* **11,** 8321–8325.

Russell, S. E., and Hall, P. A. (2005). Do septins have a role in cancer? *Br. J. Cancer* **93,** 499–503.

Russell, S. E., McIlhatton, M. A., Burrows, J. F., Donaghy, P. G., Chanduloy, S., Petty, E. M., Kalikin, L. M., Church, S. W., McIlroy, S., Harkin, D. P., *et al.* (2000). Isolation and mapping of a human septin gene to a region on chromosome 17q, commonly deleted in sporadic epithelial ovarian tumors. *Cancer Res.* **60,** 4729–4734.

Safar, A. M., Spencer, H., 3rd, Su, X., Coffey, M., Cooney, C. A., Ratnasinghe, L. D., Hutchins, L. F., and Fan, C. Y. (2005). Methylation profiling of archived non-small cell lung cancer: A promising prognostic system. *Clin. Cancer Res.* **11,** 4400–4405.

Saito, Y., Kanai, Y., Nakagawa, T., Sakamoto, M., Saito, H., Ishii, H., and Hirohashi, S. (2003). Increased protein expression of DNA methyltransferase (DNMT) 1 is significantly correlated with the malignant potential and poor prognosis of human hepatocellular carcinomas. *Int. J. Cancer* **105,** 527–532.

Saito, K., Kawakami, K., Matsumoto, I., Oda, M., Watanabe, G., and Minamoto, T. (2010). Long interspersed nuclear element 1 hypomethylation is a marker of poor prognosis in stage IA non-small cell lung cancer. *Clin. Cancer Res.* **16,** 2418–2426.

Sarkar, F. H., Li, Y., Wang, Z., Kong, D., and Ali, S. (2010). Implication of microRNAs in drug resistance for designing novel cancer therapy. *Drug Resist. Updat.* doi:10.1016/j.drup. 2010.02.001.

Sarver, A. L., French, A. J., Borralho, P. M., Thayanithy, V., Oberg, A. L., Silverstein, K. A., Morlan, B. W., Riska, S. M., Boardman, L. A., Cunningham, J. M., *et al.* (2009). Human colon cancer profiles show differential microRNA expression depending on mismatch repair status and are characteristic of undifferentiated proliferative states. *BMC Cancer* **9,** 401.

Sassaman, D. M., Dombroski, B. A., Moran, J. V., Kimberland, M. L., Naas, T. P., DeBerardinis, R. J., Gabriel, A., Swergold, G. D., and Kazazian, H. H., Jr. (1997). Many human L1 elements are capable of retrotransposition. *Nat. Genet.* **16,** 37–43.

Sathyanarayana, U. G., Padar, A., Suzuki, M., Maruyama, R., Shigematsu, H., Hsieh, J. T., Frenkel, E. P., and Gazdar, A. F. (2003). Aberrant promoter methylation of laminin-5-encoding genes in prostate cancers and its relationship to clinicopathological features. *Clin. Cancer Res.* **9,** 6395–6400.

Satoh, A., Toyota, M., Itoh, F., Sasaki, Y., Suzuki, H., Ogi, K., Kikuchi, T., Mita, H., Yamashita, T., Kojima, T., *et al.* (2003). Epigenetic inactivation of CHFR and sensitivity to microtubule inhibitors in gastric cancer. *Cancer Res.* **63,** 8606–8613.

Satzger, I., Mattern, A., Kuettler, U., Weinspach, D., Voelker, B., Kapp, A., and Gutzmer, R. (2010). MicroRNA-15b represents an independent prognostic parameter and is correlated with tumor cell proliferation and apoptosis in malignant melanoma. *Int. J. Cancer* **126,** 2553–2562.

Schaefer, A., Jung, M., Mollenkopf, H. J., Wagner, I., Stephan, C., Jentzmik, F., Miller, K., Lein, M., Kristiansen, G., and Jung, K. (2010). Diagnostic and prognostic implications of microRNA profiling in prostate carcinoma. *Int. J. Cancer* **126,** 1166–1176.

Schetter, A. J., Leung, S. Y., Sohn, J. J., Zanetti, K. A., Bowman, E. D., Yanaihara, N., Yuen, S. T., Chan, T. L., Kwong, D. L., Au, G. K., *et al.* (2008). MicroRNA expression profiles associated with prognosis and therapeutic outcome in colon adenocarcinoma. *JAMA* **299,** 425–436.

Schulmann, K., Sterian, A., Berki, A., Yin, J., Sato, F., Xu, Y., Olaru, A., Wang, S., Mori, Y., Deacu, E., *et al.* (2005). Inactivation of p16, RUNX3, and HPP1 occurs early in Barrett's-associated neoplastic progression and predicts progression risk. *Oncogene* **24,** 4138–4148.

Scott, M., Hyland, P. L., McGregor, G., Hillan, K. J., Russell, S. E. H., and Hall, P. A. (2005). Multimodality expression profiling shows SEPT9 to be overexpressed in a wide range of human tumors. *Oncogene* **24,** 4688–4700.

Seike, M., Gemma, A., Hosoya, Y., Hemmi, S., Taniguchi, Y., Fukuda, Y., Yamanaka, N., and Kudoh, S. (2000). Increase in the frequency of p16INK4 gene inactivation by hypermethylation in lung cancer during the process of metastasis and its relation to the status of p53. *Clin. Cancer Res.* **6,** 4307–4313.

Seligson, D. B., Horvath, S., Shi, T., Yu, H., Tze, S., Grunstein, M., and Kurdistani, S. K. (2005). Global histone modification patterns predict risk of prostate cancer recurrence. *Nature* **435,** 1262–1266.

Shames, D. S., Girard, L., Gao, B., Sato, M., Lewis, C. M., Shivapurkar, N., Jiang, A., Perou, C. M., Kim, Y. H., Pollack, J. R., *et al.* (2006). A genome-wide screen for promoter methylation in lung cancer identifies novel methylation markers for multiple malignancies. *PLoS Med.* **3,** e486.

Shen, W. W., Wu, J., Cai, L., Liu, B. Y., Gao, Y., Chen, G. Q., and Fu, G. H. (2007). Expression of anion exchanger 1 sequestrates p16 in the cytoplasm in gastric and colonic adenocarcinoma. *Neoplasia* **9,** 812–819.

Shen, L., Kantarjian, H., Guo, Y., Lin, E., Shan, J., Huang, X., Berry, D., Ahmed, S., Zhu, W., Pierce, S., *et al.* (2010). DNA methylation predicts survival and response to therapy in patients with myelodysplastic syndromes. *Clin. Oncol.* **28,** 605–613.

Shibata, S., Yokota, T., and Wutz, A. (2008). Synergy of Eed and Tsix in the repression of Xist gene and X-chromosome inactivation. *EMBO J.* **27,** 1816–1826.

Shim, Y. H., Kang, G. H., and Ro, J. Y. (2000). Correlation of p16 hypermethylation with p16 protein loss in sporadic gastric carcinomas. *Lab. Invest.* **80,** 689–695.

Shirahata, A., Sakata, M., Sakuraba, K., Goto, T., Mizukami, H., Saito, M., Ishibashi, K., Kigawa, G., Nemoto, H., Sanada, Y., *et al.* (2009). Vimentin methylation as a marker for advanced colorectal carcinoma. *Anticancer Res.* **29,** 279–281.

Si, M. L., Zhu, S., Wu, H., Lu, Z., Wu, F., and Mo, Y. Y. (2007). miR-21-mediated tumor growth. *Oncogene* **26,** 2799–2803.

Silver, D. P., Richardson, A. L., Eklund, A. C., Wang, Z. C., Szallasi, Z., Li, Q., Juul, N., Leong, C. O., Calogrias, D., Buraimoh, A., *et al.* (2010). Efficacy of neoadjuvant Cisplatin in triple-negative breast cancer. *J. Clin. Oncol.* **28,** 1145–1153.

Simpkins, S. B., Bocker, T., Swisher, E. M., Mutch, D. G., Gersell, D. J., Kovatich, A. J., Palazzo, J. P., Fishel, R., and Goodfellow, P. J. (1999). MLH1 promoter methylation and gene silencing is the primary cause of microsatellite instability in sporadic endometrial cancers. *Hum. Mol. Genet.* **8,** 661–666.

Slaby, O., Svoboda, M., Fabian, P., Smerdova, T., Knoflickova, D., Bednarikova, M., Nenutil, R., and Vyzula, R. (2007). Altered expression of miR-21, miR-31, miR-143 and miR-145 is related to clinicopathologic features of colorectal cancer. *Oncology* **7,** 397–402.

Sørensen, K. D., Wild, P. J., Mortezavi, A., Adolf, K., Tørring, N., Heebøll, S., Ulhøi, B. P., Ottosen, P., Sulser, T., Hermanns, T., *et al.* (2009). Genetic and epigenetic SLC18A2 silencing in prostate cancer is an independent adverse predictor of biochemical recurrence after radical prostatectomy. *Clin. Cancer Res.* **15,** 1400–1410.

Stamatopoulos, B., Meuleman, N., Haibe-Kains, B., Saussoy, P., Van Den Neste, E., Michaux, L., Heimann, P., Martiat, P., Bron, D., and Lagneaux, L. (2009). microRNA-29c and microRNA-223 down-regulation has in vivo significance in chronic lymphocytic leukemia and improves disease risk stratification. *Blood* **113,** 5237–5245.

Steensma, D. P., Baer, M. R., Slack, J. L., Buckstein, R., Godley, L. A., Garcia-Manero, G., Albitar, M., Larsen, J. S., Arora, S., Cullen, M. T., *et al.* (2010). Multicenter study of decitabine administered daily for 5 days every 4 weeks to adults with myelodysplastic syndromes: The alternative dosing for outpatient treatment (ADOPT) trial. *J. Clin. Oncol.* **27,** 3842–3848.

Stone, A. R., Bobo, W., Brat, D. J., Devi, N. S., Van Meir, E. G., and Vertino, P. M. (2004). Aberrant methylation and down-regulation of TMS1/ASC in human glioblastoma. *Am. J. Pathol.* **165,** 1151–1161.

Strathdee, G., MacKean, M. J., Illand, M., and Brown, R. (1999). A role for methylation of the hMLH1 promoter in loss of hMLH1 expression and drug resistance in ovarian cancer. *Oncogene* **18,** 2335–2341.

Suh, M. R., Lee, Y., Kim, J. Y., Kim, S. K., Moon, S. H., Lee, J. Y., Cha, K. Y., Chung, H. M., Yoon, H. S., Moon, S. Y., *et al.* (2004). Human embryonic stem cells express a unique set of microRNAs. *Dev. Biol.* **270,** 488–498.

Sun, Y., Deng, D., You, W. C., Bai, H., Zhang, L., Zhou, J., Shen, L., Ma, J. L., Xie, Y. Q., and Li, J. Y. (2004). Methylation of p16 CpG islands associated with malignant transformation of gastric dysplasia in a population-based study. *Clin. Cancer Res.* **10,** 5087–5093.

Suter, C. M., Martin, D. I. K., and Ward, R. L. (2004). Germline epimutation of MLH1 in individuals with multiple cancers. *Nat. Genet.* **36,** 497–501.

Svoboda, M., Izakovicova Holla, L., Sefr, R., Vrtkova, I., Kocakova, I., Tichy, B., and Dvorak, J. (2008). Micro-RNAs miR125b and miR137 are frequently upregulated in response to capecitabine chemoradiotherapy of rectal cancer. *Int. J. Oncol.* **33,** 541–547.

Swisher, E. M., Gonzalez, R. M., Taniguchi, T., Garcia, R. L., Walsh, T., Goff, B. A., and Welcsh, P. (2009). Methylation and protein expression of DNA repair genes: association with chemotherapy exposure and survival in sporadic ovarian and peritoneal carcinomas. *Mol. Cancer* **8,** 48.

Symer, D. E., Connelly, C., Szak, S. T., Caputo, E. M., Cost, G. J., Parmigiani, G., and Boeke, J. D. (2002). Human l1 retrotransposition is associated with genetic instability in vivo. *Cell* **110**, 327–338.

Szafranska, A. E., Davison, T. S., Shingara, J., Doleshal, M., Riggenbach, J. A., Morrison, C. D., Jewell, S., and Labourier, E. (2008). Accurate molecular characterization of formalin-fixed, paraffin-embedded tissues by microRNA expression profiling. *J. Mol. Diagn.* **10**, 415–423.

Tada, Y., Wada, M., Taguchi, K., Mochida, Y., Kinugawa, N., Tsuneyoshi, M., Naito, S., and Kuwano, M. (2002). The association of death-associated protein kinase hypermethylation with early recurrence in superficial bladder cancers. *Cancer Res.* **62**, 4048–4053.

Takai, D., Gonzales, F. A., Tsai, Y. C., Thayer, M. J., and Jones, P. A. (2001). Large scale mapping of methylcytosines in CTCF-binding sites in the human H19 promoter and aberrant hypomethylation in human bladder cancer. *Hum. Mol. Genet.* **10**, 2619–2626.

Takamizawa, J., Konishi, H., Yanagisawa, K., Tomida, S., Osada, H., Endoh, H., Harano, T., Yatabe, Y., Nagino, M., Nimura, Y., *et al.* (2004). Reduced expression of the let-7 microRNAs in human lung cancers in association with shortened postoperative survival. *Cancer Res.* **64**, 3753–3756.

Takaoka, A. S., Kakiuchi, H., Itoh, F., Hinoda, Y., Kusano, M., Ohara, M., Tsukakoshi, H., Hosokawa, M., and Imai, K. (1997). Infrequent alterations of the p16 (MTS-1) gene in human gastric cancer. *Tumour Biol.* **18**, 95–103.

Tan, A. C., Jimeno, A., Lin, S. H., Wheelhouse, J., Chan, F., Solomon, A., Rajeshkumar, N. V., Rubio-Viqueira, B., and Hidalgo, M. (2009). Characterizing DNA methylation patterns in pancreatic cancer genome. *Mol. Oncol.* **3**, 425–438.

Tanemura, A., Terando, A. M., Sim, M. S., van Hoesel, A. Q., de Maat, M. F., Morton, D. L., and Hoon, D. S. (2009). CpG island methylator phenotype predicts progression of malignant melanoma. *Clin. Cancer Res.* **15**, 1801–1807.

Tang, X., Khuri, F. R., Lee, J. J., Kemp, B. L., Liu, D., Hong, W. K., and Mao, L. (2000). Hypermethylation of the death-associated protein (DAP) kinase promoter and aggressiveness in stage-I non-small-cell lung cancer. *J. Natl. Cancer Inst.* **92**, 1511–1516.

Tänzer, M., Balluff, B., Distler, J., Hale, K., Leodolter, A., Röcken, C., Molnar, B., Schmid, R., Lofton-Day, C., Schuster, T., *et al.* (2010). Performance of epigenetic markers SEPT9 and ALX4 in plasma for detection of colorectal precancerous lesions. *PLoS ONE* **5**, e9061doi:10.1371/journal.pone.0009061.

Taylor, D. D., and Gercel-Taylor, C. (2008). MicroRNA signatures of tumor-derived exosomes as diagnostic biomarkers of ovarian cancer. *Gynecol. Oncol.* **110**, 13–21.

Thorvaldsen, J. L., Duran, K. L., and Bartolomei, M. S. (1998). Deletion of the H19 differentially methylated domain results in loss of imprinted expression of H19 and Igf2. *Genes Dev.* **12**, 3693–3702.

Tien, H. F., Tang, J. H., Tsay, W., Liu, M. C., Lee, F. Y., Wang, C. H., Chen, Y. C., and Shen, M. C. (2001). Methylation of the p15(INK4B) gene in myelodysplastic syndrome: It can be detected early at diagnosis or during disease progression and is highly associated with leukaemic transformation. *Br. J. Haematol.* **112**, 148–154.

Tomizawa, Y., Iijima, H., Nomoto, T., Iwasaki, Y., Otani, Y., Tsuchiya, S., Saito, R., Dobashi, K., Nakajima, T., and Mori, M. (2004). Clinicopathological significance of aberrant methylation of RARbeta2 at 3p24, RASSF1A at 3p21.3, and FHIT at 3p14.2 in patients with non-small cell lung cancer. *Lung Cancer* **46**, 305–312.

Toyota, M., Ahuja, N., Ohe-Toyota, M., Herman, J. G., Baylin, S. B., and Issa, J. P. (1999a). CpG island methylator phenotype in colorectal cancer. *Proc. Natl. Acad. Sci. USA* **96**, 8681–8686.

Toyota, M., Ahuja, N., Suzuki, H., Itoh, F., Ohe-Toyota, M., Imai, K., Baylin, S. B., and Issa, J. P. (1999b). Aberrant methylation in gastric cancer associated with the CpG island methylator phenotype. *Cancer Res.* **59**, 5438–5442.

Tsujiura, M., Ichikawa, D., Komatsu, S., Shiozaki, A., Takeshita, H., Kosuga, T., Konishi, H., Morimura, R., Deguchi, K., Fujiwara, H., *et al.* (2010). Circulating microRNAs in plasma of patients with gastric cancers. *Br. J. Cancer* **102**, 1174–1179.

Tufarelli, C., Stanley, J. A., Garrick, D., Sharpe, J. A., Ayyub, H., Wood, W. G., and Higgs, D. R. (2003). Transcription of antisense RNA leading to gene silencing and methylation as a novel cause of human genetic disease. *Nat. Genet.* **34**, 157–165.

Tzao, C., Tung, H. J., Jin, J. S., Sun, G. H., Hsu, H. S., Chen, B. H., Yu, C. P., and Lee, S. C. (2009). Prognostic significance of global histone modifications in resected squamous cell carcinoma of the esophagus. *Mod. Pathol.* **22**, 252–260.

Ueda, T., Volinia, S., Okumura, H., Shimizu, M., Taccioli, C., Rossi, S., Alder, H., Liu, C. G., Oue, N., Yasui, W., *et al.* (2010). Relation between microRNA expression and progression and prognosis of gastric cancer: a microRNA expression analysis. *Lancet Oncol.* **11**, 136–146.

Umetani, N., Takeuchi, H., Fujimoto, A., Shinozaki, M., Bilchik, A. J., and Hoon, D. S. (2004). Epigenetic inactivation of ID4 in colorectal carcinomas correlates with poor differentiation and unfavorable prognosis. *Clin. Cancer Res.* **10**, 7475–7483.

Usadel, H., Brabender, J., Danenberg, K. D., Jerónimo, C., Harden, S., Engles, J., Danenberg, P. V., Yang, S., and Sidransky, D. (2002). Quantitative adenomatous polyposis coli promoter methylation analysis in tumor tissue, serum, and plasma DNA of patients with lung cancer. *Cancer Res.* **62**, 371–375.

Ushijima, T. (2005). Detection and interpretation of altered methylation patterns in cancer cells. *Nat. Rev. Cancer* **5**, 223–231.

Valadi, H., Ekström, K., Bossios, A., Sjöstrand, M., Lee, J. J., and Lötvall, J. O. (2007). Exosome-mediated transfer of mRNAs and microRNAs is a novel mechanism of genetic exchange between cells. *Nat. Cell Biol.* **9**, 654–659.

Valastyan, S., Reinhardt, F., Benaich, N., Calogrias, D., Szász, A. M., Wang, Z. C., Brock, J. E., Richardson, A. L., and Weinberg, R. A. (2009). A pleiotropically acting microRNA, miR-31, inhibits breast cancer metastasis. *Cell* **137**, 1032–1046.

Vidal, D. O., Paixão, V. A., Brait, M., Souto, E. X., Caballero, O. L., Lopes, L. F., and Vettore, A. L. (2007). Aberrant methylation in pediatric myelodysplastic syndrome. *Leuk. Res.* **31**, 175–181.

Vielhauer, V., Sarafoff, M., Gais, P., and Rabes, H. M. (2001). Cell type-specific induction of O6-alkylguanine-DNA alkyltransferase mRNA expression in rat liver during regeneration, inflammation and preneoplasia. *J. Cancer Res. Clin. Oncol.* **127**, 591–602.

Viswanathan, S. R., Powers, J. T., Einhorn, W., Hoshida, Y., Ng, T. L., Toffanin, S., O'Sullivan, M., Lu, J., Phillips, L. A., Lockhart, V. L., *et al.* (2009). Lin28 promotes transformation and is associated with advanced human malignancies. *Nat. Genet.* **41**, 843–848.

Walther, A., Houlston, R., and Tomlinson, I. (2008). Association between chromosomal instability and prognosis in colorectal cancer: A meta-analysis. *Gut* **57**, 941–950.

Wang, J. S., Guo, M., Montgomery, E. A., Thompson, R. E., Cosby, H., Hicks, L., Wang, S., Herman, J. G., and Canto, M. I. (2009a). DNA promoter hypermethylation of p16 and APC predicts neoplastic progression in Barrett's esophagus. *Am. J. Gastroenterol.* **104**, 2153–2160.

Wang, J., Chen, J., Chang, P., LeBlanc, A., Li, D., Abbruzzesse, J. L., Frazier, M. L., Killary, A. M., and Sen, S. (2009b). MicroRNAs in plasma of pancreatic ductal adenocarcinoma patients as novel blood-based biomarkers of disease. *Cancer Prev. Res. (Phila Pa)* **2**, 807–813.

Watanabe, Y., Ueda, H., Etoh, T., Koike, E., Fujinami, N., Mitsuhashi, A., and Hoshiai, H. (2007). A change in promoter methylation of hMLH1 is a cause of acquired resistance to platinum-based chemotherapy in epithelial ovarian cancer. *Anticancer Res.* **27**(3B), 1449–1452.

Watts, G. S., Pieper, R. O., Costello, J. F., Peng, Y. M., Dalton, W. S., and Futscher, B. W. (1997). Methylation of discrete regions of the O6-methylguanine DNA methyltransferase (MGMT) CpG island is associated with heterochromatinization of the MGMT transcription start site and silencing of the gene. *Mol. Cell. Biol.* **17,** 5612–5619.

Weichert, W., Röske, A., Gekeler, V., Beckers, T., Ebert, M. P., Pross, M., Dietel, M., Denkert, C., and Röcken, C. (2008a). Association of patterns of class I histone deacetylase expression with patient prognosis in gastric cancer: A retrospective analysis. *Lancet Oncol.* **9,** 139–148.

Weichert, W., Röske, A., Gekeler, V., Beckers, T., Stephan, C., Jung, K., Fritzsche, F. R., Niesporek, S., Denkert, C., Dietel, M., *et al.* (2008b). Histone deacetylases 1, 2 and 3 are highly expressed in prostate cancer and HDAC2 expression is associated with shorter PSA relapse time after radical prostatectomy. *Br. J. Cancer* **98,** 604–610.

Weisenberger, D. J., Siegmund, K. D., Campan, M., Young, J., Long, T. I., Faasse, M. A., Kang, G. H., Widschwendter, M., Weener, D., Buchanan, D., *et al.* (2006). CpG island methylator phenotype underlies sporadic microsatellite instability and is tightly associated with BRAF mutation in colorectal cancer. *Nat. Genet.* **38,** 787–793.

Weksberg, R., Shuman, C., and Beckwith, J. B. (2010). Beckwith–Wiedemann syndrome. *Eur. J. Hum. Genet.* **18,** 8–14.

Woodson, K., O'Reilly, K. J., Ward, D. E., Walter, J., Hanson, J., Walk, E. L., and Tangrea, J. A. (2006). CD44 and PTGS2 methylation are independent prognostic markers for biochemical recurrence among prostate cancer patients with clinically localized disease. *Epigenetics* **1,** 183–186.

Wu, M. S., Shun, C. T., Sheu, J. C., Wang, H. P., Wang, J. T., Lee, W. J., Chen, C. J., Wang, T. H., and Lin, J. T. (1998). Overexpression of mutant p53 and c-erbB-2 proteins and mutations of the p15 and p16 genes in human gastric carcinoma: With respect to histological subtypes and stages. *J. Gastroenterol. Hepatol.* **13,** 305–310.

Wu, J., Qin, Y., Li, B., He, W. Z., and Sun, Z. L. (2008). Hypomethylated and hypermethylated profiles of H19DMR are associated with the aberrant imprinting of IGF2 and H19 in human hepatocellular carcinoma. *Genomics* **91,** 443–450.

Xi, Y., Formentini, A., Chien, M., Weir, D. B., Russo, J. J., Ju, J., Kornmann, M., and Ju, J. (2006). Prognostic values of microRNAs in colorectal cancer. *Biomark. Insights* **2,** 113–121.

Xiang, S., Liu, Z., Zhang, B., Zhou, J., Zhu, B. D., Ji, J., and Deng, D. (2010). Methylation status of individual CpG sites within Alu elements in the human genome and Alu hypomethylation in gastric carcinomas. *BMC Cancer* **10,** 44.

Xu, J., Wang, H. L., Lu, G. C., Wang, Z. J., Lin, X., and Zhou, H. W. (2007). Clinical significance of detection of tumor suppressor genes aberrant methylation in cervical carcinoma tissue. *Zhonghua Fu Chan Ke Za Zhi* **42,** 394–397.

Yan, L. X., Huang, X. F., Shao, Q., Huang, M. Y., Deng, L., Wu, Q. L., Zeng, Y. X., and Shao, J. Y. (2008). MicroRNA miR-21 overexpression in human breast cancer is associated with advanced clinical stage, lymph node metastasis and patient poor prognosis. *RNA* **14,** 2348–2360.

Yanaihara, N., Caplen, N., Bowman, E., Seike, M., Kumamoto, K., Yi, M., Stephens, R. M., Okamoto, A., Yokota, J., Tanaka, T., *et al.* (2006). Unique microRNA molecular profiles in lung cancer diagnosis and prognosis. *Cancer Cell* **9,** 189–198.

Yang, N., Kaur, S., Volinia, S., Greshock, J., Lassus, H., Hasegawa, K., Liang, S., Leminen, A., Deng, S., Smith, L., *et al.* (2008). MicroRNA microarray identifies Let-7i as a novel biomarker and therapeutic target in human epithelial ovarian cancer. *Cancer Res.* **68,** 10307–10314.

Yang, Q., Zage, P., Kagan, D., Tian, Y., Seshadri, R., Salwen, H. R., Liu, S., Chlenski, A., and Cohn, S. L. (2004). Association of epigenetic inactivation of RASSF1A with poor outcome in human neuroblastoma. *Clin. Cancer Res.* **10,** 8493–8500.

Yanokura, M., Banno, K., Kawaguchi, M., Hirao, N., Hirasawa, A., Susumu, N., Tsukazaki, K., and Aoki, D. (2007). Relationship of aberrant DNA hypermethylation of CHFR with sensitivity to taxanes in endometrial cancer. *Oncol. Rep* **17,** 41–48.

Yates, D. R., Rehman, I., Abbod, M. F., Meuth, M., Cross, S. S., Linkens, D. A., Hamdy, F. C., and Catto, J. W. (2007). Promoter hypermethylation identifies progression risk in bladder cancer. *Clin. Cancer Res.* **13,** 2046–2053.

Yegnasubramanian, S., Kowalski, J., Gonzalgo, M. L., Zahurak, M., Piantadosi, S., Walsh, P. C., Bova, G. S., De Marzo, A. M., Isaacs, W. B., and Nelson, W. G. (2004). Hypermethylation of CpG islands in primary and metastatic human prostate cancer. *Cancer Res.* **64,** 1975–1986.

Yegnasubramanian, S., Haffner, M. C., Zhang, Y., Gurel, B., Cornish, T. C., Wu, Z., Irizarry, R. A., Morgan, J., Hicks, J., DeWeese, T. L., *et al.* (2008). DNA hypomethylation arises later in prostate cancer progression than CpG island hypermethylation and contributes to metastatic tumor heterogeneity. *Cancer Res.* **68,** 8954–8967.

Yu, F., Yao, H., Zhu, P., Zhang, X., Pan, Q., Gong, C., Huang, Y., Hu, X., Su, F., Lieberman, J., and Song, E. (2007a). let-7 regulates self renewal and tumorigenicity of breast cancer cells. *Cell* **131,** 1109–1123.

Yu, J., Zhu, T., Wang, Z., Zhang, H., Qian, Z., Xu, H., Gao, B., Wang, W., Gu, L., Meng, J., *et al.* (2007b). A novel set of DNA methylation markers in urine sediments for sensitive/specific detection of bladder cancer. *Clin. Cancer Res.* **13,** 7296–7304.

Yu, S. L., Chen, H. Y., Chang, G. C., Chen, C. Y., Chen, H. W., Singh, S., Cheng, C. L., Yu, C. J., Lee, Y. C., Chen, H. S., *et al.* (2008a). MicroRNA signature predicts survival and relapse in lung cancer. *Cancer Cell* **13,** 48–57.

Yu, W., Gius, D., Onyango, P., Muldoon-Jacobs, K., Karp, J., Feinberg, A. P., and Cui, H. (2008b). Epigenetic silencing of tumour suppressor gene p15 by its antisense RNA. *Nature* **451,** 202–206.

Zenz, T., Mohr, J., Eldering, E., Kater, A. P., Bühler, A., Kienle, D., Winkler, D., Dürig, J., van Oers, M. H., Mertens, D., *et al.* (2009). miR-34a as part of the resistance network in chronic lymphocytic leukemia. *Blood* **113,** 3801–3808.

Zhang, X., Xu, H. J., Murakami, Y., Sachse, R., Yashima, K., Hirohashi, S., Hu, S. X., Benedict, W. F., and Sekiya, T. (1994). Deletions of chromosome 13q, mutations in Retinoblastoma 1, and retinoblastoma protein state in human hepatocellular carcinoma. *Cancer Res.* **54,** 4177–4182.

Zhang, L., Huang, J., Yang, N., Greshock, J., Megraw, M. S., Giannakakis, A., Liang, S., Naylor, T. L., Barchetti, A., Ward, M. R., *et al.* (2006). microRNAs exhibit high frequency genomic alterations in human cancer. *Proc. Natl. Acad. Sci. USA* **103,** 9136–9141.

Zhu, S., Si, M. L., Wu, H., and Mo, Y. Y. (2007). MicroRNA-21 targets the tumor suppressor gene tropomyosin 1 (TPM1). *J. Biol. Chem.* **282,** 14328–14336.

6

Detection of DNA Methylation Changes in Body Fluids

Triantafillos Liloglou and John K. Field
University of Liverpool Cancer Research Centre, Liverpool, United Kingdom

ABSTRACT

DNA methylation deregulation is one of the major components of epigenetic reprogramming affecting many human diseases. Due to its chemical nature and the advances in technology, DNA methylation is unique for biomarker utilization. DNA methylation changes have been detected in a variety of human body fluids associated with a large number of malignant and benign diseases. This chapter summarizes the evidence from a wide range of studies in this field and demonstrates the clinical potential of DNA methylation in cancer diagnostics.

Advances in Genetics, Vol. 71
0065-2660/10 $35.00
DOI: 10.1016/S0065-2660(10)71006-3

The technical approaches are discussed, emphasizing on quality control issues and potential pitfalls. Finally, the lack of transferring this technology to the clinic is being discussed, highlighting key study design issues that are required for biomarker validation and clinical implementation. © 2010, Elsevier Inc.

I. INTRODUCTION

Epigenetic abnormalities are among the most frequent changes in human cancer (Baylin and Ohm, 2006; Esteller, 2007; Jones and Baylin, 2007). In fact, it is considered that all human cancers contain multiple epigenetic abnormalities and this emphasizes the potential value of cancer epigenetic traits for clinical exploitation. DNA methylation is currently the most intensively studied epigenetic modification and competes for a position among the most promising of the DNA mechanisms to provide biomarkers for cancer management, including risk modeling, early detection, prediction of relapse, therapeutic stratification, and monitoring. DNA methylation is a covalent chemical modification which is robust upon removal of the clinical specimen from the body and is stable through fixation processes. In addition, DNA is the most stable human macromolecule. These characteristics provide certain advantages to DNA methylation in biomarker design and implementation as it allows for higher flexibility in clinical sample collection and preservation, fitting current clinical practice.

The concept of utilizing molecular biomarkers in human body fluids as a methodology for assisting clinical management (mainly diagnosis) is a very powerful approach and this concept was supported by a large number of research groups providing experimental evidence such as the detection of genomic instability in bronchial washings (BWs) (Field et al., 1999), sputum (Hsu et al., 2007), and ductal lavage (King et al., 2003) as well as ras mutations in sputum (Clayton et al., 2000) and urine (Fitzgerald et al., 1995; Haliassos et al., 1992).

Since the high frequency of epigenetic abnormalities in human primary tumors was made apparent, in the early 2000s, many groups have started investigating the feasibility of detecting such epigenetic aberrations in body fluids. To date the vast majority of such reports have examined DNA methylation, while there are fewer on detecting histone modifications (Deligezer et al., 2008; Paul et al., 2010). This clearly reflects the relevant literature on surgical tumor specimens but also underlines the possibly higher utility of nucleic acid changes in clinical specimens because of their stability. The detection of microRNAs in body fluids has also been reported (Gilad et al., 2008; Keller et al., 2009) and is expected to significantly contribute to cancer diagnostics. The major challenge regarding the detection of cancer-specific DNA methylation in body fluids is, as in the case of all other DNA abnormalities, that cancer cells and/or cancer circulating DNA are present in small quantities among the normal contaminating DNA.

In this chapter, evidence from a wide range of studies in this field will be presented, clearly demonstrating the clinical potential of DNA methylation in cancer diagnostics. The technical approaches will be discussed, emphasizing on quality control issues and potential pitfalls. Finally, the lack of transferring this technology to the clinic will be discussed, highlighting key study design issues that are required for biomarker validation, which currently appear to be missing from DNA methylation diagnostics.

II. DNA METHYLATION DETECTION TECHNIQUES AND METHODOLOGIES

The need for accurate detection of minute amounts of abnormally methylated DNA in body fluids inevitably leads to discussing methodology issues and the appropriateness of different approaches. As 5′-methyl cytosine pairs to guanine as effectively as unmethylated cytosine, common sequence detection techniques are inadequate for this purpose. A large number of DNA methylation detection methods and modifications exist to date, each having certain advantages and disadvantages. There are no golden, totally bulletproof methods. There are good methods which, provided the necessary quality control measures are undertaken, can successfully address the research questions in particular projects.

There are four major principles regarding the means of discrimination between 5′-methyl cytosine and cytosine:

A. Chromatin immunoprecipitation, which utilize anti-5′-methyl cytosine (Weber *et al.*, 2005) or methyl binding domain proteins (Ballestar *et al.*, 2003)
B. Methylation-specific restriction enzyme (MSRE)-based methods, which take advantage of the differential recognition of methylated DNA sequences over their unmethylated counterparts (Schumacher *et al.*, 2008)
C. Bisulfite-based methods, which exploit the conversion of cytosine—but not 5′-methyl cytosine- to uracil, converting this epigenetic modification into a sequence difference (Herman *et al.*, 1996)
D. Mixed methods utilizing combinations of these principles (Xiong and Laird, 1997).

Downstream applications following the above principles include microarrays and polymerase chain reaction (PCR)-based methods. The latter can provide a direct result, such as methylation-specific PCR (MSP) or prepare templates for subsequent detection with single-strand conformation polymorphism analysis, high-resolution melting, sequencing and pyrosequencing.

One easily recognizes the particularities in detecting abnormal DNA methylation in body fluids. These are considered to carry a tiny load of abnormal nucleic acids in the presence of high amounts of "contaminating" normal DNA. In addition, the overall amount of DNA yielded may be relatively low, frequently due to the fact that only a small part of this clinical specimen becomes available for research purposes. Exception to this is the detection of DNA methylation in the peripheral lymphocytes in leukemias or for monitoring therapeutic schemes with epigenetic drugs (Stewart et al., 2009) where the target DNA is ample.

The reliable detection of minimal residual disease associated nucleic acids in patient with solid tumors utilizing samples such as plasma/serum, urine, sputum, BWs, saliva, etc. is demanding and devious. The most widely used approach has been bisulfite conversion of DNA followed by MSP, in either its endpoint or real time (qMSP, Methylight) version. The concept here is that methylation-specific primers bearing cytosines (forward) and guanines (reverse) preferably at the 3' end of both primers will generate amplicons from only the methylated DNA copies of the target sequence under optimal conditions. The long experience of the research community in PCR amplification techniques allows, in combination with the availability of new engineered hot start Taq polymerases and reliable thermocycling hardware, to reach high levels of fidelity. Endpoint qMSP is a method which revolutionized epigenetic research (Herman et al., 1996). Its main disadvantage though is the lack of quantitation efficiency as well as problems in sensitivity and specificity. The latter can be overcome by using methylation enrichment pyrosequencing which employs pyrosequencing to confirm the status of an MSP amplicon (Shaw et al., 2006a). The real time version of MSP (qMSP or Methylight) overcomes many of the endpoint assay problems. It is highly sensitive (especially when using fluorescent probes) in visualizing minute amounts that could never be detected on gels. The use of fluorescent probes adds one more level of sequence specificity and allows for multicolor detection of internal controls to normalize for DNA input. What is currently missing from qMSP/Methylight is an internal control of bisulfite conversion.

As mentioned above, an important problem in body fluids is the low availability of DNA. To make the problem more challenging, bisulfite conversion diminishes DNA quality and the subsequent cleanup reduces significantly recovery. It is widely accepted today in the epigenetic biomarker research community that it is unlikely to find a "silver bullet." Most studies point to the discovery of panels of biomarkers, thus multiple assays. Therefore, the tiny amount of DNA recovered has to be split in different reactions. One way to overcome this problem is to multiplex four targets per reaction using probes with different fluorescent dyes (Fackler et al., 2009). Of course the level of optimization this requires to prove equal amplification factors over five concentration logs between the different amplicons is significant. The abundance of each target is different thus in the absence of appropriate calibration the high abundance target will probably consume resources disabling detection of the rare (1:1000 or less) copies. A different approach is the post bisulfite whole genome (bisulfitome) amplification (Vaissiere *et al.*, 2009) which demonstrates promising findings, but further research is required to prove the potential extent of its use.

A last important point, frequently bypassed, regarding methodology is the amount of DNA added in a qMSP reaction. qMSP users claim detection of targets in dilution with normal DNA as low as 1:10,000. In this case, one needs to ensure 10,000 genome equivalents (21 ng) are added in the reaction to have one abnormal copy keeping in mind that MSP targets only one bisulfite converted strand). This is also an important issue for consideration in whole bisulfitome amplification.

Overall, huge progress has been demonstrated in the last decade in methodological approaches for DNA methylation detection in body fluids. Many academic groups have thrown themselves into this research and the contribution of the biotech industry must also be acknowledged. Although there are still issues regarding the quality control of the used techniques, there should be no doubt that these will be overcome and a clinically useful method will be very soon available.

III. DNA METHYLATION BIOMARKER DISCOVERY AND VALIDATION

Despite the plethora of papers on potential DNA biomarkers (genetic or epigenetic) for the molecular detection of human cancer, very little is clinically utilized. The main reason for this is that the majority of the studies, although well designed for academic research purposes, lack basic principles of biomarker discovery and validation guidelines as described by the Early Detection Research Network (http://edrn.nci.nih.gov/) and Cancer Research UK (http://www.cancerresearchuk.org/). Significant obstacles in further utilizing this data are the small number of samples used, the lack of adequate follow-up time for the cancer-free controls, the lack of consistency in specimen reception, storage and preparation, the diversity of DNA methylation assays used and the lack of interlaboratory result validation. Design guidelines for creating fit-for-purpose biomarkers have been published following a major biomarkers meeting (Lee *et al.*, 2006). Hopefully in the forthcoming years researchers will team up in multigroup international consortia to perform studies of adequate size and data quality that reach clinical standards allowing for their implementation to the bedside.

The design and implementation of DNA methylation biomarkers is a long process involving discrete steps that one could group into preclinical (discovery, construction of marker panels, training retrospective sets, validation of retrospective sets) and clinical (large prospective cohorts) segments (Fig. 6.1). The discovery phase certainly involves surgical tissue of the tumor type under

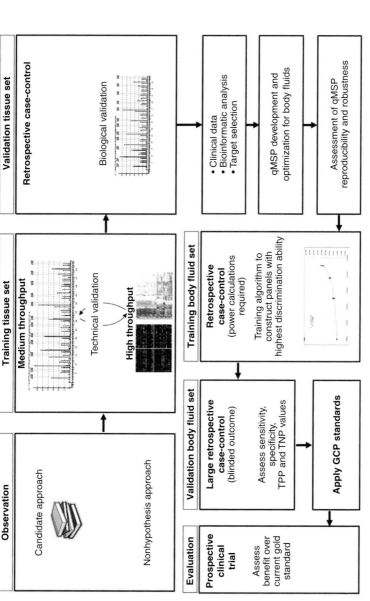

Figure 6.1. Suggested route for DNA methylation biomarker discovery and validation. Early preclinical observations are tested by examining large gene sets in a limited number of tissues. Bioinformatic analysis provides top candidate targets which are subjected to biological validation in a large independent tissue set. qMSP (or Methylight) assays are designed and optimized for the qualifying targets to achieve maximum sensitivity of detection at very high specificity. These assays are then used to screen the "training" set of body fluids with known final clinical diagnosis. Based on the latter mathematical modeling will provide the panel(s) of markers with the highest potential predictive value. This has to be tested in the

investigation. This has to be pathologically reviewed and at least grossly cleaned from normal tissue to provide > 75% tumor material. Although candidate target approaches can provide useful targets, a high-throughput genome-wide approach (such as microarrays or next generation sequencing) is considered necessary to provide potentially useful markers that are not necessarily adequately represented in the previous literature and will most likely be missed in a candidate approach. Following bioinformatic analysis, the top 20–40 markers should be validated by bisulfite sequencing or pyrosequencing in a tumor tissue set, independent to that used in the high-throughput approach. It is critical to establish a picture of the true frequency of positive biomarkers in the population being studied. This is an essential step for performing power calculations that will indicate the number of clinical samples to be assayed in the next preclinical steps in order to provide adequate power. The markers combining (a) high frequency in tumor tissue that is > 20% and (b) total absence of positives in normal adjacent tissue and peripheral blood should progress to the next phase. This is the development of the clinical assay which must combine a high sensitivity and specificity, enabling detection of methylated copies at very high dilution (1:1000–1:10,000) with normal DNA. The assay design should take into consideration the characteristics of the particular target body fluid and the potential limitations this may pose. It is also very important to have standard operating procedures (SOPs) in place to assure robustness of specimen collection, logging, storage, and DNA preparation. Finally, based on the frequency of positive biomarkers in the surgical tissue set, the undertaking of power calculations is necessary in order to indicate the number of body fluid specimens that have to be screened to provide adequate (> 90%) power.

Once the reproducibility and robustness of the biomarker assay have been tested, methylation biomarkers are used to screen the "training" set of body fluid specimens. This is a retrospective collection of patient samples with known outcome and at least 1 year (preferably 3-5 years) follow-up. Cancer-free controls should be sex and age matched to the cases and all the essential epidemiological information should be available for the cancer type under investigation to allow for appropriate adjustments. Analysis of this group will facilitate modeling the algorithm to calculate the diagnostic index. In a simple case scenario, this could be at least a number of markers positive or at least a number of *particular* markers from the panel. In a more complicated scenario, markers would contribute to an algorithm with a different coefficient. In any case, the training set will allow to clear the panel from those biomarkers of no or low potential diagnostic significance. It must be noted that the training set cannot be used to extract sensitivity and specificity figures as it is the one used to initially model the panel. Unfortunately, this concept seems to be missed in most of the relevant papers, which tend to frequently calculate sensitivity, specificity, and prediction efficiency in the training group.

The selected biomarker panel from the training group should now be used to screen the validation group. This is also a retrospective collection with known outcome, independent of the training group, and should have similar characteristics regarding the required information. Applying the algorithm on this group will provide the first figures of sensitivity, specificity, true positive ratio, false positive ratio, etc., assessing the prediction efficiency of the panel. The necessary adjustments for epidemiological risk factors must be undertaken to correct for confounding factors. An ideal biomarker panel should provide >80% sensitivity and >95% specificity prior to any clinical implementation. In addition, a number of conditions must be satisfied to confirm good clinical practice. The clinic set up must facilitate uniform sampling, handling, and storage. The sample processing and the assays must be brought to good clinical laboratory practice standards and the lab should receive appropriate accreditation. An important characteristic of a biomarker is that it must provide patient benefit by improving the current gold standards. Thus, the biomarker panel has to prove its diagnostic efficiency over the current clinical examinations.

IV. DNA METHYLATION DETECTION IN SPUTUM AND BRONCHIAL WASHINGS

Lung cancer is the first cause of cancer-related deaths in the western world and a clear example of the lack of early diagnostics and screening programs (Field and Duffy, 2008; Smith et al., 2008). The cytological examination of BWs (also referred to as bronchial or bronchoalveolar lavage) and sputum (induced or spontaneous) is the standard practice in assisting in lung cancer diagnosis, following suspicious symptoms and/or imaging outcome (Smith and Berg, 2008). However, cytology alone has generally low efficiency in diagnosing lung tumors (Dobler and Crawford, 2009). Studies on DNA methylation aberration in lung tumors have produced a long list of genes that may serve as candidate biomarkers for lung cancer diagnosis (Divine et al., 2005; Ehrich et al., 2006; Field et al., 2005; Tessema et al., 2009). Many of these genes have been tested in sputum and BWs providing diverse results.

The feasibility of detecting p16 promoter hypermethylation in bronchial lavage was shown in an early study by Ahrendt et al. (1999) while subsequent studies investigated panels of promoters (Kim et al., 2004; Topaloglu et al., 2004). DNA methylation analysis of APC, p16, RASSF1, and RARβ in bronchial lavage correctly detected lung cancer in 67% of cytologically positive but also 35% of cytologically negative cases (Schmiemann et al., 2005) while the same group reported good specificity for RASSF1 (Grote et al., 2006), as well as for p16 and RARβ but lack of specificity for SEMAB promoter (Grote et al., 2005). Analysis of DNA methylation in bronchial lavage appears to achieve higher sensitivity and specificity in detecting central rather than peripheral tumors (de Fraipont *et al.*, 2005) which is consistent with the nature of the bronchoscopic examination.

DNA methylation in sputum of multiple genes (among p16, MGMT, DAPK, RASSF1A, PAX5 *β*, and GATA5) was shown to be a promising predictor for lung cancer (Belinsky *et al.*, 2005, 2006; Palmisano *et al.*, 2000). Increased methylation in sputum from lung cancer patients has been shown for TCF21 (Shivapurkar *et al.*, 2008) and RASSF1 (van der Drift *et al.*, 2008). ACS/TMS1 methylation was demonstrated in the sputum of 41% of patients with stage III NSCLC, 15% of patients with stage I NSCLC, and 2% of cancer-free smokers (Machida *et al.*, 2006). A panel of four genes (p16, DAPK, PAX5beta, and GATA5) assayed in sputum was reported to reflect DNA methylation in biopsies, with the highest specificity shown for p16 (Belinsky *et al.*, 2007). Specificity and the positive and negative predictive values of DNA methylation biomarkers are still unclear. Although the prevalence of DNA methylation of specific genes is higher in the sputum/lavage of lung cancer patients, it is also reported at diverse frequencies in the sputum/lavage of cancer-free controls (Baryshnikova *et al.*, 2008; Cirincione *et al.*, 2006; Georgiou *et al.*, 2007; Hsu *et al.*, 2007; van der Drift *et al.*, 2008). This probably is indicative of a field cancerization effect and a manifestation of early preneoplastic foci in smokers (Russo *et al.*, 2005; Verri *et al.*, 2009; Zochbauer-Muller *et al.*, 2003). This smoking-related methylation "noise" poses certain challenges in biomarker implementation as it may lead to overdiagnosis. Thus the landscape of methylation marker specificity is far from being clear, keeping in mind that additional to tobacco carcinogens may also cause similar molecular noise. As previously discussed, different studies utilize different techniques, promoter regions, primers, and enzymes. Also, the quality control implemented in academic studies is most usually inferior to that required for clinical diagnostics. A significant issue mentioned in most studies is that the cancer-free smokers utilized as controls are not followed up to 5 years, which is required to establish a real clinical gold standard for estimating lung cancer risk (Baryshnikova *et al.*, 2008). Thus, the frequency of real positives in the sputum and BWs of genuinely confirmed cancer-free individuals is not yet clear. Molecular data on sputum and BW available in research papers are not always followed by cytological reports, thus the relationship between adequacy (presence of lung macrophages) and positive DNA methylation results is not totally clear. It has been reported that DNA methylation of particular genes is found in samples independent of their adequacy report (Belinsky *et al.*, 2005, 2006). This can be interpreted twofold. It may suggest the presence of circulating cancer DNA in the specimen and/or it may be indicative of a field cancerization effect. In the latter case, one is not detecting an abnormality from the tumor cells themselves but abnormalities from histologically normal bronchial or oral cells. Although this is considered as background or noise, it can significantly assist diagnosis; instead of a direct indicator of a cancer lesion, this could be considered as a surrogate marker of increased risk prompting patient follow-up over a definitive time period.

There are a number of issues one should consider when designing biomarker studies utilizing BWs and sputum. Even assuming that molecular assays used for the methylation detection are "bulletproof", the basic limitations come from the very nature of these specimens. The BW is a representative specimen of a particular bronchus. Thus, the molecular detection of a tumor in BW depends on the proximity of the bronchus examined to the tumor. BWs are usually rich in bronchial cells and frequently contain traces of blood but lack contaminating oral squamous epithelial cells. The latter are frequently present in sputum, which theoretically represents a wider area of the lung; however, its cell content strongly depends on the training given to the patient of how to produce a good sputum specimen. Sputum can be induced or spontaneous; for example, three early morning sputum collections in order to increase the number of cells in the specimen. It is common understanding for researchers using BW and sputum that, as both specimens originate from the main bronchi, they will provide higher sensitivity in detection of central (usually squamous carcinomas) than peripheral (frequently adenocarcinomas) tumors. This has also experimentally been confirmed (Field et al., 1999). It is also known that there is low consistency of the cell profile in the two specimen types among different patients in the same clinical setting. This variability increases between different clinical settings. Sputum samples contain a large number of oral squamous cells while one cannot exclude the presence of cells exfoliating from the esophagus. These issues pose a limitation in the molecular diagnosis of lung cancer and are frequently overlooked by researchers, who often concentrate more on the efficiency of the chemistry of their PCR assays, commonly tested on standard DNA dilutions.

V. DNA METHYLATION DETECTION IN SALIVA

Oral tumors are no exception regarding their epigenetic deregulation and hypermethylation of many promoters has been demonstrated (Shaw et al., 2006b, 2007a,b, 2008). It has also been shown that DNA methylation might be a predictor of malignant transformation in oral dysplasias (Hall et al., 2008). Saliva specimens carry cells from the oral mucosa and are thus considered appropriate in detecting molecular abnormalities that may assist in oral tumor diagnosis. The DNA methylation status of p16, MGMT, and DAPK in sputum was reported to match 65% of corresponding methylation positive tumors while only one of 30 cancer-free controls demonstrated hypermethylation in saliva (Rosas et al., 2001). DNA methylation of a panel consisting of TIMP3, ECAD, p16, MGMT, DAPK, and RASSF1 was shown to serve as a very useful marker subgroup in predicting oral cancer relapse allowing for curable surgery (Righini *et al.*, 2007). A microarray-based study utilizing preoperative and postoperative saliva from patients with oral tumors demonstrated gene panels with variable sensitivity and specificity in detecting the tumor (Viet and Schmidt, 2008). The utility of methylation marker panels over single markers has also been shown in saliva from patients with squamous carcinoma of the head and neck, also pointing out the different profiles detected between saliva and serum (Carvalho *et al.*, 2008).

VI. DNA METHYLATION DETECTION IN BLOOD, PLASMA, AND SERUM

Plasma and serum are commonly used in clinical research as potential sources of minimally invasive specimen acquisition for DNA methylation-related diagnostics. The relative suitability between these two sample types is not totally clear, as very few studies have used both types of specimens in parallel; however they appear to be comparable (Wong *et al.*, 2003). Plasma seems to be used in preference to serum, most probably because of the concern that cancer circulating DNA might be trapped in the coagulating clot. The great advantage of plasma/serum is that it probably presents with the highest uniformity of specimen collection and preparation in comparison to any other clinical sample. It is less subject to site- or operator-dependent variability and can nevertheless be easily standardized among hospitals. The main disadvantage of plasma /serum, at least for as long as site-specific markers do not exist, is that the detected methylation abnormalities may have originated from anywhere in the body. Thus, although the feasibility of detecting tumor-specific epigenetic abnormalities has been demonstrated, the sources of potential contaminating signal are increased, skewing the results and escalating false positives. It is currently difficult to envisage how a plasma/serum positive screening assay would point the clinician toward the site of malignancy.

There is extensive literature on DNA methylation abnormalities in plasma related to many types of malignances. As in the case of other body fluids, different genes have been tested providing diverse and sometimes conflicting results. DNA methylation abnormalities have been reported in the plasma of patients with a variety of cancers such as lung (Belinsky *et al.*, 2005), breast (Bae *et al.*, 2005; Hoque *et al.*, 2006b; Ling *et al.*, 2010; Papadopoulou *et al.*, 2006; Yazici *et al.*, 2009), prostate (Bryzgunova *et al.*, 2008; Chuang *et al.*, 2007; Papadopoulou *et al.*, 2006; Payne *et al.*, 2009), gastric (Bernal *et al.*, 2008; Kolesnikova *et al.*, 2008), pancreatic (Jiao *et al.*, 2007; Liggett *et al.*, 2010b; Melnikov *et al.*, 2009b), colorectal (Grutzmann *et al.*, 2008; Lee *et al.*, 2009;

Lofton-Day et al., 2008; Tanzer et al., 2010), hepatocellular (Chang et al., 2008; Iyer et al., 2010; Zhang et al., 2006), ovarian (Melnikov et al., 2009a), and glioma (Weaver et al., 2006).

Similarly, a long list of reports exists for hypermethylation of various promoters in the serum of patient with lung (Belinsky et al., 2007; Hoffmann et al., 2009a; Umemura et al., 2008), breast (Dulaimi et al., 2004; Sharma et al., 2007, 2010; Shukla et al., 2006; Van der Auwera et al., 2009), neuroblastoma (Misawa et al., 2009), glioma (Lavon et al., 2010; Liu et al., 2010; Wakabayashi et al., 2009), gastric (Abbaszadegan et al., 2008; Sakakura et al., 2009; Wang et al., 2008), colorectal (Taback et al., 2006; Wang et al., 2008), hepatocellular (Zhang et al., 2007), testicular (Ellinger et al., 2009), and bladder (Ellinger et al., 2008) cancer.

The studies mentioned above mainly focus on early diagnosis of disease, however, the use of these samples have also demonstrated a benefit in other aspects of cancer management, such as establishing associations with risk factor exposure (Brait et al., 2009), prediction of relapse (Hoffmann et al., 2009b), and therapeutic monitoring (Aparicio et al., 2009; Friso et al., 2007; Sonpavde et al., 2009).

While plasma and serum samples are used to target cell-free circulating DNA from solid tumors, white blood cells are the apparent target sample for leukemia-related studies (Bullinger et al., 2010; Dunwell et al., 2010; Figueroa et al., 2010; Milani et al., 2010; Schafer et al., 2010). In addition, methylation profiling of WBCs has been shown to provide important information on risk prediction for breast (Widschwendter et al., 2008), bladder (Wilhelm et al., 2010), and gastric (Hou et al., 2010) cancer, as chronic exposures to carcinogens most probably leave their epigenetic imprint in blood cells.

In addition to its involvement in cancer biology, DNA methylation is known for its critical developmental role (Geiman and Muegge, 2010). There is also an increasing line of evidence implicating DNA methylation deregulation in many human nonmalignant diseases including neurodegenerative (Kronenberg et al., 2009; Wang et al., 2010), cardiovascular (Gluckman et al., 2009), and diabetes (Ling and Groop, 2009). Thus the use of DNA methylation biomarkers in blood/plasma/serum for the clinical management of other conditions grows continuously. There are reports demonstrating the use of DNA methylation biomarkers in blood products to diagnose fragile X syndrome (Godler et al., 2010), multiple sclerosis (Liggett et al., 2010a), insulin resistance (Gemma et al., 2010), underweight state in anorexia nervosa (Ehrlich et al., 2010), coronary artery disease (Sharma et al., 2008), and hypertension (Smolarek et al., 2010). In addition, hypermethylation of RASSF1 (Tsui et al., 2007) and Maspin (Chim et al., 2008) in the maternal plasma has been associated with pre-eclampsia although there are contradicting reports (Bellido et al., 2010). Hyper-methylated RASSF1 is a fetal DNA marker that can be readily detectable in maternal plasma (Chan et al., 2006) while detection of placental DNA methylation patterns of chromosome 21 in maternal plasma have been suggested as a noninvasive means of prenatal diagnosis for trisomy 21 (Chim et al., 2008). DNA methylation in both cord blood and plasma have also been suggested as potential biomarkers for skewed X chromosome inactivation in IVF-conceived infants (King et al., 2010) and as a biological measure of maternal folic acid intake (Fryer et al., 2009).

VII. DNA METHYLATION DETECTION IN URINE

Urine is a specimen widely used in clinical practice in order to assist diagnosis of various conditions. It is noninvasive, low cost, and its acquisition has minimal variability. Urine is a very useful body fluid for potentially detecting cancer of the bladder and probably kidney as well as prostate in males, who have a common genitourinary tract. Normally, cells from urine are collected by centrifugation and the DNA extracted from the pellet is subjected to the molecular assays of preference. The feasibility of detecting abnormal DNA methylation in the urine of bladder cancer patients has been demonstrated. A four marker panel (p16, p14ARF, MGMT, GSTP1) out of nine genes examined was shown to provide 69% sensitivity at 100% specificity while the modeling of the remaining five genes led to a two-stage prediction ROC curve with 82% sensitivity, 96% specificity (Hoque et al., 2006a). Diverse results come from another study in which CCND2 and APC methylation in urine provide 100% specificity for malignant over benign diagnosis for bladder lesions while p16, p14ARF, RASSF1, GSTP1, and E-Cad demonstrated reduced specificity (Pu et al., 2006). Evidence from another study demonstrates that RASSF1 and APC methylation provided higher specificity (Yates et al., 2006). DNA methylation of TWIST1 and NID2 was reported to provide 90% sensitivity/93% specificity for bladder cancer detection while the corresponding figures for cytology of these specimens were 48% and 96%, respectively (Renard et al., 2009).

Urine specimens have been used to assess DNA methylation in patients with prostate cancer. Concordance between digital rectal examination and post biopsy urine sample was shown for GSTP1 and APC (94%) as well as EDNRB (82%); however, 60% of urine samples demonstrated hypermethylation in at least one of three markers among controls (Rogers et al., 2006). Urine collected after prostate massage was tested for hypermethylation in 10 genes, resulting in a panel (GSTP1, RASSF1, RARβ2, and APC) that provided 86% sensitivity and 89% specificity (Roupret et al., 2007). The utilization of GSTP1 hypermethylation in urine has provided encouraging results in additional studies (Bryzgunova et al., 2008; Payne et al., 2009). Interestingly, DNA methylation detection has been tested alongside HPV detection with promising results in the urine of

Table 6.1. Compilation of representative studies on DNA methylation in body fluids

Target disease	Body fluid	Method	Genes	Reference
Anorexia	PBMC	Sequencing	POCM	Ehrlich et al. (2010)
Bladder Ca	Blood	Pyrosequencing	LINE1	Wilhelm et al. (2010)
Bladder Ca	Serum	qPCR+MSRE	APC, DAPK, GSTP1, PTGS2, TIG1, Reprimo	Ellinger et al. (2008)
Bladder Ca	Urine	qMSP	APC, ARF, CDH1, GSTP1, MGMT, CDKN2A, RARbeta2, RASSF1A, and TIMP3	Hoque et al. (2006)
Bladder Ca	Urine	MSP	p14ARF, p16INK4a, RASSF1, GSTP, APC, and E-Cad	Pu et al. (2006)
Bladder Ca	Urine	MSP	TWIST, NID2	Renard et al. (2009)
Bladder Ca	Urine	qMSP	RASSF1a, E-cad, and APC	Yates et al. (2006)
Breast Ca	Blood	qMSP	PITX2, TITF1, GDNF, MYOD1, ZNF217	Widschwendter et al. (2008)
Breast Ca	Ductal fluid	MSP	p16, RASSF1A, twist, RARb	Antill et al. (2010)
Breast Ca	Ductal fluid	qMSP	cyclin D2, APC, HIN1, RASSF1A, and RAR-beta2	Euhus et al. (2007)
Breast Ca	Ductal fluid	MSP	RASSF1, TWIST1, HIN1	Fackler et al. (2006)
Breast Ca	Ductal fluid	qMSP	RASSF1A, RARb, TWIST, HIN1, and Cyclin D2	Fackler et al. (2006)
Breast Ca	Ductal fluid	MSP	RARb, HIN1, twist, CDKD2	Locke et al. (2007)
Breast Ca	Ductal fluid	qMSP	RASSF1	Suijkerbuijk et al. (2008)
Breast Ca	Ductal fluid	qMSP	CCND-2, p16, RARb	Zhu et al. (2010)
Breast Ca	Plasma	MSP	CDKD2, RARb, twist, HIN-1	Bae et al. (2005)
Breast Ca	Plasma	MSP	APC, GSTP1, RASSF1, RARb	Hoque et al. (2006a,b)
Breast Ca	Plasma	MSP	THRB	Ling et al. (2010)
Breast Ca	Plasma	MSP	RASSF1	Yazici et al. (2009)
Breast Ca	Serum	MSP	RASSF1A,APC, DAPK	Dulaimi et al. (2004).
Breast Ca	Serum	MSP	CyclinD2,p16INK16a, p14ARF, Slit2	Sharma et al. (2007)
Breast Ca	Serum	MSP	MDR1	Sharma et al. (2010)
Breast Ca	Serum	MSP	RASSF1A,RAR-beta	Shukla et al. (2006)

Disease	Sample	Method	Genes	Reference
Breast Ca	Serum	qMSP	APC, RASSF1AA, ESR1	Van der Auwera et al. (2009)
Cervical Ca	Urine	MethyLight	DAPK1, RARB, TWIST1, and CDH13	Feng et al. (2007)
Colorectal Ca	Peritoneal lavage	qMSP	CDH1, p16, MGMT, APC	Kamiyama et al. (2009)
Colorectal Ca	Plasma	MSP	SEPT9	Grutzmann et al. (2008)
Colorectal Ca	Plasma	MSP	APC, MGMT, RASSF1, Wif-1	Lee et al. (2009)
Colorectal Ca	Plasma	MSP	TMEFF, NGFR, SEPT9	Lofton-Day et al. (2008)
Colorectal Ca	Plasma	MSP	SEPT9, ALX4	Tanzer et al. (2010)
Colorectal Ca	Serum	MSP	MGMT,P16,RAR-beta2, RASSF1A, APC	Taback et al. (2006)
Colorectal Ca	Serum	MSP	RASSF1AA	Wang et al. (2008)
Coronary artery disease	Blood	MSP	ApoE	Sharma et al. (2008)
Esophageal Ca	Blood	qMSP	DAPK, APC	Hoffmann et al. (2009a,b)
Fragile X syndrome	Blood	MALDI-TOF	FMR1	Godler et al. (2010)
Gastric Ca	Blood	Pyrosequencing	Alu,LINE1	Hou et al. (2010)
Gastric Ca	Gastric juice	MSP	CDH1	Muretto et al. (2008)
Gastric Ca	Plasma	MSP	19 genes	Bernal et al. (2008)
Gastric Ca	Plasma	MSP	MGMT, p15, hMLH1	Kolesnikova et al. (2008)
Gastric Ca	Serum	MSP	p16	Abbaszadegan et al. (2008)
Gastric Ca	Serum	qMSP	RUNX3	Sakakura et al. (2009)
Glioma	Plasma	MSP	p16, MGMT, p73, RARb	Weaver et al. (2006)
Glioma	Serum	MSP	MGMT	Lavon et al. (2010)
Glioma	Serum	qMSP	MGMT, p16, TIMP-3, THBS1	Liu et al. (2010)
Glioma	Serum	MSP	p16	Wakabayashi et al. (2009)
Head & Neck Ca	Saliva	MSP	TIMP3,ECAD,p16,MGMT,DAPK,RASSF1A	Righini et al. (2007)
Head & Neck Ca	Saliva	MSP	p16, MGMT, DAPK	Rosas et al. (2001)
Head & Neck Ca	Saliva	Microarray	RASSF1A, RARB, and RIZ1	Viet and Schmidt (2008)
Head & Neck Ca	Serum and salivary rinses	qMSP	CDH1, MGMT, DAPK,	Carvalho et al. (2008)
Hepatocellular Ca	Serum	MSP	p15,p16,RASSF1AA	Zhang et al. (2007)

(Continues)

Table 6.1. (*Continued*)

Target disease	Body fluid	Method	Genes	Reference
Hypertension	Blood	TLC	Global methylation	Smolarek et al. (2010)
Insulin resistance	Blood	MSP	TFAM	Gemma et al. (2010)
Late pregnancy test	Cord blood	Pyrosequencing	LINE-1	Fryer et al. (2009)
Leukemia	Blood	MALDI-TOF	92 genomic regions	Bullinger et al. (2010)
Leukemia	Blood	Microarray	genome-wide	Dunwell et al. (2010)
Leukemia	Blood	Microarray	Genome-wide	Figueroa et al. (2010)
Leukemia	Blood	Microarray	1320 CpG sites	Milani et al. (2010)
Leukemia	Blood	qMSP	MLL	Schafer et al. (2010)
Leukemia	Plasma	MSP	p15	Paul et al. (2010)
Liver Ca	Plasma	MSP	APC, E-cad, p16	Chang et al. (2008)
Liver Ca	Plasma	MSP	APC, FHIT, p15, p16, E-cad	Iyer et al. (2010)
Liver Ca	Plasma	MSP	p16	Zhang et al. (2006)
Liver Ca	Plasma, serum	qMSP	p16	Wong et al. (2003)
Lung Ca	BAL	MSP	p16	Ahrendt et al. (1999)
Lung Ca	BAL	MSP	p16, DAPK, MGMT,FHIT,APC	de Fraipont et al. (2005)
Lung Ca	BAL	qMSP	p16, RARB2, SEMA3B	Grote et al. (2005)
Lung Ca	BAL	qMSP	RASSF1	Grote et al. (2006)
Lung Ca	BAL	qMSP	APC, p16INK4a, and RASSF1A	Grote et al. (2006)
Lung Ca	BAL	MSP	p16, RASSSF1, FHIT RARb	Kim et al. (2004)
Lung Ca	BAL	qMSP	APC, p16INK4a, RASSF1A	Schmiemann et al. (2005)
Lung Ca	BAL	qMSP	CDH1, APC,MGMT, RASSF1A, GSTP1, p16, ARF, RARbeta-2	Topaloglu et al. (2004)
Lung Ca	Plasma	MSP	p16	Belinsky et al. (2005)
Lung Ca	Serum	MSP	DAPK, MGMT, and GSTPI	Hoffman et al. (2009)
Lung Ca	Serum	MSP	MGMT, p16, RASSF1A, DAPK, RARb	Umemura et al. (2008)
Lung Ca	Serum, Sputum	MSP	p16, DAPK, PAX5 beta, and GATA5	Belinsky et al. (2007)
Lung Ca	Sputum	MSP	p16, RASSF1, NORE1A	Baryshnikova et al. (2008)

Lung Ca	Sputum	MSP	p16, RASSF1	Belinsky et al. (2006)
Lung Ca	Sputum	MSP	p16, DAPK, PAX5 beta, and GATA5	Belinsky et al. (2007)
Lung Ca	Sputum	MSP	RARbeta2, p16INK4A, and RASSF1A	Cirincione et al. (2006)
Lung Ca	Sputum	MSP	p16	Georgiou et al. (2007)
Lung Ca	Sputum	Nested qMSP	RAR beta,p16INK4a	Hsu et al. (2007)
Lung Ca	Sputum	MSP	ASC/TMS	Machida et al. (2006)
Lung Ca	Sputum	MSP	p16 MGMT	Palmisano et al. (2000)
Lung Ca	Sputum	qMSP	TCF21	Shivapurkar et al. (2008)
Lung Ca	Sputum	R-T beta-globin PCR	RASSF1	van der Drift et al. (2008)
Lung Ca	Sputum	MSP	FHIT	Verri et al. (2009)
Lung Ca	Sputum, BAL	MSP	RARbeta-2, CDH13, p16, RASSF1A	Zochbauer-Muller et al. (2003)
Lung Ca	Sputum, plasma	MSP	p16, MGMT, RASSF1A, PAX5alpha, DAPK	Belinsky et al. (2005)
Multiple sclerosis	Plasma	MethDet-56	56 promoters	Liggett et al. (2010a,b)
Neural tube defect risk	Blood	MALDI-TOF	global methylation	Wang et al. (2010)
Neuroblastoma	Serum	MSP	RASSF1Aa	Misawa et al. (2009)
Oestrogen-related carcinogenesis	PBMC	LC/MS	ER-alpha, Erbeta, p16	Friso et al. (2007)
Ovarian Ca	Plasma	MSP	BRCA1, HIC1, PAX5, PGR, THBS1	Melnikov et al. (2009a,b)
Pancreatic Ca	Duodenal juice	MSP	26 genes	Matsubayashi et al. (2006)
Pancreatic Ca	Pancreatic juice	MSP	ppENK, pi6	Fukushima et al. (2003)
Pancreatic Ca	Pancreatic juice	qMSP	TFPI-2	Jiang et al. (2006)
Pancreatic Ca	Pancreatic juice	qMSP	Cyclin D2, FOXE1, NPTX2, ppENK, p16, and TFPI2	Matsubayashi et al. (2006)
Pancreatic Ca	Pancreatic Juice	qMSP	TFPI2	Ohtsubo et al. (2008)
Pancreatic Ca	Pancreatic juice	MSP	UCHL1, NPTX2, SARP2, CLDN5, reprimo, LHX1, WNT7A, FOXE1, TJP2, CDH3, ST14	Sato et al. (2003).

(Continues)

Table 6.1. (*Continued*)

Target disease	Body fluid	Method	Genes	Reference
Pancreatic Ca	Pancreatic juice	qMSP	SARP2	Watanabe et al. (2006)
Pancreatic Ca	Pancreatic juice	qMSP	p16	Yan et al. (2005)
Pancreatic Ca	Plasma	MSP	ppENK, p16	Jiao et al. (2007)
Pancreatic Ca	Plasma	MethDet56	BRCA1, CCND2, HMLH1, CDKN1C, PGR,SYK, VHL,	Liggett et al. (2010a,b)
Pancreatic Ca	Plasma	MSP	CCND2, SOCS1, THBS1, PLAU, VHL	Melnikov et al. (2009a,b)
Pre-eclampsia	Plasma	MALDI-TOF	RASSF1, SERPINB5	Bellido et al. (2010)
Pre-eclampsia	Plasma	qMSP	RASSF1	Chan et al. (2006)
Pre-eclampsia	Plasma	Sequencing	Maspin	Chim et al. (2005)
Pre-eclampsia	Plasma	qMSP	RASSF1	Tsui et al. (2007)
Prostate and Breast Ca	Plasma	MSP, MethyLight	GSTP1, RASSF1, ATM	Papadopoulou et al. (2006)
Prostate Ca	Plasma	qMSP	GSTP1	Chuang et al. (2007)
Prostate Ca	Plasma	Pyrosequencing	LINE-1	Sonpavde et al. (2009)
Prostate Ca	Plasma, Urine	MS sequencing	GSTP1	Bryzgunova et al. (2008)
Prostate Ca	Plasma, Urine	qMSP	GSTP1, RASSF2, HIST1H4K, TFAP2E	Payne et al. (2009)
Prostate Ca	Urine	MSP	GSTP1, EDNRB, and APC	Rogers et al. (2006)
Prostate Ca	Urine	qMSP	GSTP1, RASSF1a, ECDH1, APC, DAPK, MGMT, p14, p16, RARb, TIMP3	Roupret et al. (2007)
Risk factor exposure	Plasma	qMSP	RARbeta2,APC, CCND2, GSTP1, MGMT, p16	Brait et al. (2009)
Solid tumors	Plasma	Pyrosequencing	LINE-1,MAGE-1	Aparicio et al. (2009)
Testicular Ca	Serum	qPCR+MSRE	APC, p16, P14arf,RASSF1AA, PTGS2, GSTP1	Ellinger et al. (2009)
Trisomy 21	Plasma	Sequencing	Chr 21	Chim et al. (2008)
X-inactivation	Cord blood	MSP	Chr X	King et al. (2010)

females with cervical neoplasia (Feng *et al.*, 2007). Although it is not clear how cervical cancer cells or free DNA end up in the urine, these results certainly point to the need for further investigation.

VIII. DNA METHYLATION DETECTION IN PANCREATIC, DUODENAL, AND GASTRIC JUICE

Pancreatic ductal adenocarcinoma is the most frequent form of pancreatic cancer and despite its low incidence it is the deadliest human cancer (Jemal *et al.*, 2009). The clinical management of this disease is hampered by late detection (Greenhalf *et al.*, 2009). Endoluminal Ultrasound and Computerized Tomography are used in the screening for, and diagnosis of, pancreatic cancer (Langer *et al.*, 2009; Poley *et al.*, 2009), but neither imaging modality is adequate in itself to give the necessary sensitivity and specificity to allow reliable early detection while avoiding unacceptable numbers of false positives (Greenhalf *et al.*, 2009). Therefore, a number of groups have looked at molecular analysis of pancreatic juice as a possible adjunct to conventional imaging (Carlson *et al.*, 2008). Pancreatic juice can be taken endoscopically either directly from the pancreatic duct (Yan *et al.*, 2005) or with the use of secretin (to increase juice flow) from the duodenum (Matsubayashi *et al.*, 2005). Use of pancreatic juice is considered particularly important in addressing the issue of accurately differentiate chronic pancreatitis from malignant disease. This is a pressing need as the symptoms of the diseases are very similar and distinguishing fibrosis and a small malignant mass is practically impossible with conventional imaging, particularly as pancreatic cancer inevitably results in some level of pancreatitis. To further complicate this problem chronic pancreatitis is a risk factor for pancreatic cancer, particularly the inherited form of the disease, so patients with pancreatitis are prime candidates for screening (Vitone *et al.*, 2005). DNA methylation of various genes is frequent in pancreatic cancers (Sato *et al.*, 2003), however, and lower frequencies are also reported for cancer-free patients with inflammatory conditions (Yan *et al.*, 2005).

A methylation microarray approach on pancreatic tumors resulted in a panel of three genes (NPTX2, SARP2, and CLDN5) that demonstrated methylation in all pancreatic tumors and 75% of related pancreatic juice while none of the pancreatic juice specimens of controls showed any methylation for these three genes (Sato *et al.*, 2003). ppENK and pi6 methylation was reported to be very frequent in pancreatic juice of cases but absent in controls. The high frequency, however, of ppENK and pi6 methylation in duodenal aspirates led to the suggestion that direct cannulation of pancreatic duct during sampling should be used to avoid contamination by duodenal secretions (Fukushima *et al.*, 2003). A similar suggestion comes from an independent study (Matsubayashi *et al.*, 2005). The same group published evidence on a 5 (of 17 tested) gene panel which provided 26/55 positive epigenotypes in the pancreatic juice of patients with pancreatic cancer and only 12/195 epigenotypes in the high risk groups (Matsubayashi et al., 2006). The potential diagnostic utilization of pure pancreatic juice has been also shown by investigating DNA methylation of TFPI-2 (Jiang et al., 2006; Ohtsubo et al., 2008) and SARP (Watanabe et al., 2006).

The feasibility of DNA methylation detection in the gastric juice has been demonstrated (Muretto et al., 2008); however, there is still not enough evidence to report on sensitivity and specificity.

IX. DNA METHYLATION DETECTION IN BREAST DUCTAL FLUID

Breast ductal fluid (also referred to as ductal lavage or nipple aspirate) can be collected by vacuum aspiration following oxytocin induction. Ductal lavage contains cells, the epigenetic analysis of which can potentially give away the presence of a cancerous growth in the breast (Suijkerbuijk et al., 2008). qMSP analysis of nine genes in ductal lavage of breast cancer patients and controls was shown to improve cancer detection efficiency when used in parallel to cytology (Fackler et al., 2006). DNA methylation of a panel of 4 genes is also reported in breast tumors and matching ductal lavage (Zhu et al., 2010). Another study suggests that both atypia and DNA methylation index from APC, CCND2, HIN1, RASSF1, and RARβ2 independently predict abundant cellularity, and risk group classification (Euhus et al., 2007). Particularly interesting are also the associations between promoter methylation in ductal lavage and the presence of germline BRCA1/2 mutations (Antill et al., 2010; Locke et al., 2007), suggesting the potential of surrogate epigenetic biomarkers to predict breast cancer risk.

X. CONCLUSIONS

To date, there is considerable literature demonstrating the feasibility of detecting abnormal DNA methylation in body fluids from patients with a variety of neoplastic and non-neoplastic diseases. Most studies agree in that methylation biomarkers may be used to assist in clinical management of disease, however, very few have progressed to rigorous clinical evaluation. One of the reasons is that, with the exception of few studies, the number of specimens screened is usually small and power calculations are often not included. The design of the biomarker studies is frequently elusive omitting training and validation sets. In many cases, the sensitivity and specificity are calculated in the same sample set which was used to "train" the construction of panels, resulting in misleading figures. The possibility of meta-analyses to increase statistical significance on certain biomarkers is hampered by the diversity of methodological approaches used, with different protocols and certainly different cut-off points. Even though MSP is used in the majority of the studies, there is a lot of diversity among the assays. Older studies utilized endpoint MSP which is not quantitative and is recognized today to have specificity issues. Real-time assays (qMSP or Methylight) employed either intercalator dyes (SYBR) or Taqman probes and most of them different primers. The quality control of assays is occasionally limited and cross validation by external laboratories is rare.

A second set of problems arises from the differences in study design. Clinical sampling, storage, and preparation of body fluids are subject to the local circumstances and practices; the documentation of SOPs is not always clear. There are different biases in the populations studied, especially regarding the controls utilized in each study. Cancer-free controls range from healthy donors to hospital cohorts with nonproliferative disorders, patients with chronic inflammation, and patients with preneoplasia. Case-control matching and adjustment for known risk factors is not always a standard practice. Thus, it becomes hard to assess the level of agreement/disagreement between studies as it is difficult to normalize the data sets.

A third issue is the diversity of genes and sequences examined and the organization in different panels, contributing to the patchy landscape. It is encouraging, however, that similar findings are reported for many gene promoters including p16, RASSF1, MGMT, DAPK, and GSTP1. Overall, despite the invaluable evidence regarding the potential of DNA methylation biomarkers in body fluids for diagnostic purposes, these fundamental problems currently impair further exploitation.

It has to be acknowledged that most of the previously listed issues are very difficult to overcome outside large consortia. Researchers have to face various problems due to everyday practicalities associated with availability of funding, local codes for sample collection, availability of and accessibility to clinical information. Biomarker discovery and validation must follow a lengthy, laborious, and very expensive route prior to clinical establishment.

Currently there is a large gap between the literature on DNA methylation detection feasibility and clinical implementation, suggesting that appropriate validation of DNA methylation biomarkers in body fluids will be facilitated by large consortia where associated members share resources and information, consent on SOPs and common methodology and cross validate methods between different labs. The existing studies on epigenetic alterations in body fluids have certainly paved the way. It is now time for implementation.

Acknowledgments

The authors would like to thank Dr Bill Greenhalf, Mr George Nikolaidis, Benjamin Brown, and Miss Sevasti Giakoumelou for their critical review and constructive comments. Research on DNA methylation detection in body fluids in the University of Liverpool Cancer Research Centre is funded by the Roy Castle Lung Cancer Foundation and Cancer Research UK.

References

Abbaszadegan, M. R., Moaven, O., Sima, H. R., Ghafarzadegan, K., A'rabi, A., Forghani, M. N., Raziee, H. R., Mashhadinejad, A., Jafarzadeh, M., Esmaili-Shandiz, E., and Dadkhah, E. (2008). p16 promoter hypermethylation: a useful serum marker for early detection of gastric cancer. *World J. Gastroenterol.* **14**, 2055–2060.

Ahrendt, S. A., Chow, J. T., Xu, L. H., Yang, S. C., Eisenberger, C. F., Esteller, M., Herman, J. G., Wu, L., Decker, P. A., Jen, J., *et al.* (1999). Molecular detection of tumor cells in bronchoalveolar lavage fluid from patients with early stage lung cancer. *J. Natl. Cancer Inst.* **91**, 332–339.

Antill, Y. C., Mitchell, G., Johnson, S. A., Devereux, L., Milner, A., Di Iulio, J., Lindeman, G. J., Kirk, J., Phillips, K. A., and Campbell, I. G. (2010). Gene methylation in breast ductal fluid from BRCA1 and BRCA2 mutation carriers. *Cancer Epidemiol. Biomarkers Prev.* **19**, 265–274.

Aparicio, A., North, B., Barske, L., Wang, X., Bollati, V., Weisenberger, D., Yoo, C., Tannir, N., Horne, E., Groshen, S., *et al.* (2009). LINE-1 methylation in plasma DNA as a biomarker of activity of DNA methylation inhibitors in patients with solid tumors. *Epigenetics* **4**, 176–184.

Bae, Y. K., Shim, Y. R., Choi, J. H., Kim, M. J., Gabrielson, E., Lee, S. J., Hwang, T. Y., and Shin, S. O. (2005). Gene promoter hypermethylation in tumors and plasma of breast cancer patients. *Cancer Res. Treat.* **37**, 233–240.

Ballestar, E., Paz, M. F., Valle, L., Wei, S., Fraga, M. F., Espada, J., Cigudosa, J. C., Huang, T. H., and Esteller, M. (2003). Methyl-CpG binding proteins identify novel sites of epigenetic inactivation in human cancer. *EMBO J.* **22**, 6335–6345.

Baryshnikova, E., Destro, A., Infante, M. V., Cavuto, S., Cariboni, U., Alloisio, M., Ceresoli, G. L., Lutman, R., Brambilla, G., Chiesa, G., *et al.* (2008). Molecular alterations in spontaneous sputum of cancer-free heavy smokers: results from a large screening program. *Clin. Cancer Res.* **14**, 1913–1919.

Baylin, S. B., and Ohm, J. E. (2006). Epigenetic gene silencing in cancer - a mechanism for early oncogenic pathway addiction? *Nat. Rev. Cancer* **6**, 107–116.

Belinsky, S. A., Klinge, D. M., Dekker, J. D., Smith, M. W., Bocklage, T. J., Gilliland, F. D., Crowell, R. E., Karp, D. D., Stidley, C. A., and Picchi, M. A. (2005). Gene promoter methylation in plasma and sputum increases with lung cancer risk. *Clin. Cancer Res.* **11**, 6505–6511.

Belinsky, S. A., Liechty, K. C., Gentry, F. D., Wolf, H. J., Rogers, J., Vu, K., Haney, J., Kennedy, T. C., Hirsch, F. R., Miller, Y., *et al.* (2006). Promoter hypermethylation of multiple genes in sputum precedes lung cancer incidence in a high-risk cohort. *Cancer Res.* **66**, 3338–3344.

Belinsky, S. A., Grimes, M. J., Casas, E., Stidley, C. A., Franklin, W. A., Bocklage, T. J., Johnson, D. H., and Schiller, J. H. (2007). Predicting gene promoter methylation in non-small-cell lung cancer by evaluating sputum and serum. *Br. J. Cancer* **96**, 1278–1283.

Bellido, M. L., Radpour, R., Lapaire, O., De Bie, I., Hosli, I., Bitzer, J., Hmadcha, A., Zhong, X. Y., and Holzgreve, W. (2010). MALDI-TOF Mass Array Analysis of RASSF1A and SERPINB5 Methylation Patterns in Human Placenta and Plasma. *Biol. Reprod.* **82**, 745–750.

Bernal, C., Aguayo, F., Villarroel, C., Vargas, M., Diaz, I., Ossandon, F. J., Santibanez, E., Palma, M., Aravena, E., Barrientos, C., *et al.* (2008). Reprimo as a potential biomarker for early detection in gastric cancer. *Clin. Cancer Res.* **14**, 6264–6269.

Brait, M., Ford, J. G., Papaiahgari, S., Garza, M. A., Lee, J. I., Loyo, M., Maldonado, L., Begum, S., McCaffrey, L., Howerton, M., *et al.* (2009). Association between lifestyle factors and CpG island methylation in a cancer-free population. *Cancer Epidemiol. Biomarkers Prev.* **18**, 2984–2991.

Bryzgunova, O. E., Morozkin, E. S., Yarmoschuk, S. V., Vlassov, V. V., and Laktionov, P. P. (2008). Methylation-specific sequencing of GSTP1 gene promoter in circulating/extracellular DNA from blood and urine of healthy donors and prostate cancer patients. *Ann. NY Acad. Sci.* **1137**, 222–225.

Bullinger, L., Ehrich, M., Dohner, K., Schlenk, R. F., Dohner, H., Nelson, M. R., and van den Boom, D. (2010). Quantitative DNA methylation predicts survival in adult acute myeloid leukemia. *Blood* **115**, 636–642.

Carlson, C., Greenhalf, W., and Brentall, T. A. (eds.) (2008). Screening of hereditary pancreatic cancer families. Blackwell, Oxford.

Carvalho, A. L., Jeronimo, C., Kim, M. M., Henrique, R., Zhang, Z., Hoque, M. O., Chang, S., Brait, M., Nayak, C. S., Jiang, W. W., *et al.* (2008). Evaluation of promoter hypermethylation detection in body fluids as a screening/diagnosis tool for head and neck squamous cell carcinoma. *Clin. Cancer Res.* **14**, 97–107.

Chan, K. C., Ding, C., Gerovassili, A., Yeung, S. W., Chiu, R. W., Leung, T. N., Lau, T. K., Chim, S. S., Chung, G. T., Nicolaides, K. H., *et al.* (2006). Hypermethylated RASSF1A in maternal plasma: A universal fetal DNA marker that improves the reliability of noninvasive prenatal diagnosis. *Clin. Chem.* **52**, 2211–2218.

Chang, H., Yi, B., Li, L., Zhang, H. Y., Sun, F., Dong, S. Q., and Cao, Y. (2008). Methylation of tumor associated genes in tissue and plasma samples from liver disease patients. *Exp. Mol. Pathol.* **85**, 96–100.

Chim, S. S., Tong, Y. K., Chiu, R. W., Lau, T. K., Leung, T. N., Chan, L. Y., Oudejans, C. B., Ding, C., and Lo, Y. M. (2005). Detection of the placental epigenetic signature of the maspin gene in maternal plasma. *Proc. Natl. Acad. Sci. USA* **102**, 14753–14758.

Chim, S. S., Jin, S., Lee, T. Y., Lun, F. M., Lee, W. S., Chan, L. Y., Jin, Y., Yang, N., Tong, Y. K., Leung, T. Y., *et al.* (2008). Systematic search for placental DNA-methylation markers on chromosome 21: toward a maternal plasma-based epigenetic test for fetal trisomy 21. *Clin. Chem.* **54**, 500–511.

Chuang, C. K., Chu, D. C., Tzou, R. D., Liou, S. I., Chia, J. H., and Sun, C. F. (2007). Hypermethylation of the CpG islands in the promoter region flanking GSTP1 gene is a potential plasma DNA biomarker for detecting prostate carcinoma. *Cancer Detect. Prev.* **31**, 59–63.

Cirincione, R., Lintas, C., Conte, D., Mariani, L., Roz, L., Vignola, A. M., Pastorino, U., and Sozzi, G. (2006). Methylation profile in tumor and sputum samples of lung cancer patients detected by spiral computed tomography: a nested case-control study. *Int. J. Cancer* **118**, 1248–1253.

Clayton, S. J., Scott, F. M., Walker, J., Callaghan, K., Haque, K., Liloglou, T., Xinarianos, G., Shawcross, S., Ceuppens, P., Field, J. K., *et al.* (2000). K-ras point mutation detection in lung cancer: comparison of two approaches to somatic mutation detection using ARMS allele-specific amplification. *Clin. Chem.* **46**, 1929–1938.

de Fraipont, F., Moro-Sibilot, D., Michelland, S., Brambilla, E., Brambilla, C., and Favrot, M. C. (2005). Promoter methylation of genes in bronchial lavages: a marker for early diagnosis of primary and relapsing non-small cell lung cancer? *Lung Cancer* **50**, 199–209.

Deligezer, U., Akisik, E. E., Erten, N., and Dalay, N. (2008). Sequence-specific histone methylation is detectable on circulating nucleosomes in plasma. *Clin. Chem.* **54**, 1125–1131.

Divine, K. K., Pulling, L. C., Marron-Terada, P. G., Liechty, K. C., Kang, T., Schwartz, A. G., Bocklage, T. J., Coons, T. A., Gilliland, F. D., and Belinsky, S. A. (2005). Multiplicity of abnormal promoter methylation in lung adenocarcinomas from smokers and never smokers. *Int. J. Cancer* **114**, 400–405.

Dobler, C. C., and Crawford, A. B. (2009). Bronchoscopic diagnosis of endoscopically visible lung malignancies: should cytological examinations be carried out routinely? *Intern. Med. J.* **39**, 806–811.

Dulaimi, E., Hillinck, J., Ibanez de Caceres, I., Al-Saleem, T., and Cairns, P. (2004). Tumor suppressor gene promoter hypermethylation in serum of breast cancer patients. *Clin. Cancer Res.* **10**, 6189–6193.

Dunwell, T., Hesson, L., Rauch, T. A., Wang, L., Clark, R. E., Dallol, A., Gentle, D., Catchpoole, D., Maher, E. R., Pfeifer, G. P., et al. (2010). A genome-wide screen identifies frequently methylated genes in haematological and epithelial cancers. Mol. Cancer **9,** 44.

Ehrich, M., Field, J. K., Liloglou, T., Xinarianos, G., Oeth, P., Nelson, M. R., Cantor, C. R., and van den Boom, D. (2006). Cytosine methylation profiles as a molecular marker in non-small cell lung cancer. Cancer Res. **66,** 10911–10918.

Ehrlich, S., Weiss, D., Burghardt, R., Infante-Duarte, C., Brockhaus, S., Muschler, M. A., Bleich, S., Lehmkuhl, U., and Frieling, H. (2010). Promoter specific DNA methylation and gene expression of POMC in acutely underweight and recovered patients with anorexia nervosa. J. Psychiatr. Res. Epub ahead of print.

Ellinger, J., El Kassem, N., Heukamp, L. C., Matthews, S., Cubukluoz, F., Kahl, P., Perabo, F. G., Muller, S. C., von Ruecker, A., and Bastian, P. J. (2008). Hypermethylation of cell-free serum DNA indicates worse outcome in patients with bladder cancer. J. Urol. **179,** 346–352.

Ellinger, J., Albers, P., Perabo, F. G., Muller, S. C., von Ruecker, A., and Bastian, P. J. (2009). CpG island hypermethylation of cell-free circulating serum DNA in patients with testicular cancer. J. Urol. **182,** 324–329.

Esteller, M. (2007). Cancer epigenomics: DNA methylomes and histone-modification maps. Nat. Rev. Genet. **8,** 286–298.

Euhus, D. M., Bu, D., Ashfaq, R., Xie, X. J., Bian, A., Leitch, A. M., and Lewis, C. M. (2007). Atypia and DNA methylation in nipple duct lavage in relation to predicted breast cancer risk. Cancer Epidemiol. Biomarkers Prev. **16,** 1812–1821.

Fackler, M. J., Malone, K., Zhang, Z., Schilling, E., Garrett-Mayer, E., Swift-Scanlan, T., Lange, J., Nayar, R., Davidson, N. E., Khan, S. A., et al. (2006). Quantitative multiplex methylation-specific PCR analysis doubles detection of tumor cells in breast ductal fluid. Clin. Cancer Res. **12,** 3306–3310.

Fackler, M. J., Rivers, A., Teo, W. W., Mangat, A., Taylor, E., Zhang, Z., Goodman, S., Argani, P., Nayar, R., Susnik, B., et al. (2009). Hypermethylated genes as biomarkers of cancer in women with pathologic nipple discharge. Clin. Cancer Res. **15,** 3802–3811.

Feng, Q., Hawes, S. E., Stern, J. E., Dem, A., Sow, P. S., Dembele, B., Toure, P., Sova, P., Laird, P. W., and Kiviat, N. B. (2007). Promoter hypermethylation of tumor suppressor genes in urine from patients with cervical neoplasia. Cancer Epidemiol. Biomarkers Prev. **16,** 1178–1184.

Field, J. K., and Duffy, S. W. (2008). Lung cancer screening: the way forward. Br. J. Cancer **99,** 557–562.

Field, J. K., Liloglou, T., Xinarianos, G., Prime, W., Fielding, P., Walshaw, M. J., and Turnbull, L. (1999). Genetic alterations in bronchial lavage as a potential marker for individuals with a high risk of developing lung cancer. Cancer Res. **59,** 2690–2695.

Field, J. K., Liloglou, T., Warrak, S., Burger, M., Becker, E., Berlin, K., Nimmrich, I., and Maier, S. (2005). Methylation discriminators in NSCLC identified by a microarray based approach. Int. J. Oncol. **27,** 105–111.

Figueroa, M. E., Lugthart, S., Li, Y., Erpelinck-Verschueren, C., Deng, X., Christos, P. J., Schifano, E., Booth, J., van Putten, W., Skrabanek, L., et al. (2010). DNA methylation signatures identify biologically distinct subtypes in acute myeloid leukemia. Cancer Cell **17,** 13–27.

Fitzgerald, J. M., Ramchurren, N., Rieger, K., Levesque, P., Silverman, M., Libertino, J. A., and Summerhayes, I. C. (1995). Identification of H-ras mutations in urine sediments complements cytology in the detection of bladder tumors. J. Natl. Cancer Inst. **87,** 129–133.

Friso, S., Lamon-Fava, S., Jang, H., Schaefer, E. J., Corrocher, R., and Choi, S. W. (2007). Oestrogen replacement therapy reduces total plasma homocysteine and enhances genomic DNA methylation in postmenopausal women. Br. J. Nutr. **97,** 617–621.

Fryer, A. A., Nafee, T. M., Ismail, K. M., Carroll, W. D., Emes, R. D., and Farrell, W. E. (2009). LINE-1 DNA methylation is inversely correlated with cord plasma homocysteine in man: a preliminary study. Epigenetics **4,** 394–398.

Fukushima, N., Walter, K. M., Uek, T., Sato, N., Matsubayashi, H., Cameron, J. L., Hruban, R. H., Canto, M., Yeo, C. J., and Goggins, M. (2003). Diagnosing pancreatic cancer using methylation specific PCR analysis of pancreatic juice. *Cancer Biol. Ther.* **2,** 78–83.

Geiman, T. M., and Muegge, K. (2010). DNA methylation in early development. *Mol. Reprod. Dev.* **77,** 105–113.

Gemma, C., Sookoian, S., Dieuzeide, G., Garcia, S. I., Gianotti, T. F., Gonzalez, C. D., and Pirola, C. J. (2010). Methylation of TFAM gene promoter in peripheral white blood cells is associated with insulin resistance in adolescents. *Mol. Genet. Metab.* **100,** 83–87.

Georgiou, E., Valeri, R., Tzimagiorgis, G., Anzel, J., Krikelis, D., Tsilikas, C., Sarikos, G., Destouni, C., Dimitriadou, A., and Kouidou, S. (2007). Aberrant p16 promoter methylation among Greek lung cancer patients and smokers: correlation with smoking. *Eur. J. Cancer Prev.* **16,** 396–402.

Gilad, S., Meiri, E., Yogev, Y., Benjamin, S., Lebanony, D., Yerushalmi, N., Benjamin, H., Kushnir, M., Cholakh, H., Melamed, N., Bentwich, Z., Hod, M., *et al.* (2008). Serum microRNAs are promising novel biomarkers. *PLoS ONE* **3**(9), e3148.

Gluckman, P. D., Hanson, M. A., Buklijas, T., Low, F. M., and Beedle, A. S. (2009). Epigenetic mechanisms that underpin metabolic and cardiovascular diseases. *Nat. Rev. Endocrinol.* **5,** 401–408.

Godler, D. E., Tassone, F., Loesch, D. Z., Taylor, A. K., Gehling, F., Hagerman, R. J., Burgess, T., Ganesamoorthy, D., Hennerich, D., Gordon, L., *et al.* (2010). Methylation of novel markers of fragile X alleles is inversely correlated with FMRP expression and FMR1 activation ratio. *Hum. Mol. Genet.* **19,** 1618–1632.

Greenhalf, W., Grocock, C., Harcus, M., and Neoptolemos, J. (2009). Screening of high-risk families for pancreatic cancer. *Pancreatology* **9,** 215–222.

Grote, H. J., Schmiemann, V., Geddert, H., Rohr, U. P., Kappes, R., Gabbert, H. E., and Bocking, A. (2005). Aberrant promoter methylation of p16(INK4a), RARB2 and SEMA3B in bronchial aspirates from patients with suspected lung cancer. *Int. J. Cancer* **116,** 720–725.

Grote, H. J., Schmiemann, V., Geddert, H., Bocking, A., Kappes, R., Gabbert, H. E., and Sarbia, M. (2006). Methylation of RAS association domain family protein 1A as a biomarker of lung cancer. *Cancer* **108,** 129–134.

Grutzmann, R., Molnar, B., Pilarsky, C., Habermann, J. K., Schlag, P. M., Saeger, H. D., Miehlke, S., Stolz, T., Model, F., Roblick, U. J., *et al.* (2008). Sensitive detection of colorectal cancer in peripheral blood by septin 9 DNA methylation assay. *PLoS ONE* **3,** e3759.

Haliassos, A., Liloglou, T., Licourinas, M., Doumas, C., Ricci, N., and Spandidos, D. A. (1992). H ras oncogene mutations in the urine of patients with bladder tumors: description of a non invasive method for the detection of neoplasia. *Int. J. Oncol.* **1,** 4.

Hall, G. L., Shaw, R. J., Field, E. A., Rogers, S. N., Sutton, D. N., Woolgar, J. A., Lowe, D., Liloglou, T., Field, J. K., and Risk, J. M. (2008). p16 Promoter methylation is a potential predictor of malignant transformation in oral epithelial dysplasia. *Cancer Epidemiol. Biomarkers Prev.* **17,** 2174–2179.

Herman, J. G., Graff, J. R., Myohanen, S., Nelkin, B. D., and Baylin, S. B. (1996). Methylation-specific PCR: a novel PCR assay for methylation status of CpG islands. *Proc. Natl. Acad. Sci. USA* **93,** 9821–9826.

Hoffmann, A. C., Kaifi, J. T., Vallbohmer, D., Yekebas, E., Grimminger, P., Leers, J. M., Izbicki, J. R., Holscher, A. H., Schneider, P. M., Metzger, R., *et al.* (2009a). Lack of prognostic significance of serum DNA methylation of DAPK, MGMT, and GSTPI in patients with non-small cell lung cancer. *J. Surg. Oncol.* **100,** 414–417.

Hoffmann, A. C., Vallbohmer, D., Prenzel, K., Metzger, R., Heitmann, M., Neiss, S., Ling, F., Holscher, A. H., Schneider, P. M., and Brabender, J. (2009b). Methylated DAPK and APC promoter DNA detection in peripheral blood is significantly associated with apparent residual tumor and outcome. *J. Cancer Res. Clin. Oncol.* **135,** 1231–1237.

Hoque, M. O., Begum, S., Topaloglu, O., Chatterjee, A., Rosenbaum, E., Van Criekinge, W., Westra, W. H., Schoenberg, M., Zahurak, M., Goodman, S. N., et al. (2006a). Quantitation of promoter methylation of multiple genes in urine DNA and bladder cancer detection. J. Natl. Cancer Inst. 98, 996–1004.

Hoque, M. O., Feng, Q., Toure, P., Dem, A., Critchlow, C. W., Hawes, S. E., Wood, T., Jeronimo, C., Rosenbaum, E., Stern, J., et al. (2006b). Detection of aberrant methylation of four genes in plasma DNA for the detection of breast cancer. J. Clin. Oncol. 24, 4262–4269.

Hou, L., Wang, H., Sartori, S., Gawron, A., Lissowska, J., Bollati, V., Tarantini, L., Zhang, F. F., Zatonski, W., Chow, W. H., et al. (2010). Blood leukocyte DNA hypomethylation and gastric cancer risk in a high-risk Polish population. Int. J. Cancer. Epub ahead of print.

Hsu, H. S., Chen, T. P., Wen, C. K., Hung, C. H., Chen, C. Y., Chen, J. T., and Wang, Y. C. (2007). Multiple genetic and epigenetic biomarkers for lung cancer detection in cytologically negative sputum and a nested case-control study for risk assessment. J. Pathol. 213, 412–419.

Iyer, P., Zekri, A. R., Hung, C. W., Schiefelbein, E., Ismail, K., Hablas, A., Seifeldin, I. A., and Soliman, A. S. (2010). Concordance of DNA methylation pattern in plasma and tumor DNA of Egyptian hepatocellular carcinoma patients. Exp. Mol. Pathol. 88, 107–111.

Jemal, A., Center, M. M., Ward, E., and Thun, M. J. (2009). Cancer occurrence. Methods Mol. Biol. 471, 3–29.

Jiang, P., Watanabe, H., Okada, G., Ohtsubo, K., Mouri, H., Tsuchiyama, T., Yao, F., and Sawabu, N. (2006). Diagnostic utility of aberrant methylation of tissue factor pathway inhibitor 2 in pure pancreatic juice for pancreatic carcinoma. Cancer Sci. 97, 1267–1273.

Jiao, L., Zhu, J., Hassan, M. M., Evans, D. B., Abbruzzese, J. L., and Li, D. (2007). K-ras mutation and p16 and preproenkephalin promoter hypermethylation in plasma DNA of pancreatic cancer patients: in relation to cigarette smoking. Pancreas 34, 55–62.

Jones, P. A., and Baylin, S. B. (2007). The epigenomics of cancer. Cell 128, 683–692.

Keller, A., Leidinger, P., Borries, A., Wendschlag, A., Wucherpfennig, F., Scheffler, M., Huwer, H., Lenhof, H. P., and Meese, E. (2009). miRNAs in lung cancer - studying complex fingerprints in patient's blood cells by microarray experiments. BMC Cancer 9, 353.

Kim, H., Kwon, Y. M., Kim, J. S., Lee, H., Park, J. H., Shim, Y. M., Han, J., Park, J., and Kim, D. H. (2004). Tumor-specific methylation in bronchial lavage for the early detection of non-small-cell lung cancer. J. Clin. Oncol. 22, 2363–2370.

King, B. L., Tsai, S. C., Gryga, M. E., D'Aquila, T. G., Seelig, S. A., Morrison, L. E., Jacobson, K. K., Legator, M. S., Ward, D. C., Rimm, D. L., et al. (2003). Detection of chromosomal instability in paired breast surgery and ductal lavage specimens by interphase fluorescence in situ hybridization. Clin. Cancer Res. 9, 1509–1516.

King, J. L., Yang, B., Sparks, A. E., Mains, L. M., Murray, J. C., and Van Voorhis, B. J. (2010). Skewed X inactivation and IVF-conceived infants. Reprod. Biomed. Online 20, 660–663.

Kolesnikova, E. V., Tamkovich, S. N., Bryzgunova, O. E., Shelestyuk, P. I., Permyakova, V. I., Vlassov, V. V., Tuzikov, A. S., Laktionov, P. P., and Rykova, E. Y. (2008). Circulating DNA in the blood of gastric cancer patients. Ann. NY Acad. Sci. 1137, 226–231.

Kronenberg, G., Colla, M., and Endres, M. (2009). Folic acid, neurodegenerative and neuropsychiatric disease. Curr. Mol. Med. 9, 315–323.

Langer, P., Kann, P. H., Fendrich, V., Habbe, N., Schneider, M., Sina, M., Slater, E. P., Heverhagen, J. T., Gress, T. M., Rothmund, M., et al. (2009). Five years of prospective screening of high-risk individuals from families with familial pancreatic cancer. Gut 58, 1410–1418.

Lavon, I., Refael, M., Zelikovitch, B., Shalom, E., and Siegal, T. (2010). Serum DNA can define tumor-specific genetic and epigenetic markers in gliomas of various grades. Neuro. Oncol. 12, 173–180.

Lee, J. W., Devanarayan, V., Barrett, Y. C., Weiner, R., Allinson, J., Fountain, S., Keller, S., Weinryb, I., Green, M., Duan, L., *et al.* (2006). Fit-for-purpose method development and validation for successful biomarker measurement. *Pharm. Res.* **23,** 312–328.

Lee, B. B., Lee, E. J., Jung, E. H., Chun, H. K., Chang, D. K., Song, S. Y., Park, J., and Kim, D. H. (2009). Aberrant methylation of APC, MGMT, RASSF2A, and Wif-1 genes in plasma as a biomarker for early detection of colorectal cancer. *Clin. Cancer Res.* **15,** 6185–6191.

Liggett, T., Melnikov, A., Tilwalli, S., Yi, Q., Chen, H., Replogle, C., Feng, X., Reder, A., Stefoski, D., Balabanov, R., *et al.* (2010a). Methylation patterns of cell-free plasma DNA in relapsing-remitting multiple sclerosis. *J. Neurol. Sci.* **290,** 16–21.

Liggett, T., Melnikov, A., Yi, Q. L., Replogle, C., Brand, R., Kaul, K., Talamonti, M., Abrams, R. A., and Levenson, V. (2010b). Differential methylation of cell-free circulating DNA among patients with pancreatic cancer versus chronic pancreatitis. *Cancer* **116,** 1674–1680.

Ling, C., and Groop, L. (2009). Epigenetics: a molecular link between environmental factors and type 2 diabetes. *Diabetes* **58,** 2718–2725.

Ling, Y., Xu, X., Hao, J., Ling, X., Du, X., Liu, X., and Zhao, X. (2010). Aberrant methylation of the THRB gene in tissue and plasma of breast cancer patients. *Cancer Genet. Cytogenet.* **196,** 140–145.

Liu, B. L., Cheng, J. X., Zhang, W., Zhang, X., Wang, R., Lin, H., Huo, J. L., and Cheng, H. (2010). Quantitative detection of multiple gene promoter hypermethylation in tumor tissue, serum, and cerebrospinal fluid predicts prognosis of malignant gliomas. *Neuro. Oncol.* **12,** 540–548.

Locke, I., Kote-Jarai, Z., Fackler, M. J., Bancroft, E., Osin, P., Nerurkar, A., Izatt, L., Pichert, G., Gui, G. P., and Eeles, R. A. (2007). Gene promoter hypermethylation in ductal lavage fluid from healthy BRCA gene mutation carriers and mutation-negative controls. *Breast Cancer Res.* **9,** R20.

Lofton-Day, C., Model, F., Devos, T., Tetzner, R., Distler, J., Schuster, M., Song, X., Lesche, R., Liebenberg, V., Ebert, M., *et al.* (2008). DNA methylation biomarkers for blood-based colorectal cancer screening. *Clin. Chem.* **54,** 414–423.

Machida, E. O., Brock, M. V., Hooker, C. M., Nakayama, J., Ishida, A., Amano, J., Picchi, M. A., Belinsky, S. A., Herman, J. G., Taniguchi, S., *et al.* (2006). Hypermethylation of ASC/TMS1 is a sputum marker for late-stage lung cancer. *Cancer Res.* **66,** 6210–6218.

Matsubayashi, H., Sato, N., Brune, K., Blackford, A. L., Hruban, R. H., Canto, M., Yeo, C. J., and Goggins, M. (2005). Age- and disease-related methylation of multiple genes in nonneoplastic duodenum and in duodenal juice. *Clin. Cancer Res.* **11,** 573–583.

Matsubayashi, H., Canto, M., Sato, N., Klein, A., Abe, T., Yamashita, K., Yeo, C. J., Kalloo, A., Hruban, R., and Goggins, M. (2006). DNA methylation alterations in the pancreatic juice of patients with suspected pancreatic disease. *Cancer Res.* **66,** 1208–1217.

Melnikov, A., Scholtens, D., Godwin, A., and Levenson, V. (2009a). Differential methylation profile of ovarian cancer in tissues and plasma. *J. Mol. Diagn.* **11,** 60–65.

Melnikov, A. A., Scholtens, D., Talamonti, M. S., Bentrem, D. J., and Levenson, V. V. (2009b). Methylation profile of circulating plasma DNA in patients with pancreatic cancer. *J. Surg. Oncol.* **99,** 119–122.

Milani, L., Lundmark, A., Kiialainen, A., Nordlund, J., Flaegstad, T., Forestier, E., Heyman, M., Jonmundsson, G., Kanerva, J., Schmiegelow, K., *et al.* (2010). DNA methylation for subtype classification and prediction of treatment outcome in patients with childhood acute lymphoblastic leukemia. *Blood* **115,** 1214–1225.

Misawa, A., Tanaka, S., Yagyu, S., Tsuchiya, K., Iehara, T., Sugimoto, T., and Hosoi, H. (2009). RASSF1A hypermethylation in pretreatment serum DNA of neuroblastoma patients: a prognostic marker. *Br. J. Cancer* **100,** 399–404.

Muretto, P., Ruzzo, A., Pizzagalli, F., Graziano, F., Maltese, P., Zingaretti, C., Berselli, E., Donnarumma, N., and Magnani, M. (2008). Endogastric capsule for E-cadherin gene (CDH1) promoter hypermethylation assessment in DNA from gastric juice of diffuse gastric cancer patients. *Ann. Oncol.* **19,** 516–519.

Ohtsubo, K., Watanabe, H., Okada, G., Tsuchiyama, T., Mouri, H., Yamaguchi, Y., Motoo, Y., Okai, T., Amaya, K., Kitagawa, H., et al. (2008). A case of pancreatic cancer with formation of a mass mimicking alcoholic or autoimmune pancreatitis in a young man. Possibility of diagnosis by hypermethylation of pure pancreatic juice. JOP 9, 37–45.

Palmisano, W. A., Divine, K. K., Saccomanno, G., Gilliland, F. D., Baylin, S. B., Herman, J. G., and Belinsky, S. A. (2000). Predicting lung cancer by detecting aberrant promoter methylation in sputum. Cancer Res. 60, 5954–5958.

Papadopoulou, E., Davilas, E., Sotiriou, V., Georgakopoulos, E., Georgakopoulou, S., Koliopanos, A., Aggelakis, F., Dardoufas, K., Agnanti, N. J., Karydas, I., et al. (2006). Cell-free DNA and RNA in plasma as a new molecular marker for prostate and breast cancer. Ann. NY Acad. Sci. 1075, 235–243.

Paul, T. A., Bies, J., Small, D., and Wolff, L. (2010). Signatures of polycomb repression and reduced H3K4 trimethylation are associated with p15INK4b DNA methylation in AML. Blood 115, 3098–3108.

Payne, S. R., Serth, J., Schostak, M., Kamradt, J., Strauss, A., Thelen, P., Model, F., Day, J. K., Liebenberg, V., Morotti, A., et al. (2009). DNA methylation biomarkers of prostate cancer: confirmation of candidates and evidence urine is the most sensitive body fluid for non-invasive detection. Prostate 69, 1257–1269.

Poley, J. W., Kluijt, I., Gouma, D. J., Harinck, F., Wagner, A., Aalfs, C., van Eijck, C. H., Cats, A., Kuipers, E. J., Nio, Y., et al. (2009). The yield of first-time endoscopic ultrasonography in screening individuals at a high risk of developing pancreatic cancer. Am. J. Gastroenterol. 104, 2175–2181.

Pu, R. T., Laitala, L. E., and Clark, D. P. (2006). Methylation profiling of urothelial carcinoma in bladder biopsy and urine. Acta Cytol. 50, 499–506.

Renard, I., Joniau, S., van Cleynenbreugel, B., Collette, C., Naome, C., Vlassenbroeck, I., Nicolas, H., de Leval, J., Straub, J., Van Criekinge, W., et al. (2009). Identification and validation of the methylated TWIST1 and NID2 genes through real-time methylation-specific polymerase chain reaction assays for the noninvasive detection of primary bladder cancer in urine samples. Eur. Urol. 58, 96–104.

Righini, C. A., de Fraipont, F., Timsit, J. F., Faure, C., Brambilla, E., Reyt, E., and Favrot, M. C. (2007). Tumor-specific methylation in saliva: a promising biomarker for early detection of head and neck cancer recurrence. Clin. Cancer Res. 13, 1179–1185.

Rogers, C. G., Gonzalgo, M. L., Yan, G., Bastian, P. J., Chan, D. Y., Nelson, W. G., and Pavlovich, C. P. (2006). High concordance of gene methylation in post-digital rectal examination and post-biopsy urine samples for prostate cancer detection. J. Urol. 176, 2280–2284.

Rosas, S. L., Koch, W., da Costa Carvalho, M. G., Wu, L., Califano, J., Westra, W., Jen, J., and Sidransky, D. (2001). Promoter hypermethylation patterns of p16, O6-methylguanine-DNA-methyltransferase, and death-associated protein kinase in tumors and saliva of head and neck cancer patients. Cancer Res. 61, 939–942.

Roupret, M., Hupertan, V., Yates, D. R., Catto, J. W., Rehman, I., Meuth, M., Ricci, S., Lacave, R., Cancel-Tassin, G., de la Taille, A., et al. (2007). Molecular detection of localized prostate cancer using quantitative methylation-specific PCR on urinary cells obtained following prostate massage. Clin. Cancer Res. 13, 1720–1725.

Russo, A. L., Thiagalingam, A., Pan, H., Califano, J., Cheng, K. H., Ponte, J. F., Chinnappan, D., Nemani, P., Sidransky, D., and Thiagalingam, S. (2005). Differential DNA hypermethylation of critical genes mediates the stage-specific tobacco smoke-induced neoplastic progression of lung cancer. Clin. Cancer Res. 11, 2466–2470.

Sakakura, C., Hamada, T., Miyagawa, K., Nishio, M., Miyashita, A., Nagata, H., Ida, H., Yazumi, S., Otsuji, E., Chiba, T., et al. (2009). Quantitative analysis of tumor-derived methylated RUNX3 sequences in the serum of gastric cancer patients. Anticancer Res. 29, 2619–2625.

Sato, N., Fukushima, N., Maitra, A., Matsubayashi, H., Yeo, C. J., Cameron, J. L., Hruban, R. H., and Goggins, M. (2003). Discovery of novel targets for aberrant methylation in pancreatic carcinoma using high-throughput microarrays. *Cancer Res.* **63**, 3735–3742.

Schafer, E., Irizarry, R., Negi, S., McIntyre, E., Small, D., Figueroa, M. E., Melnick, A., and Brown, P. (2010). Promoter hypermethylation in MLL-r infant acute lymphoblastic leukemia: biology and therapeutic targeting. *Blood.* **115**, 4627–4628.

Schmiemann, V., Bocking, A., Kazimirek, M., Onofre, A. S., Gabbert, H. E., Kappes, R., Gerharz, C. D., and Grote, H. J. (2005). Methylation assay for the diagnosis of lung cancer on bronchial aspirates: a cohort study. *Clin. Cancer Res.* **11**, 7728–7734.

Schumacher, A., Weinhäusel, A., and Petronis, A. (2008). Application of microarrays for DNA methylation profiling. *Methods Mol. Biol.* **439**, 109–129.

Sharma, G., Mirza, S., Prasad, C. P., Srivastava, A., Gupta, S. D., and Ralhan, R. (2007). Promoter hypermethylation of p16INK4A, p14ARF, CyclinD2 and Slit2 in serum and tumor DNA from breast cancer patients. *Life Sci.* **80**, 1873–1881.

Sharma, P., Kumar, J., Garg, G., Kumar, A., Patowary, A., Karthikeyan, G., Ramakrishnan, L., Brahmachari, V., and Sengupta, S. (2008). Detection of altered global DNA methylation in coronary artery disease patients. *DNA Cell Biol.* **27**, 357–365.

Sharma, G., Mirza, S., Parshad, R., Srivastava, A., Datta Gupta, S., Pandya, P., and Ralhan, R. (2010). CpG hypomethylation of MDR1 gene in tumor and serum of invasive ductal breast carcinoma patients. *Clin. Biochem.* **43**, 373–379.

Shaw, R. J., Akufo-Tetteh, E. K., Risk, J. M., Field, J. K., and Liloglou, T. (2006a). Methylation enrichment pyrosequencing: combining the specificity of MSP with validation by pyrosequencing. *Nucleic Acids Res.* **34**, e78.

Shaw, R. J., Liloglou, T., Rogers, S. N., Brown, J. S., Vaughan, E. D., Lowe, D., Field, J. K., and Risk, J. M. (2006b). Promoter methylation of P16, RARbeta, E-cadherin, cyclin A1 and cytoglobin in oral cancer: quantitative evaluation using pyrosequencing. *Br. J. Cancer* **94**, 561–568.

Shaw, R. J., Hall, G. L., Lowe, D., Bowers, N. L., Liloglou, T., Field, J. K., Woolgar, J. A., and Risk, J. M. (2007a). CpG island methylation phenotype (CIMP) in oral cancer: associated with a marked inflammatory response and less aggressive tumour biology. *Oral Oncol.* **43**, 878–886.

Shaw, R. J., Hall, G. L., Woolgar, J. A., Lowe, D., Rogers, S. N., Field, J. K., Liloglou, T., and Risk, J. M. (2007b). Quantitative methylation analysis of resection margins and lymph nodes in oral squamous cell carcinoma. *Br. J. Oral Maxillofac. Surg.* **45**, 617–622.

Shaw, R. J., Hall, G. L., Lowe, D., Liloglou, T., Field, J. K., Sloan, P., and Risk, J. M. (2008). The role of pyrosequencing in head and neck cancer epigenetics: correlation of quantitative methylation data with gene expression. *Arch. Otolaryngol. Head Neck Surg.* **134**, 251–256.

Shivapurkar, N., Stastny, V., Xie, Y., Prinsen, C., Frenkel, E., Czerniak, B., Thunnissen, F. B., Minna, J. D., and Gazdar, A. F. (2008). Differential methylation of a short CpG-rich sequence within exon 1 of TCF21 gene: a promising cancer biomarker assay. *Cancer Epidemiol. Biomarkers Prev.* **17**, 995–1000.

Shukla, S., Mirza, S., Sharma, G., Parshad, R., Gupta, S. D., and Ralhan, R. (2006). Detection of RASSF1A and RARbeta hypermethylation in serum DNA from breast cancer patients. *Epigenetics* **1**, 88–93.

Smith, J. J., and Berg, C. D. (2008). Lung cancer screening: promise and pitfalls. *Semin. Oncol. Nurs.* **24**, 9–15.

Smith, R. A., Field, J. K., and Duffy, S. W. (2008). A global approach to cancer-screening trials. *Lancet Oncol.* **9**, 908–909.

Smolarek, I., Wyszko, E., Barciszewska, A. M., Nowak, S., Gawronska, I., Jablecka, A., and Barciszewska, M. Z. (2010). Global DNA methylation changes in blood of patients with essential hypertension. *Med. Sci. Monit.* **16**, CR149–CR155.

Sonpavde, G., Aparicio, A. M., Zhan, F., North, B., Delaune, R., Garbo, L. E., Rousey, S. R., Weinstein, R. E., Xiao, L., Boehm, K. A., et al. (2009). Azacitidine favorably modulates PSA kinetics correlating with plasma DNA LINE-1 hypomethylation in men with chemonaive castration-resistant prostate cancer. *Urol. Oncol.* Epub ahead of print.

Stewart, D. J., Issa, J. P., Kurzrock, R., Nunez, M. I., Jelinek, J., Hong, D., Oki, Y., Guo, Z., Gupta, S., and Wistuba, I. I. (2009). Decitabine effect on tumor global DNA methylation and other parameters in a phase I trial in refractory solid tumors and lymphomas. *Clin. Cancer Res.* **15**, 3881–3888.

Suijkerbuijk, K. P., van der Wall, E., Vooijs, M., and van Diest, P. J. (2008). Molecular analysis of nipple fluid for breast cancer screening. *Pathobiology* **75**, 149–152.

Taback, B., Saha, S., and Hoon, D. S. (2006). Comparative analysis of mesenteric and peripheral blood circulating tumor DNA in colorectal cancer patients. *Ann. NY Acad. Sci.* **1075**, 197–203.

Tanzer, M., Balluff, B., Distler, J., Hale, K., Leodolter, A., Rocken, C., Molnar, B., Schmid, R., Lofton-Day, C., Schuster, T., et al. (2010). Performance of epigenetic markers SEPT9 and ALX4 in plasma for detection of colorectal precancerous lesions. *PLoS ONE* **5**, e9061.

Tessema, M., Yu, Y. Y., Stidley, C. A., Machida, E. O., Schuebel, K. E., Baylin, S. B., and Belinsky, S. A. (2009). Concomitant promoter methylation of multiple genes in lung adenocarcinomas from current, former and never smokers. *Carcinogenesis* **30**, 1132–1138.

Topaloglu, O., Hoque, M. O., Tokumaru, Y., Lee, J., Ratovitski, E., Sidransky, D., and Moon, C. S. (2004). Detection of promoter hypermethylation of multiple genes in the tumor and bronchoalveolar lavage of patients with lung cancer. *Clin. Cancer Res.* **10**, 2284–2288.

Tsui, D. W., Chan, K. C., Chim, S. S., Chan, L. W., Leung, T. Y., Lau, T. K., Lo, Y. M., and Chiu, R. W. (2007). Quantitative aberrations of hypermethylated RASSF1A gene sequences in maternal plasma in pre-eclampsia. *Prenat. Diagn.* **27**, 1212–1218.

Umemura, S., Fujimoto, N., Hiraki, A., Gemba, K., Takigawa, N., Fujiwara, K., Fujii, M., Umemura, H., Satoh, M., Tabata, M., et al. (2008). Aberrant promoter hypermethylation in serum DNA from patients with silicosis. *Carcinogenesis* **29**, 1845–1849.

Vaissiere, T., Cuenin, C., Paliwal, A., Vineis, P., Hoek, G., Krzyzanowski, M., Airoldi, L., Dunning, A., Garte, S., Hainaut, P., et al. (2009). Quantitative analysis of DNA methylation after whole bisulfitome amplification of a minute amount of DNA from body fluids. *Epigenetics* **4**, 221–230.

Van der Auwera, I., Elst, H. J., Van Laere, S. J., Maes, H., Huget, P., van Dam, P., Van Marck, E. A., Vermeulen, P. B., and Dirix, L. Y. (2009). The presence of circulating total DNA and methylated genes is associated with circulating tumour cells in blood from breast cancer patients. *Br. J. Cancer* **100**, 1277–1286.

van der Drift, M. A., Prinsen, C. F., Hol, B. E., Bolijn, A. S., Jeunink, M. A., Dekhuijzen, P. N., and Thunnissen, F. B. (2008). Can free DNA be detected in sputum of lung cancer patients? *Lung Cancer* **61**, 385–390.

Verri, C., Roz, L., Conte, D., Liloglou, T., Livio, A., Vesin, A., Fabbri, A., Andriani, F., Brambilla, C., Tavecchio, L., et al. (2009). Fragile histidine triad gene inactivation in lung cancer: the European Early Lung Cancer project. *Am. J. Respir. Crit. Care Med.* **179**, 396–401.

Viet, C. T., and Schmidt, B. L. (2008). Methylation array analysis of preoperative and postoperative saliva DNA in oral cancer patients. *Cancer Epidemiol. Biomarkers Prev.* **17**, 3603–3611.

Vitone, L. J., Greenhalf, W., Howes, N. R., and Neoptolemos, J. P. (2005). Hereditary pancreatitis and secondary screening for early pancreatic cancer. *Rocz. Akad. Med. Bialymst.* **50**, 73–84.

Wakabayashi, T., Natsume, A., Hatano, H., Fujii, M., Shimato, S., Ito, M., Ohno, M., Ito, S., Ogura, M., and Yoshida, J. (2009). p16 promoter methylation in the serum as a basis for the molecular diagnosis of gliomas. *Neurosurgery* **64**, 455–461, discussion 461-452.

Wang, Y. C., Yu, Z. H., Liu, C., Xu, L. Z., Yu, W., Lu, J., Zhu, R. M., Li, G. L., Xia, X. Y., Wei, X. W., *et al.* (2008). Detection of RASSF1A promoter hypermethylation in serum from gastric and colorectal adenocarcinoma patients. *World J. Gastroenterol.* **14,** 3074–3080.

Wang, L., Wang, F., Guan, J., Le, J., Wu, L., Zou, J., Zhao, H., Pei, L., Zheng, X., and Zhang, T. (2010). Relation between hypomethylation of long interspersed nucleotide elements and risk of neural tube defects. *Am. J. Clin. Nutr.* **91,** 1359–1367.

Watanabe, H., Okada, G., Ohtsubo, K., Yao, F., Jiang, P. H., Mouri, H., Wakabayashi, T., and Sawabu, N. (2006). Aberrant methylation of secreted apoptosis-related protein 2 (SARP2) in pure pancreatic juice in diagnosis of pancreatic neoplasms. *Pancreas* **32,** 382–389.

Weaver, K. D., Grossman, S. A., and Herman, J. G. (2006). Methylated tumor-specific DNA as a plasma biomarker in patients with glioma. *Cancer Invest.* **24,** 35–40.

Weber, M., Davies, J. J., Wittig, D., Oakeley, E. J., Haase, M., Lam, W. L., and Schubeler, D. (2005). Chromosome-wide and promoter-specific analyses identify sites of differential DNA methylation in normal and transformed human cells. *Nat. Genet.* **37,** 853–862.

Widschwendter, M., Apostolidou, S., Raum, E., Rothenbacher, D., Fiegl, H., Menon, U., Stegmaier, C., Jacobs, I. J., and Brenner, H. (2008). Epigenotyping in peripheral blood cell DNA and breast cancer risk: a proof of principle study. *PLoS ONE* **3,** e2656.

Wilhelm, C. S., Kelsey, K. T., Butler, R., Plaza, S., Gagne, L., Zens, M. S., Andrew, A. S., Morris, S., Nelson, H. H., Schned, A. R., *et al.* (2010). Implications of LINE1 Methylation for Bladder Cancer Risk in Women. *Clin. Cancer Res.* **16,** 1682–1689.

Wong, I. H., Zhang, J., Lai, P. B., Lau, W. Y., and Lo, Y. M. (2003). Quantitative analysis of tumor-derived methylated p16INK4a sequences in plasma, serum, and blood cells of hepatocellular carcinoma patients. *Clin. Cancer Res.* **9,** 1047–1052.

Xiong, Z., and Laird, P. W. (1997). COBRA: a sensitive and quantitative DNA methylation assay. *Nucleic Acids Res.* **25,** 2532–2534.

Yan, L., McFaul, C., Howes, N., Leslie, J., Lancaster, G., Wong, T., Threadgold, J., Evans, J., Gilmore, I., Smart, H., *et al.* (2005). Molecular analysis to detect pancreatic ductal adenocarcinoma in high-risk groups. *Gastroenterology* **128,** 2124–2130.

Yates, D. R., Rehman, I., Meuth, M., Cross, S. S., Hamdy, F. C., and Catto, J. W. (2006). Methylational urinalysis: a prospective study of bladder cancer patients and age stratified benign controls. *Oncogene* **25,** 1984–1988.

Yazici, H., Terry, M. B., Cho, Y. H., Senie, R. T., Liao, Y., Andrulis, I., and Santella, R. M. (2009). Aberrant methylation of RASSF1A in plasma DNA before breast cancer diagnosis in the Breast Cancer Family Registry. *Cancer Epidemiol. Biomarkers Prev.* **18,** 2723–2725.

Zhang, Y. J., Rossner, P., Jr., Chen, Y., Agrawal, M., Wang, Q., Wang, L., Ahsan, H., Yu, M. W., Lee, P. H., and Santella, R. M. (2006). Aflatoxin B1 and polycyclic aromatic hydrocarbon adducts, p53 mutations and p16 methylation in liver tissue and plasma of hepatocellular carcinoma patients. *Int. J. Cancer* **119,** 985–991.

Zhang, Y. J., Wu, H. C., Shen, J., Ahsan, H., Tsai, W. Y., Yang, H. I., Wang, L. Y., Chen, S. Y., Chen, C. J., and Santella, R. M. (2007). Predicting hepatocellular carcinoma by detection of aberrant promoter methylation in serum DNA. *Clin. Cancer Res.* **13,** 2378–2384.

Zhu, W., Qin, W., Hewett, J. E., and Sauter, E. R. (2010). Quantitative evaluation of DNA hypermethylation in malignant and benign breast tissue and fluids. *Int. J. Cancer* **126,** 474–482.

Zochbauer-Muller, S., Lam, S., Toyooka, S., Virmani, A. K., Toyooka, K. O., Seidl, S., Minna, J. D., and Gazdar, A. F. (2003). Aberrant methylation of multiple genes in the upper aerodigestive tract epithelium of heavy smokers. *Int. J. Cancer* **107,** 612–616.

Application of Epigenetics in Molecular Epidemiology and Epigenetic Cancer Prevention

7

Epigenetics in Molecular Epidemiology of Cancer: A New Scope

Yasuhito Yuasa

Department of Molecular Oncology, Graduate School of Medical and Dental Sciences, Tokyo Medical and Dental University, Tokyo, Japan

ABSTRACT

Classical epidemiologic studies have made important contributions to identifying the etiology of most common cancers and have had substantive public health impact. Molecular epidemiology is an extension of classical epidemiologic research to incorporate biochemical and molecular biomarkers with questionnaire data to advance our understanding of mechanisms of carcinogenesis and of events between exposure and cancer development. Risk prediction, prognostication, and therapy prediction are the clinical uses of molecular epidemiology in cancer

Advances in Genetics, Vol. 71
0065-2660/10 $35.00
DOI: 10.1016/S0065-2660(10)71008-7

management. Lifestyle and environmental factors associated with carcinogenesis also strongly affect epigenetic statuses, and thus epigenetic mechanisms may mediate environmental influences on gene expression and even diseases, resulting in a focus of epidemiologic investigation. DNA methylation can be studied on candidate genes or on a genome-wide scale, although the genotype is fixed at conception but the epigenome is dynamic. Unlike simple genotyping, the levels and patterns of epigenetic changes differ in different tissues and cell types, and may not reflect events in target tissues. Still as a possible risk factor and surrogate marker for liability to cancer, the methylation statuses of blood leukocyte DNA seem to be ideal for the analysis and are emerging as a new scope. Epigenetic changes in comparison with genetic ones are reversible and are acquired in a gradual manner. These epigenetic features offer a huge potential for prevention strategies. © 2010, Elsevier Inc.

I. INTRODUCTION

Classical epidemiologic studies have made important contributions to identifying the etiology of most common cancers and have had substantive public health impact. The smoking and lung cancer association is best known and has promoted successful cancer prevention initiatives and policy changes. Molecular epidemiology was conceived of as an extension of traditional (classical) epidemiologic research to incorporate biomarkers (biochemical and molecular) with questionnaire data to advance our understanding of mechanisms of carcinogenesis and of events between exposure and cancer development.

Genetic variation in the human genome is an emerging resource for studying cancer. One of the largest types of inherited genetic variation is the single nucleotide polymorphism (SNP). A SNP is defined as a genomic locus where two or more alternative bases occur with appreciable frequency ($>1\%$). SNPs have been studied extensively for defining disease candidate gene regions and establishing functional relationships between genotypic and phenotypic differences. Early molecular epidemiologic studies were constrained by existing technology to use the traditional hypothesis-driven candidate gene approach to identify genetic variants that conferred susceptibility to cancer. These studies used current knowledge of cancer biology and prior experimental or *in vitro* data to explore genes of interest. However, this candidate gene approach is a simplistic one and cannot address the role of multiple genetic loci in complex diseases with complex environmental backgrounds like cancer.

According to recent technological advances, particularly DNA sequencing, genotyping several hundred thousand to more than a million SNPs at one time has been possible. Such genome-wide association studies (GWAS) using population-based designs have identified many genetic loci associated with risk of a range of complex diseases, including cancer (Galvan *et al.*, 2010). Risk prediction, prognostication, and therapy prediction are the clinical uses of molecular epidemiology in cancer management.

Epigenetic mechanisms may mediate environmental influences on gene expression and are therefore also becoming a focus of epidemiologic investigation. The term "Epigenetic Epidemiology" has been a framework for studies that seek to understand the joint influences of epigenetic changes and environmental exposures on cancer risk. DNA methylation can be studied on candidate genes or on a genome-wide scale, although unlike the genotype that is fixed at conception, the epigenome is dynamic. Unlike simple genotyping, the levels and patterns of epigenetic changes differ in different tissues and cell types, and may not reflect events in target tissues.

Lifestyle and environmental factors associated with carcinogenesis affect epigenetic statuses and there have been many studies on them (Arasaradnam *et al.*, 2008; Baccarelli and Bollati, 2009; Herceg, 2007). As a possible risk factor and surrogate marker for liability to cancer, the methylation statuses of blood leukocyte DNA seem to be ideal for the analysis and are emerging as a new scope. These important issues on epigenetic epidemiology will be discussed in this chapter.

II. FACTORS AFFECTING EPIGENETIC STATUSES

A. Aging

There is a significant association between advanced age and an increased incidence of cancer. The cancer-prone phenotype of aged people may be caused by the combined effects of a cumulative mutational load, telomere dysfunction, deterioration of immunity, an altered stromal milieu, and increased epigenetic gene silencing. The studies on epigenetic statuses in humans and animals indicate global hypomethylation and promoter region hypermethylation as follows (Table 7.1).

Long interspersed repeated sequence (L1Md) of mouse was examined in genomic DNA of neonatal and adult mice. L1Md was fully methylated in young animals but demethylated in 27-month-old mice (Mays-Hoopes *et al.*, 1986). The 5-methyldeoxycytidine content of DNA, isolated from the tissues of two rodent species of various ages, were determined. The rate of loss of DNA 5-methyldeoxycytidine residues appears to be inversely related to life span (Wilson *et al.*, 1987). Issa *et al.* reported that CpG island methylation of the estrogen receptor (*ER*) gene occurred in a subpopulation of cells, which increased as a direct function of age in the normal human colonic mucosa (Ahuja *et al.*, 1998;

Table 7.1. Factors Affecting Epigenetic Statuses

Agent/exposure	DNA or histone	Type	Up or Down	Species	Tissue	References
Aging	Long interspersed repeated sequence	Global methylation	D	Mouse	Liver, brain	Mays-Hoopes et al. (1986)
	5-methyldeoxycytidine	Global methylation	D	Mouse	Brain, liver, small intestine	Wilson et al. (1987)
	Estrogen receptor	Promoter methylation	U	Human	Colon	Issa et al. (1994)
	MyoD, N33, ER	Promoter methylation	U	Human	Colon	Ahuja et al. (1998)
Sex	CALCA, MGMT, MTHFR	Promoter methylation	U in male	Human	Blood leukocyte	Sarter et al. (2005)
	RASSF1A	Promoter methylation	U in male	Human	Lung cancer	Vaissière et al. (2009)
Lifestyle factors						
Diet						
Folate	Agouti	CpG methylation	U	Mouse	Coat color	Waterland et al., (2003)
	[3H] methyl incorporation	Global methylation	U	Human	Blood leukocyte	Pufulete et al. (2006)
Early life energy restriction	IGF-2	Promoter methylation	D	Human	Blood leukocyte	Heijmans et al. (2008)
	CACNA1G, IGF2, NEUROG1, RUNX3, SOCS1	Promoter methylation	D	Human	Colorecatal cancer	Hughes et al. (2009)
Genistein	CpG islands	Global methylation	U	Mouse	Prostate	Day et al. (2002)
	Agouti	CpG methylation	U	Mouse	Coat color	Dolinoy et al. (2006)
Green tea	CDX2, BMP2	Promoter methylation	D	Human	Gastric cancer	Yuasa et al. (2009)
Tobacco smoking	D17S5, p16/INK4a	Promoter methylation	U	Human	Lung cancer	Eguchi et al. (1997), Kim et al. (2001)
	p53	Binding to methylated CpGs	U	Human	Lung cancer	Chen et al. (1998), Yoon et al. (2001)
Alcohol	[3H] methyl incorporation	Global methylation	D	Rat	Colon	Choi et al. (1999)
	APC, p14/ARF, p16/INK4a, hMLH1, MGMT, RASSF1A	Promoter methylation	U	Human	Colorecatal cancer	van Engelsand et al. (2003)
Physical activity	MGMT, SFRP1	Promoter methylation	U	Human	Head and neck cancer	Puri et al. (2005), Marsit et al. (2006)
	CACNA2D3	Promoter methylation	D	Human	Gastric cancer	Yuasa et al. (2009)
Environment						
Metal						
Arsenic	Genome DNA	Global methylation	D	Rat	Liver cells	Zhao et al. (1997)

Agent	Gene/Target	Mechanism		Species	Tissue/Cancer	References
Chromium	*p16*	Promoter methylation	U	Human	Lung cancer	Kondo *et al.* (2006)
	Histone	linhibit remodeling		Mouse	Hepatoma cells	Schnekenburger *et al.* (2007)
Nickel	*p16*	Promoter methylation	U	Mouse	Histiocytoma	Govindarajan *et al.* (2002)
	Histone	Reduced acetylation, increased dimethylation of H3K9, Increased ubiquitination of H2A and H2B		Human	Hepatoma cells	Ke *et al.* (2006)
Asbestos	GPC3, *p16*	Promoter methylation	U	Human	Mesothelioma	Murthy *et al.* (2000), Hirao *et al.* (2002)
Chemicals						
Diethylstilbestrol	*Lactoferrin*	Promoter methylation	D	Mouse	Uterus	Li *et al.* (1997)
Bisphenol A	*Agouti*	CpG methylation	D	Mouse	Coat color	Dolinoy *et al.* (2007)
Vinclozolin	CpG-rich regions	Global methylation	U and D	Rat	Testis	Anway *et al.* (2005)
Benzene	LINE-1, AluI	Global methylation	D	Human	Blood leukocyte	Bollati *et al.* (2007)
	p15	Promoter methylation	U			
Aflatoxin B1	*RASSF1A*	Promoter methylation	U	Human	Hepatocellular carcinoma	Zhang *et al.* (2002)
Infection						
Helicobacter pylori	HAND1, LOX, THBD, GPR37, NPTX2, CXCL2, IL-1β	Promoter methylation	U	Human, Mongolian gerbil	Gastric mucosae	Maekita *et al.* (2006), Niwa *et al.* (2010)
EBV	PTEN, RASSF1, *p16*	Promoter methylation	U	Human	Gastric cancer	Kang *et al.* (2002), Kusano *et al.* (2006)
HBV	*p16*	Promoter methylation	U	Human	Hepatocellular carcinoma	Baek *et al.* (2000), Azechi *et al.* (2001)
HPV	*p16, FHIT*	Promoter methylation	U	Human	Cervical cancer	Virmani *et al.* (2001), Flatley *et al.* (2009)
Ionizing radiation	5-methyldeoxycytidine	Global methylation	D	Mouse	Liver	Tawa *et al.* (1998)
	p16	Promoter methylation	U	Human	Lung adenocarcinoma	Belinsky *et al.* (2004)
UV radiation	5-methylcytosine	Global methylation	D	Mouse	Skin	Mittal *et al.* (2003)

Issa *et al.*, 1994). Such *ER* gene methylation was also found in all colonic tumors examined, including small adenomas, suggesting that methylation-associated inactivation of the *ER* gene in the aging colorectal mucosa could be one of the earliest events in colorectal tumorigenesis.

Age-related methylation increases at the population level in a linear manner, although the rate of increase may differ among individuals. Because methylation has the potential to change cellular physiology, the risk of neoplastic transformation of affected cells may increase, providing a mechanistic link between aging and cancer.

B. Sex

For the *CALCA*, *MGMT*, and *MTHFR* genes, male gender was associated with higher gene-specific percentage of methylated reference values in human leukocytes (Sarter *et al.*, 2005). A methylation level in *RASSF1A* in lung cancer was influenced by sex, with males showing higher levels of methylation (Vaissière *et al.*, 2009). Since the frequencies of many human cancers, such as lung and gastric cancers, are predominant in male, sex differences in methylation levels may be an important factor for carcinogenesis.

C. Lifestyle factors

1. Diet

a. Folate

Folate has been one of the nutrients most strongly implicated in terms of protection against colorectal cancer (CRC). An inverse relationship between folate intake and risk of colorectal adenoma (CRA) was demonstrated. Folate through its role as a methyl donor in the one carbon metabolism is crucial for DNA synthesis as well as methylation (Arasaradnam *et al.*, 2008).

With viable yellow agouti (A*vy*) mice, which harbor a transposable element in the *agouti* gene, dietary methyl supplementation of *a/a* dams with extra folic acid, vitamin B12, choline, and betaine altered the phenotype of their A*vy/a* offspring via increased CpG methylation at the A*vy* locus (Waterland and Jirtle, 2003). Folic acid supplementation increased serum and erythrocyte folate concentrations and increased DNA methylation in leukocytes and colonic mucosa in 31 patients with histologically confirmed CRA (Pufulete *et al.*, 2005). These results suggest that DNA hypomethylation can be reversed by physiological intake of folic acid.

b. Early life energy restriction

CpG methylation patterns are largely erased in the early embryos and then reestablished in a tissue-specific manner. Therefore, early embryonic development may represent a sensitive stage, and dietary and environmental factors that affect DNA methylation reaction and the activity of DNA methyltransferases may result in permanent fixation of aberrant methylation patterns. One of the rare opportunities for studying the relevance of such findings to humans is presented by individuals who were prenatally exposed to famine during the Dutch Hunger Winter of 1944–1945 toward the end of World War II. The famine-exposed individual had, six decades later, less DNA methylation of the imprinted *IGF2* gene compared with their unexposed, same-sex siblings. The association was specific for periconceptional exposure, reinforcing that very early mammalian development is a crucial period for establishing and maintaining epigenetic marks (Heijmans *et al.*, 2008). In a large Dutch population, it has also been observed that individuals who experienced severe caloric restriction as adolescents through exposure to the Hunger Winter have a lower risk of developing CRC later in life. Residing in a western famine area during the Hunger Winter was inversely associated with the degree of promoter hypermethylation compared to individuals who did not reside in a western area (Hughes *et al.*, 2009). These data contribute empirical support for the hypothesis that early-life environmental conditions can cause epigenetic changes in humans that persist throughout life.

c. Phytoestrogen

Phytoestrogens which include coumestans, isoflavones, and lignans are plant-derived estrogen-like compounds. They are naturally occurring in many foods including fruits, legumes such as soy and rice. Phytoestrogens have several biological actions including antiestrogenic, anti-inflammatory, and anticarcinogenic effects (Arasaradnam *et al.*, 2008).

Male mice were fed a control diet or the same diet containing soy phytoestrogen genistein. DNA from prostate were then screened with differential methylation hybridization arrays to test for changes in the methylation status of the CpG islands in the mouse genome. Consumption of genistein diet was positively correlated with changes in prostate DNA methylation at CpG islands of specific mouse genes (Day *et al.*, 2002).

Maternal dietary genistein supplementation of mice during gestation, at levels comparable with humans consuming high soy diets, shifted the coat color of heterozygous viable yellow agouti (*A*vy/a) offspring toward pseudoagouti. This marked phenotypic change was significantly associated with increased methylation of six cytosine–guanine sites in a retrotransposon upstream of the transcription start site of the *Agouti* gene. The extent of this DNA methylation was similar in endodermal, mesodermal, and ectodermal tissues, indicating that genistein acts during early embryonic development. Moreover, this genistein-induced hypermethylation persisted into adulthood, decreasing ectopic *Agouti* expression. Thus, *in utero* dietary genistein affects gene expression in adulthood by permanently altering the epigenome (Dolinoy *et al.*, 2006).

d. Green tea

Yuasa *et al.* examined the methylation statuses of six tumor-related genes including *CDX2* and *BMP-2* in 106 primary gastric carcinomas by methylation-specific PCR (MSP) and compared them with the past lifestyles of the patients. Significant association was found between a decreased intake of green tea and methylation of *CDX2* and *BMP-2*. When the intake of green tea was stratified, the prevalence of aberrant methylation of *CDX2* and *BMP-2* decreased significantly with a higher intake of green tea (Fig. 7.1A and B). Thus, some epidemiological factors, such as green tea intake, could be important as to determination of the methylation statuses of selected genes and may influence the development of cancer, including that of the stomach (Yuasa *et al.*, 2009).

Green tea contains several polyphenolic compounds, such as epigallocatechin gallate (EGCG). Significant inhibitory effects of EGCG or green tea extracts on carcinogenesis of rodents in various organs including the stomach have been demonstrated. Most epidemiological studies in Japan revealed cancer-preventive effects of drinking green tea. Although it has been reported that EGCG dose-dependently inhibited DNA methyltransferase activity in several cancer cells, resulting in reactivation of methylation-silenced genes, such as *retinoic acid receptor β*, *p16*, and *hMLH1* (Fang *et al.*, 2003), the *in vitro* activity of EGCG on DNA methylation is controversial (Chuang *et al.*, 2005).

2. Tobacco smoking

Tobacco smoking is strongly carcinogenic in many cancers including lung cancer. Tobacco smoke contains many carcinogens inducing genetic changes, of which polycyclic aromatic hydrocarbons (PAHs), such as benzo[a]pyrene, are considered to be most carcinogenic. On the other hand, little is known about the epigenetic events in tobacco-associated cancer. The incidences of hypermethylation of the *D17S5* locus and *p16/INK4A* in lung cancers are significantly higher in cigarette smokers than those who have never smoked (Eguchi *et al.*, 1997; Kim *et al.*, 2001). It is noteworthy that methylation levels of several genes had significant correlations with smoking duration in the background mucosae of esophageal squamous cell carcinoma patients, suggesting an epigenetic field for cancerization (Oka *et al.*, 2009).

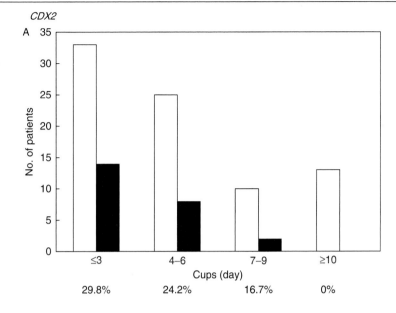

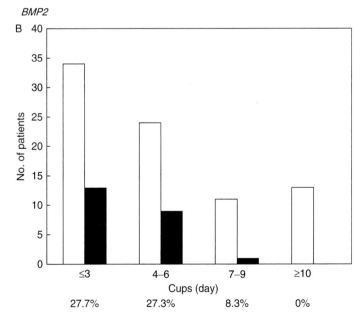

Figure 7.1. Frequencies of the presence (closed bars) or absence (open bars) of *CDX2* and *BMP-2* methylation in gastric cancers stratified as to intake of green tea.

Epigenetic targets of PAHs from tobacco smoke may also be involved in carcinogenesis in a different way (Chen et al., 1998; Yoon et al., 2001). A large fraction of the p53 mutations in lung cancers from smokers are G-to-T transversions, a type of mutation that is infrequent in lung cancers from nonsmokers and in most other tumors. There is an association between G-to-T transversion hotspots in lung cancers and sites of preferential formation of PAHs adducts along the p53 gene. The occurrence of transversion hotspots at methylated CpGs correlated with high levels of benzo[a]pyrene diol epoxide adducts formed at such sites. These findings provide evidence that G-to-T mutational processes can occur selectively at methylated CpG sequences and strengthen a link between PAHs present in cigarette smoke and lung cancer mutations.

These data suggest that stopping aberrant DNA methylation and mutations by cessation of cigarette smoking and by other interventions may be potential new strategies for the prevention of, or therapy for, smoking-related cancers, such as lung and esophageal cancers.

3. Alcohol

Alcoholic drinks are a cause of cancer of the mouth, pharynx, and larynx, esophagus, colorectum in men, and breast, and probably a cause of liver cancer, and of CRC in women (Marmot et al., 2007). Alcohol is known to enhance the effects of environmental carcinogens directly and by contributing to nutritional deficiency.

Chronic alcohol consumption induced genomic DNA hypomethylation in rat colon but not p53-specific methylation change. In the hepatic tissue of the alcohol-fed rats, concentrations of S-adenosylmethionine (SAM), the compound that provides methyl groups for DNA methylation, was also significantly diminished (Choi et al., 1999). van Engeland et al. (2003) revealed that the prevalence of promoter hypermethylation of several genes including APC-1A, p14/ARF, p16/INK4A, hMLH1, MGMT, and RASSF1A was higher in CRC patients with high alcohol (and low folate) intake than in those with low alcohol/ high folate intake. These data suggest that the mechanistic link between high alcohol intake and increased CRC risk seems to be mediated through low folate status. Significant correlation was also reported between a positive alcohol use history and promoter hypermethylation of the MGMT and SFRP1 genes in head and neck cell carcinomas (Marsit et al., 2006; Puri et al., 2005).

4. Physical activity

Physical activity of all types protects against cancers of the colon, breast (post-menopause), and end metrium (Marmot et al., 2007). CACNA2D3 methylation was more frequently found in gastric carcinoma patients with no physical activity than in those with physical activity (Yuasa et al., 2009). The mechanism underlying this result remains unclear.

D. Environmental factors

1. Metal

a. Arsenic

Inorganic arsenic in drinking water is a cause of lung cancer and probably a cause of skin cancer (Marmot *et al.*, 2007). Inorganic arsenic is enzymatically methylated for detoxication, consuming SAM in the process. The observation that DNA methyltransferases also require SAM as their methyl donor suggests a role for DNA methylation in arsenic carcinogenesis.

Chronic exposure of arsenic induced transformation of a rat liver epithelial cell line. Global DNA hypomethylation occurred concurrently with malignant transformation and in the presence of depressed levels of SAM. Whereas, transcription of DNA methyltransferase was elevated, the methyltransferase enzymatic activity was reduced with arsenic transformation (Zhao *et al.*, 1997). Arsenite produced significant increases in the steady-state expression of c-*myc* during the malignant transformation process using a rat liver epithelial cell line. A prominent overexpression of c-*myc*, a gene frequently activated during hepatocarcinogenesis, is strongly correlated with several events possibly associated with arsenic-induced malignant transformation, including hyperproliferation, DNA hypomethylation, and tumor formation upon inoculation into nude mice (Chen *et al.*, 2001). These results indicate arsenic can act as a carcinogen by inducing DNA hypomethylation, which in turn facilitates aberrant gene expression, and they constitute a reasonable theory of mechanism in arsenic carcinogenesis.

b. Cadmium

Epidemiological studies have provided evidence that occupational and/or environmental exposure to cadmium is associated with human lung cancers and possibly cancers in other tissues, but the mechanisms underlying cadmium carcinogenesis remain poorly understood. In rat liver cells, cadmium was an effective inhibitor of DNA methyltransferase and kinetic studies indicated the likely mechanism is via interference with enzyme–DNA interactions. Rat liver cells exposed to cadmium initially exhibited reduced DNA methyltransferase activity and decreased genomic DNA methylation, but with longer exposures, these cells showed increases in both DNA methyltransferase activity and DNA methylation (Takiguchi *et al.*, 2003).

c. Chromium

Chromium is an inhaled carcinogen that causes lung cancer. Intracellular Cr(VI) reduction, which generates reactive oxygen species, has been proposed as the most probable cause of chromium-induced tissue damage, underlying its toxicity and carcinogenicity. As to the effects of chromium on epigenetics, one-third of lung cancers in chromate workers showed *p16* hypermethylation, most of which repressed p16 protein (Kondo *et al.*, 2006). It has also been reported that chromium cross-linked histone deacetylase 1–DNA methyltransferase 1 complexes to chromatin and inhibited histone-remodeling marks critical for transcriptional activation in mouse hepatoma cells (Schnekenburger *et al.*, 2007). These data suggest that not only genetic, but also epigenetic alterations are involved in the carcinogenesis due to chromate.

d. Nickel

Occupational exposure to nickel compounds has been linked to lung and nasal cancers. Nickel sulfide is an insoluble carcinogen which is thought to act through the production of free radicals. Implanted nickel sulfide into C57BL/6 mice induced malignant fibrous histiocytomas and all tumors demonstrated hypermethylation of the tumor suppressor gene *p16* (Govindarajan *et al.*, 2002). Nickel also affects histone modification. For example, soluble $NiCl_2$ was delivered into the cells and caused transgene silencing, and this was associated with loss of histone acetylation. Exposure of cells to $NiCl_2$ also led to increased dimethylation of H3K9 and increased ubiquitination of H2A and H2B (Ke *et al.*, 2006).

2. Asbestos

Exposure to asbestos, occurring as a result of direct exposure from industrial sources or indirect exposure from household is a cause of mesothelioma and lung cancer. Malignant mesothelioma is a rare neoplasm arising primarily from the surface serosal cells of the pleural, peritoneal, and pericardial cavities. Downregulation of GPC3 commonly occurred in malignant mesothelioma and GPC3, an X-linked recessive overgrowth gene, may encode a negative regulator of mesothelial cell growth (Murthy *et al.*, 2000). Gene changes of *p16/INK4A*, a cell cycle inhibitor, are relatively common in malignant mesothelioma. The changes consist of deletion as well as promoter methylation (Hirao *et al.*, 2002).

3. Chemicals

a. Diethylstilbestrol

Prenatal exposure to diethylstilbestrol (DES), a synthetic estrogen, was shown to increase the risk of cervical and vaginal cancer and pregnancy-related problems in women and testicular abnormalities in men. DES exposure in neonatal mice decreased promoter DNA methylation in CpG sites of the *lactoferrin* gene.

Moreover, the demethylation was maintained in uterine tumors of the neonatally DES-treated mice, indicating the importance of epigenetic changes in DES treatment (Li *et al.*, 1997).

b. Bisphenol A

Bisphenol A, one of endocrine disruptors, is a high production volume chemical used in the manufacture of polycarbonate plastic and epoxy resins. It is present in many commonly used products including food and beverage containers, baby bottles, and dental composites. Rodent studies have associated pre- or perinatal bisphenol A exposure with higher body weight, increased breast and prostate cancers, altered reproductive function, and other chronic health effects. Maternal exposure to bisphenol A shifted the coat color distribution of viable yellow agouti (A*vy*) mouse offspring toward yellow by decreasing CpG methylation in an intracisternal A particle retrotransposon upstream of the *Agouti* gene (Dolinoy *et al.*, 2007).

c. Vinclozolin

Transgenerational effects of environmental toxins require either a genetic or epigenetic alteration in the germ line. The endocrine disruptor vinclozolin, an antiandrogenic compound, induced transgenerational pathogenesis in rats, resulting in spermatogenic defects, male infertility, breast cancer, kidney disease, prostate disease, and immune abnormalities. These transgenerational disease phenotypes are transmitted to the majority of progeny for four generations and, although the transgenerational effect is transmitted through only the male germ line, both males and females are affected. The epigenetic alterations observed involve both hypermethylation and hypomethylation events in the germ line (Anway *et al.*, 2005).

d. Benzene

Exposure to benzene, a widespread airborne pollutant emitted from traffic exhaust fumes and cigarette smoking, has been consistently associated with acute myelogenous leukemia. Blood DNA samples were obtained from subjects with different levels of benzene exposure, including gas station attendants, traffic police officers, and unexposed referents. Airborne benzene exposure was associated with a significant reduction in global methylation measured in LINE-1 and Alu, and was also associated with hypermethylation in *p15*. These findings indicate that low-level benzene exposure may induce altered DNA methylation (Bollati *et al.*, 2007).

e. Aflatoxin B1

Foods, such as peanuts and cereals, contaminated with aflatoxins, strong carcinogenic mycotoxins, are a cause of liver cancer is convincing (Marmot *et al.*, 2007). *RASSF1A* and *p16* undergo epigenetic silencing by methylation of promoter region CpG islands in hepatocellular carcinoma (HCC) tissues and may play an important role in the development of HCC in Taiwan. There was also a significant association between *RASSF1A* methylation and the aflatoxin B1-DNA adduct level in HCC. These results suggest the hypothesis that exposure to aflatoxins may be involved in altered methylation of genes involved in cancer development (Zhang *et al.*, 2002).

4. Infection

a. *Helicobacter pylori*

Helicobacter pylori infection is a major etiologic risk factor for gastric cancers. *H. pylori* infection was shown to induce various degrees of methylation of CpG islands in promoters of several genes in human gastric mucosa, which appears to reflect gastric cancer risk (Maekita *et al.*, 2006). Methylation levels of several genes, such as *GPR37* and *NPTX2*, in gastric epithelial cells infected with *H. pylori* in Mongolian gerbils (*Meriones unguiculatus*) started to increase at several weeks after infection. When *H. pylori* was eradicated with antibiotics, methylation levels markedly decreased, but they remained higher than those in gerbils that were not infected by *H. pylori*. Expression levels of several inflammation-related genes (*CXCL2*, *IL-1β*, *NOS2*, and *TNF-α*) paralleled the temporal changes of methylation levels. Significantly suppressing inflammation with the immunosuppressive drug cyclosporin A did not affect colonization by *H. pylori* but blocked the induction of altered DNA methylation. These data suggest that DNA methylation alterations are the infection-associated inflammatory response, rather than *H. pylori* itself (Niwa *et al.*, 2010).

b. Epstein–Barr virus

Epstein–Barr virus (EBV) is a ubiquitous herpes virus that infects most children during early childhood and causes few, if any, symptoms. However, EBV is involved in several types of lymphomas and a subset (about 9%) of gastric carcinomas, although its specific role in gastric carcinogenesis remains unclear. EBV-positive gastric carcinoma demonstrates more frequent aberrant methylation of multiple genes, such as *PTEN*, *RASSF1*, *GSTP1*, and *p16*, than EBV-negative gastric carcinoma and constitutes high CpG island methylator phenotype gastric carcinoma (Kang *et al.*, 2002; Kusano *et al.*, 2006).

c. Hepatitis B virus

Hepatitis B virus (HBV) is a significant risk factor for HCC in the developing world. Although most persons will not develop any illness related to this infection, some will develop chronic liver disease, which can lead to cirrhosis or HCC or both. *p16/INK4A* is known as a major target for human hepatocellular carcinogenesis and *p16* alterations in HCCs include hypermethylation and homozygous deletions (Azechi *et al.*, 2001; Baek *et al.*, 2000).

d. Human papilloma virus

Virtually all cervical cancers are associated with human papilloma virus (HPV) infection, but the majority of women with HPV do not develop cervical cancer. Thus, HPV infection is a necessary but not a sufficient cause of cervical cancer. The methylation analyses on cervical cancers exhibited following results. Aberrant methylation was found early during multistage pathogenesis and an increasing trend for methylation was seen with increasing pathological change. Methylation of *RAR-β* and *GSTP1* were early events, *p16* and *MGMT* methylation were intermediate, and *FHIT* methylation was a late, tumor-associated event. Methylation occurred independently of HPV infection (Virmani *et al.*, 2001). On the other hand, a recent study reported that lower folate status was associated with high-risk HPV infection (Flatley *et al.*, 2009).

5. Ionizing radiation

The molecular mechanisms leading to lung cancer from high linear energy transfer radiation have largely been linked to α-particles that cause DNA damage primarily through large deletions and, to a lesser extent, point mutations. As to epigenetic changes by radiation, effects of ionizing radiation on the genomic DNA methylation level in various organs of mouse were examined. Global hypomethylation was observed in liver, but not in brain or spleen (Tawa *et al.*, 1998). Inversely, the prevalence for methylation of *p16* was increased significantly in adenocarcinomas from plutonium-exposed workers compared with nonworker controls (Belinsky *et al.*, 2004).

6. Ultraviolet radiation

Ultraviolet (UV) light from sunlight or sunlamps is divided into three bands of wavelengths, UVA, UVB, and UVC. UVB is the most effective carcinogen and is absorbed by bases in the DNA, causing characteristic patterns of DNA damage through generation of reactive oxygen species. UV radiation causes both malignant melanoma and nonmelanoma skin cancer. Chronic UVB irradiation resulted in the hypomethylation pattern of DNA (5-methyl cytosine-specific) in mouse skin (Mittal *et al.*, 2003).

III. BLOOD LEUKOCYTE DNA METHYLATION STATUSES IN CANCER CASES AND CONTROL

Since epigenetic change including aberrant DNA methylation is one of important mechanisms underlying carcinogenesis, it would be useful to determine the DNA methylation status in individuals for diagnosis of liability to cancer. Aberrant DNA methylation is found in most tumors of all organs, indicating that it may occur everywhere in the body including blood leukocyte. Because blood leukocyte is one of the best accessible cells of the body, methylation statuses of blood leukocyte DNA have been analyzed for a possible risk factor and surrogate marker for liability to cancer as well as a marker for environmental exposure (Table 7.2).

A. Global methylation level

Global DNA hypomethylation may result in chromosomal instability and oncogene activation, and may be associated with many types of cancer as a surrogate of systemic methylation activity.

1. Methylcytosine and 5-methyldeoxycytidine

Lim et al. (2008) examined genomic methylation of leukocyte DNA in relation to CRA among asymptomatic women. Genomic methylation of leukocyte DNA was determined as a methylcytosine level by liquid chromatography mass spectrometry. Compared with women in the lowest tertile of genomic methylation, women in the second and third tertiles had lower risk of CRA, suggesting systemic genomic hypomethylation as a potential etiologic factor for an early stage of CRA.

%Cytosine methylation (5-mC) was measured in leukocyte DNA from 775 Spanish bladder cancer cases and 397 controls. Median %5-mC DNA was significantly lower in cases than in controls. The lowest cancer risk was noted in never smokers in the highest methylation quartile. By comparison with never smokers in the highest quartile, current smokers in the lowest methylation quartile had the highest risk of bladder cancer. In analyses stratified by smoking, hypomethylation was a strong risk factor in never smokers. These results indicate that leukocyte DNA hypomethylation is associated with increased risk of developing bladder cancer, and this association is independent of smoking (Moore et al., 2008).

Samples and data were obtained from women with incident early-stage breast cancer (I–IIIa) and women who were cancer free. 5-methyldeoxycytosine (5-mdC) level in leukocyte DNA was significantly lower in breast cancer cases than healthy controls (Choi et al., 2009).

Table 7.2. Blood Leukocyte DNA Methylation Statuses in Cancer Cases and Control

DNA type	Tissue	No. of Cases	Change	No. of Control	References
Global methylation level					
methylcytosine	Colorectal adenoma	115	Hypomethylation	115	Lim et al. (2008)
methylcytosine	Bladder cancer	775	Hypomethylation	397	Moore et al. (2008)
5-methyldeoxycytidine	Breast cancer	179	Hypomethylation	180	Choi et al. (2009)
LRE1	Head and neck squamous cell carcinoma	278	Hypomethylation	526	Hsiung et al. (2007)
LINE-1, Alu	Gastric cancer	300	Hypomethylation	419	Hou et al. (2010)
Gene					
IGF2	Colorectal neoplasia	65	More LOI	107	Cui et al. (2003)
	CRC	225	No difference	435	Ito et al. (2008)
	Breast Cancer	228	No difference	460	
	CRC	97	No difference	190	Kaaks et al. (2009)
DAPK, ECAD, MGMT, p16	Lung cancer, smoker	49, 22	Hypermethylation	5	Russo et al. (2005)
ATM	Bilateral breast cancer	190	Hypermethylation	190	Flanagan et al. (2009)
NUP155, ZNF217, TITF1, NEUROD1, SFRP1	Breast Cancer	320	Hypomethylation	676	Widschwendter et al. (2008)
Estrogen receptor alpha	Colorectal lesion	73	No difference	57	Ally et al. (2009)

2. Repeated sequences

Genome-wide (global) hypomethylation seems to occur early and is a feature of genomic DNA derived from head and neck squamous cell carcinoma (HNSCC). To determine whether global methylation in DNA derived from whole blood is associated with HNSCC and to assess potential modification of this property by environmental or behavioral risk factors, global DNA methylation levels were assessed. Hypomethylation was associated with a significant 1.6-fold increased risk for disease in models controlled for other HNSCC risk factors. Smoking showed a significant differential effect on blood relative methylation between cases and controls (Hsiung *et al.*, 2007).

Hou *et al.* (2010) determined whether global methylation in blood leukocyte DNA was associated with gastric cancer in a population-based study on 302 gastric cancer cases and 421 age- and sex-matched controls in Warsaw, Poland, between 1994 and 1996. Using PCR-pyrosequencing, they analyzed methylation levels of Alu and LINE-1, 2 CG-rich repetitive elements, to measure global methylation levels. Gastric cancer risk was highest among those with lowest level of methylation in either Alu or LINE-1 relative to those with the highest levels, although the trends were not statistically significant. For LINE-1, the association tended to be stronger among individuals with a family history of cancer, current alcohol drinkers, current smokers, those who rarely or never consumed fruit, CC carriers for the MTRR Ex51123C>T polymorphism and TT carriers for the MTRR Ex151572T>C polymorphism. The association was not different by sex, *H. pylori* infection, intake of folate, vitamin B6, and total protein and the remaining polymorphisms examined. These results indicate that interactions between blood leukocyte DNA hypomethylation and host characteristics may determine gastric cancer risk.

B. Promoter methylation

1. IGF-2

Genomic imprinting is a form of gene silencing that is epigenetic in origin. Loss of imprinting (LOI) of the insulin-like growth factor II gene (*IGF2*) was discovered in embryonal tumors of childhood, such as Wilms' tumor, but it is found commonly in many types of cancer, including ovarian, lung, liver, and colon. In Wilms' tumors, especially those with late onset, LOI resulted in the increased expression of *IGF2*, an important autocrine growth factor for many cancers, including CRC. LOI was found in normal colonic mucosa of about 30% of CRC patients, but it was found in only 10% of healthy individuals. In a study to investigate the utility of LOI as a marker of CRC risk, 172 patients were evaluated at a colonoscopy clinic. The adjusted odds ratio for LOI in lymphocytes was 5.15 for patients with a positive family history, 3.46 for patients with adenomas, and 21.7 for patients with CRC. Thus, LOI of lymphocyte DNA may be a valuable predictive marker of an individual's risk for CRC (Cui et al., 2003).

The above data is intriguing, but there have been two reports against the first one. DNA samples from tumor tissues and matched nontumor tissues from 22 breast and 42 CRC patients as well as peripheral blood samples obtained from colorectal and breast cancer patients were analyzed. IGF2 methylation levels in tumors were lower than matched nontumor tissue. Hypomethylation of IGF2 was detected in breast (33%) and colorectal (80%) tumor tissues with a higher frequency than LOI indicating that methylation levels are a better indicator of cancer than LOI. The prevalence of IGF2 hypomethylation was 9.5% and this correlated with increased age but not cancer risk. Thus, IGF2 hypomethylation occurs as an acquired tissue-specific somatic event rather than a constitutive innate epimutation. These results indicate that IGF2 hypomethylation has diagnostic potential for colon cancer rather than value as a surrogate biomarker for constitutive LOI (Ito et al., 2008).

Kaaks et al. (2009) performed a case-control study of 97 colon cancer cases and 190 age-matched and gender-matched controls, nested within the prospective Northern Sweden Health and Disease Study cohort. Mean fractions of IGF2 CpG methylation in lymphocyte DNA were identical for cases and controls, and logistic regression analyses showed no relationship between colon cancer risk and quartile levels of CpG methylation. The results from this study population do not support the hypothesis that colon cancer can be predicted from the different degrees of methylation of IGF2 from lymphocyte DNA.

2. Other genes

Promoter DNA methylation levels of six genes in samples derived from 27 bronchial epithelial cells and matching blood samples from 22 former/current smokers as well as 49 primary non small cell lung cancer samples with corresponding blood controls were determined using MSP. p16 promoter DNA methylation in tumors was observed at consistently higher levels when compared with all the other samples analyzed. Interestingly, similar levels of methylation were observed in bronchial epithelial cells and corresponding blood from smokers for all four genes (ECAD, p16, MGMT, and DAPK) that showed smoking/lung cancer-associated methylation changes. These data provide preliminary evidence that peripheral lymphocytes could potentially be used as a surrogate for bronchial epithelial cells to detect altered DNA methylation in smokers (Russo et al., 2005).

Bilaterality of breast cancer is an indicator of constitutional cancer susceptibility. Significant hypermethylation of one intragenic repetitive element around the *ATM* gene in breast cancer cases was shown compared with controls, with the highest quartile of methylation associated with a threefold increased risk of breast cancer. Increased methylation of this locus is associated with lower steady-state *ATM* mRNA level and correlates with age of cancer patients but not controls, suggesting a combined age–phenotype-related association (Flanagan *et al.*, 2009).

Using quantitative methylation analysis in a case-control study, it was found that DNA hypomethylation of peripheral blood cell DNA indicated good prediction of breast cancer risk. Invasive ductal and invasive lobular breast cancers were characterized by two different sets of genes, the latter particular by genes involved in the differentiation of the mesenchyme (*PITX2*, *TITF1*, *GDNF*, and *MYOD1*). Only ER-α target genes predicted ER positive BC; lack of peripheral blood cell DNA methylation of ZNF217 predicted BC independent of age and family history and was associated with ER-α bioactivity in the corresponding serum (Widschwendter *et al.*, 2008).

DNA was extracted from frozen stored whole blood samples of 27 subjects with CRC, 30 with adenoma, 16 with hyperplastic polyps, and 57 disease free subjects, and DNA methylation of the *estrogen receptor* α (*ER-α*) gene was quantitated. *ER-α* was partially methylated in leukocyte DNA in all subjects, with no significant difference between disease groups. *ER-α* methylation in leukocytes was 60% lower than that in normal colonic tissue. There was a positive relationship between *ER-α* methylation in leukocytes and colonic tissue in subjects with and without colorectal tumors. However, unlike in colonic tissue, *ER-α* methylation in leukocytes was unable to distinguish between disease groups, although this may be because of the narrower range of methylation values in leukocytes compared with colon (0–12.6 vs. 0.8–38.7, respectively; Ally *et al.*, 2009).

IV. CONCLUSION

Molecular epidemiologic studies on cancer using epigenetic signatures are still in its infancy. It is essential to enhance statistical power by conducting studies with larger sample size, and developing consortia to coordinate research efforts. Similar to GWAS on SNPs, a genome-wide search for epigenetic risk factors may clarify associations between a phenotype and thousands of biologic variables, and provides important information for the discovery of new epigenetic biomarkers.

Understanding the similarities and differences in genetic versus epigenetic data will help with planning the next step of epidemiologic studies that prospectively incorporate both. Any peripheral tissue samples can be used for genotyping an individual, while the level and pattern of epigenetic marks vary across different tissue and cell types, posing a formidable challenge for epigenetic analysis. Epigenetic marks in readily accessible cells, such as blood, buccal, and skin, may not reflect those in generally inaccessible tissues like brain. Analysis of postmortem tissue will provide valuable insight into the epigenetic profile of inaccessible tissues, but the identification and validation of peripheral epigenetic marker will be required for epidemiologic studies to benefit from incorporation of epigenetic data. Several human studies have already investigated the effects of environmental exposures on blood leukocytes. However, since effects of environmental factors may be tissue- or even cell-specific, the data on blood cell DNA do not necessarily represent epigenetic patterns in the target tissues.

There might be two types of effects of each environmental factor on epigenetic statuses, that is, systemic and local. Aging and smoking might cause systemic epigenetic effects, because these affect epigenetic changes even in blood leukocyte DNA. On the other hand, infection may induce only local epigenetic changes in certain susceptible tissues, such as the liver and stomach. Blood leukocyte DNA may be used as biomarkers representing systemic epigenetic changes, but not the local ones. Further investigations are necessary to fully understand extents of epigenetic changes induced by each environmental factor.

Epigenetic alterations in comparison with genetic changes are reversible and are typically acquired in a gradual manner. These features offer an enormous potential for prevention strategies. For example, the impact of certain micronutrients on DNA methylation adds to our current understanding of possible mechanisms of how diet is linked with cancer. However, the specific mechanistic effects of diet on DNA methylation requires further study, for example, the effect on promoter methylation of tumor suppressor genes and alteration in transcript levels of DNA methyltransferases by varying concentrations of a particular micronutrient. Once these mechanisms have been elucidated, the next step would be human interventional trials using these substrates as nutraceuticals. Subsequent findings would then result in a paradigm shift moving us into an era of targeted modification of methylation patterns using diet.

Future prospective investigations are also needed to determine whether exposed individuals develop epigenetic alterations over time and whether such alterations consequently increase the risk of disease including cancer.

Acknowledgment

This work was supported by JSPS A3 Foresight Program.

References

Ahuja, N., Li, Q., Mohan, A. L., Baylin, S. B., and Issa, J. P. (1998). Aging and DNA methylation in colorectal mucosa and cancer. *Cancer Res.* **58**, 5489–5494.

Ally, M. S., Al-Ghnaniem, R., and Pufulete, M. (2009). The relationship between gene-specific DNA methylation in leukocytes and normal colorectal mucosa in subjects with and without colorectal tumors. *Cancer Epidemiol. Biomarkers Prev.* **18**, 922–928.

Anway, M. D., Cupp, A. S., Uzumcu, M., and Skinner, M. K. (2005). Epigenetic transgenerational actions of endocrine disruptors and male fertility. *Science* **308**, 1466–1469.

Arasaradnam, R. P., Commane, D. M., Bradburn, D., and Mathers, J. C. (2008). A review of dietary factors and its influence on DNA methylation in colorectal carcinogenesis. *Epigenetics* **3**, 193–198.

Azechi, H., Nishida, N., Fukuda, Y., Nishimura, T., Minata, M., Katsuma, H., Kuno, M., Ito, T., Komeda, T., Kita, R., *et al.* (2001). Disruption of the p16/cyclin D1/retinoblastoma protein pathway in the majority of human hepatocellular carcinomas. *Oncology* **60**, 346–354.

Baccarelli, A., and Bollati, V. (2009). Epigenetics and environmental chemicals. *Curr. Opin. Pediatr.* **21**, 243–251.

Baek, M. J., Piao, Z., Kim, N. G., Park, C., Shin, E. C., Park, J. H., Jung, H. J., Kim, C. G., and Kim, H. (2000). p16 is a major inactivation target in hepatocellular carcinoma. *Cancer* **89**, 60–68.

Belinsky, S. A., Klinge, D. M., Liechty, K. C., March, T. H., Kang, T., Gilliland, F. D., Sotnic, N., Adamova, G., Rusinova, G., and Telnov, V. (2004). Plutonium targets the p16 gene for inactivation by promoter hypermethylation in human lung adenocarcinoma. *Carcinogenesis* **25**, 1063–1067.

Bollati, V., Baccarelli, A., Hou, L., Bonzini, M., Fustinoni, S., Cavallo, D., Byun, H. M., Jiang, J., Marinelli, B., Pesatori, A. C., Bertazzi, P. A., and Yang, A. S. (2007). Changes in DNA methylation patterns in subjects exposed to low-dose benzene. *Cancer Res.* **67**, 876–880.

Chen, J. X., Zheng, Y., West, M., and Tang, M. S. (1998). Carcinogens preferentially bind at methylated CpG in the p53 mutational hot spots. *Cancer Res.* **58**, 2070–2075.

Chen, H., Liu, J., Zhao, C. Q., Diwan, B. A., Merrick, B. A., and Waalkes, M. P. (2001). Association of c-myc overexpression and hyperproliferation with arsenite-induced malignant transformation. *Toxicol. Appl. Pharmacol.* **175**, 260–268.

Choi, S. W., Stickel, F., Baik, H. W., Kim, Y. I., Seitz, H. K., and Mason, J. B. (1999). Chronic alcohol consumption induces genomic but not p53-specific DNA hypomethylation in rat colon. *J. Nutr.* **129**, 1945–1950.

Choi, J. Y., James, S. R., Link, P. A., McCann, S. E., Hong, C. C., Davis, W., Nesline, M. K., Ambrosone, C. B., and Karpf, A. R. (2009). Association between global DNA hypomethylation in leukocytes and risk of breast cancer. *Carcinogenesis* **30**, 1889–1897.

Chuang, J. C., Yoo, C. B., Kwan, J. M., Li, T. W., Liang, G., Yang, A. S., and Jones, P. A. (2005). Comparison of biological effects of non-nucleoside DNA methylation inhibitors versus 5-aza-2′-deoxycytidine. *Mol. Cancer Ther.* **4**, 1515–1520.

Cui, H., Cruz-Correa, M., Giardiello, F. M., Hutcheon, D. F., Kafonek, D. R., Brandenburg, S., Wu, Y., He, X., Powe, N. R., and Feinberg, A. P. (2003). Loss of IGF2 imprinting: A potential marker of colorectal cancer risk. *Science* **299**, 1753–1755.

Day, J. K., Bauer, A. M., DesBordes, C., Zhuang, Y., Kim, B. E., Newton, L. G., Nehram, V., Forsee, K. M., MacDonald, R. S., Besch-Williford, C., Huang, T. H., and Lubahnm, D. B. (2002). Genistein alters methylation patterns in mice. *J. Nutr. Suppl.* **132**, 2419S–2423S.

Dolinoy, D. C., Weidmanm, J. R., Waterland, R. A., and Jirtle, R. L. (2006). Maternal genistein alters coat color and protects Avy mouse offspring from obesity by modifying the fetal epigenome. *Environ. Health Perspect.* **114**, 567–572.

Dolinoy, D. C., Huang, D., and Jirtle, R. L. (2007). Maternal nutrient supplementation counteracts bisphenol A-induced DNA hypomethylation in early development. *Proc. Natl. Acad. Sci. USA* **104**, 13056–13061.

Eguchi, K., Kanai, Y., Kobayashi, K., and Hirohashi, S. (1997). DNA hypermethylation at the D17S5 locus in non-small cell lung cancers: Its association with smoking history. *Cancer Res.* **57,** 4913–4915.

Fang, M. Z., Wang, Y., Ai, N., Hou, Z., Sun, Y., Lu, H., Welsh, W., and Yang, C. S. (2003). Tea polyphenol (-)-epigallocatechin-3-gallate inhibits DNA methyltransferase and reactivates methylation-silenced genes in cancer cell lines. *Cancer Res.* **63,** 7563–7570.

Flanagan, J. M., Munoz-Alegre, M., Henderson, S., Tang, T., Sun, P., Johnson, N., Fletcher, O., Dos Santos Silva, I., Peto, J., Boshoff, C., *et al.* (2009). Gene-body hypermethylation of ATM in peripheral blood DNA of bilateral breast cancer patients. *Hum. Mol. Genet.* **8,** 1332–1342.

Flatley, J. E., McNeir, K., Balasubramani, L., Tidy, J., Stuart, E. L., Young, T. A., and Powers, H. J. (2009). Folate status and aberrant DNA methylation are associated with HPV infection and cervical pathogenesis. *Cancer Epidemiol. Biomarkers Prev.* **18,** 2782–2789.

Galvan, A., Ioannidis, J. P. A., and Dragani, T. A. (2010). Beyond genome-wide association studies: Genetic heterogeneity and individual predisposition to cancer. *Trends Genet.* **26,** 132–141.

Govindarajan, B., Klafter, R., Miller, M. S., Mansur, C., Mizesko, M., Bai, X., LaMontagne, K., Jr., and Arbiser, J. L. (2002). Reactive oxygen-induced carcinogenesis causes hypermethylation of p16^{Ink4a} and activation of MAP kinase. *Mol. Med.* **8,** 1–8.

Heijmans, B. T., Tobi, E. W., Stein, A. D., Putter, H., Blauw, G. J., Susser, E. S., Slagboom, P. E., and Lumey, L. H. (2008). Persistent epigenetic differences associated with prenatal exposure to famine in humans. *Proc. Natl. Acad. Sci. USA* **105,** 17046–17049.

Herceg, Z. (2007). Epigenetics and cancer: Towards an evaluation of the impact of environmental and dietary factors. *Mutagenesis* **22,** 91–103.

Hirao, T., Bueno, R., Chen, C. J., Gordon, G. J., Heilig, E., and Kelsey, K. T. (2002). Alterations of the p16INK4 locus in human malignant mesothelial tumors. *Carcinogenesis* **23,** 1127–1130.

Hou, L., Wang, H., Sartori, S., Gawron, A., Lissowska, J., Bollati, V., Tarantini, L., Zhang, F. F., Zatonski, W., Chow, W. H., *et al.* (2010). Blood leukocyte DNA hypomethylation and gastric cancer risk in a high-risk Polish population. *Int. J. Cancer* DOI: 10.1002/ijc.25190.

Hsiung, D. T., Marsit, C. J., Houseman, E. A., Eddy, K., Furniss, C. S., McClean, M. D., and Kelsey, K. T. (2007). Global DNA methylation level in whole blood as a biomarker in head and neck squamous cell carcinoma. *Cancer Epidemiol. Biomarkers Prev.* **16,** 108–114.

Hughes, L. A., van den Brandt, P. A., de Bruïne, A. P., Wouters, K. A., Hulsmans, S., Spiertz, A., Goldbohm, R. A., de Goeij, A. F., Herman, J. G., Weijenberg, M. P., *et al.* (2009). Early life exposure to famine and colorectal cancer risk: A role for epigenetic mechanisms. *PLoS ONE* **4,** e7951.

Issa, J. P., Ottaviano, Y. L., Celano, P., Hamilton, S. R., Davidson, N. E., and Baylinm, S. B. (1994). Methylation of the oestrogen receptor CpG island links ageing and neoplasia in human colon. *Nat. Genet.* **7,** 536–540.

Ito, Y., Koessler, T., Ibrahim, A. E., Raim, S., Vowler, S. L., Abu-Amerom, S., Silvam, A. L., Maia, A. T., Huddleston, J. E., Uribe-Lewis, S., *et al.* (2008). Somatically acquired hypomethylation of IGF2 in breast and colorectal cancer. *Hum. Mol. Genet.* **17,** 2633–2643.

Kaaks, R., Stattin, P., Villar, S., Poetsch, A. R., Dossusm, L., Nieters, A., Riboli, E., Palmqvist, R., Hallmans, G., Plass, C., *et al.* (2009). Insulin-like growth factor-II methylation status in lymphocyte DNA and colon cancer risk in the Northern Sweden Health and Disease cohort. *Cancer Res.* **69,** 5400–5405.

Kang, G. H., Lee, S., Kim, W. H., Lee, H. W., Kim, J. C., Rhyu, M. G., and Ro, J. Y. (2002). Epstein-Barr virus-positive gastric carcinoma demonstrates frequent aberrant methylation of multiple genes and constitutes CpG island methylator phenotype-positive gastric carcinoma. *Am. J. Pathol.* **160,** 787–794.

Ke, Q., Davidson, T., Chen, H., Kluz, T., and Costa, M. (2006). Alterations of histone modifications and transgene silencing by nickel chloride. *Carcinogenesis* **27,** 1481–1488.

Kim, D. H., Nelson, H. H., Wiencke, J. K., Zheng, S., Christiani, D. C., Wain, J. C., Mark, E. J., and Kelsey, K. T. (2001). p16INK4a and histology-specific methylation of CpG islands by exposure to tobacco smoke in non-small cell lung cancer. *Cancer Res.* **61,** 3419–3424.

Kondo, K., Takahashi, Y., Hirose, Y., Nagao, T., Tsuyuguchi, M., Hashimoto, M., Ochiai, A., Monden, Y., and Tangoku, A. (2006). The reduced expression and aberrant methylation of p16INK4a in chromate workers with lung cancer. *Lung Cancer* **53,** 295–302.

Kusano, M., Toyota, M., Suzuki, H., Akino, K., Aoki, F., Fujita, M., Hosokawa, M., Shinomura, Y., Imai, K., and Tokino, T. (2006). Genetic, epigenetic, and clinicopathologic features of gastric carcinomas with the CpG island methylator phenotype and an association with Epstein-Barr virus. *Cancer* **106,** 1467–1479.

Li, S., Washburn, K. A., Moore, R., Uno, T., Teng, C., Newbold, R. R., McLachlan, J. A., and Negishi, M. (1997). Developmental exposure to diethylstilbestrol elicits demethylation of estrogen-responsive lactoferrin gene in mouse uterus. *Cancer Res.* **57,** 4356–4359.

Lim, U., Flood, A., Choi, S. W., Albanes, D., Cross, A. J., Schatzkin, A., Sinha, R., Katki, H. A., Cash, B., Schoenfeld, P., *et al.* (2008). Genomic methylation of leukocyte DNA in relation to colorectal adenoma among asymptomatic women. *Gastroenterology* **134,** 47–55.

Maekita, T., Nakazawa, K., Mihara, M., Nakajima, T., Yanaoka, K., Iguchi, M., Arii, K., Kaneda, A., Tsukamoto, T., Tatematsu, M., *et al.* (2006). High levels of aberrant DNA methylation in *Helicobacter pylori*-infected gastric mucosae and its possible association with gastric cancer risk. *Clin. Cancer Res.* **12,** 989–995.

Marmot, M., Atinmo, T., Byers, T., Chen, J., Hirohata, T., Jackson, A., James, W. P. T., Kolonel, L. N., Kumanyika, S., and Leitzmann, C. *et al.* (eds.) (2007). *In* "Food, nutrition, physical activity, and the prevention of cancer: A global perspective". World Cancer Research Fund/ American Institute for Cancer Res, Washington, DC.

Marsit, C. J., McClean, M. D., Furniss, C. S., and Kelsey, K. T. (2006). Epigenetic inactivation of the SFRP genes is associated with drinking, smoking and HPV in head and neck squamous cell carcinoma. *Int. J. Cancer* **119,** 1761–1766.

Mays-Hoopes, L., Chao, W., Butcher, H. C., and Huang, R. C. (1986). Decreased methylation of the major mouse long interspersed repeated DNA during aging and in myeloma cells. *Dev. Genet.* **7,** 65–73.

Mittal, A., Piyathilake, C., Hara, Y., and Katiyar, S. K. (2003). Exceptionally high protection of photocarcinogenesis by topical application of (-)-epigallocatechin-3-gallate in hydrophilic cream in SKH-1 hairless mouse model: Relationship to inhibition of UVB-induced global DNA hypomethylation. *Neoplasia* **5,** 555–565.

Moore, L. E., Pfeiffer, R. M., Poscablo, C., Real, F. X., Kogevinas, M., Silverman, D., García-Closas, R., Chanock, S., Tardón, A., Serra, C., *et al.* (2008). Genomic DNA hypomethylation as a biomarker for bladder cancer susceptibility in the Spanish Bladder Cancer Study: A case-control study. *Lancet Oncol.* **9,** 359–366.

Murthy, S. S., Shen, T., De Rienzo, A., Lee, W. C., Ferriola, P. C., Jhanwar, S. C., Mossman, B. T., Filmus, J., and Testa, J. R. (2000). Expression of GPC3, an X-linked recessive overgrowth gene, is silenced in malignant mesothelioma. *Oncogene* **19,** 410–416.

Niwa, T., Tsukamoto, T., Toyoda, T., Mori, A., Tanaka, H., Maekita, T., Ichinose, M., Tatematsu, M., and Ushijima, T. (2010). Inflammatory processes triggered by *Helicobacter pylori* infection cause aberrant DNA methylation in gastric epithelial cells. *Cancer Res.* **70,** 1430–1440.

Oka, D., Yamashita, S., Tomioka, T., Nakanishi, Y., Kato, H., Kaminishi, M., and Ushijima, T. (2009). The presence of aberrant DNA methylation in noncancerous esophageal mucosae in association with smoking history. *Cancer* **115,** 3412–3426.

Pufulete, M., Al-Ghnaniem, R., Khushal, A., Appleby, P., Harris, N., Gout, S., Emery, P. W., and Sanders, T. A. (2005). Effect of folic acid supplementation on genomic DNA methylation in patients with colorectal adenoma. *Gut* **54,** 648–653.

Puri, S. K., Si, L., Fan, C. Y., and Hanna, E. (2005). Aberrant promoter hypermethylation of multiple genes in head and neck squamous cell carcinoma. *Am. J. Otolaryngol.* **26,** 12–17.

Russo, A. L., Thiagalingam, A., Pan, H., Califano, J., Cheng, K. H., Ponte, J. F., Chinnappan, D., Nemani, P., Sidransky, D., and Thiagalingam, S. (2005). Differential DNA hypermethylation of critical genes mediates the stage-specific tobacco smoke-induced neoplastic progression of lung cancer. *Clin. Cancer Res.* **11,** 2466–2470.

Sarter, B., Long, T. I., Tsong, W. H., Koh, W. P., Yu, M. C., and Laird, P. W. (2005). Sex differential in methylation patterns of selected genes in Singapore Chinese. *Hum. Genet.* **117,** 402–403.

Schnekenburger, M., Talaska, G., and Puga, A. (2007). Chromium cross-links histone deacetylase 1-DNA methyltransferase 1 complexes to chromatin, inhibiting histone-remodeling marks critical for transcriptional activation. *Mol. Cell. Biol.* **27,** 7089–7101.

Takiguchi, M., Achanzar, W. E., Qu, W., Li, G., and Waalkes, M. P. (2003). Effects of cadmium on DNA-(Cytosine-5) methyltransferase activity and DNA methylation status during cadmium-induced cellular transformation. *Exp. Cell Res.* **286,** 355–365.

Tawa, R., Kimura, Y., Komura, J., Miyamura, Y., Kurishita, A., Sasaki, M. S., Sakurai, H., and Ono, T. (1998). Effects of X-ray irradiation on genomic DNA methylation levels in mouse tissues. *J. Radiat. Res.* **39,** 271–278.

Vaissière, T., Hung, R. J., Zaridze, D., Moukeria, A., Cuenin, C., Fasolo, V., Ferro, G., Paliwal, A., Hainaut, P., Brennan, P., *et al.* (2009). Quantitative analysis of DNA methylation profiles in lung cancer identifies aberrant DNA methylation of specific genes and its association with gender and cancer risk factors. *Cancer Res.* **69,** 243–252.

van Engeland, M., Weijenberg, M. P., Roemen, G. M., Brink, M., de Bruïne, A. P., Goldbohm, R. A., van den Brandt, P. A., Baylin, S. B., de Goeij, A. F., and Herman, J. G. (2003). Effects of dietary folate and alcohol intake on promoter methylation in sporadic colorectal cancer: The Netherlands cohort study on diet and cancer. *Cancer Res.* **63,** 3133–3137.

Virmani, A. K., Muller, C., Rathi, A., Zoechbauer-Mueller, S., Mathis, M., and Gazdar, A. F. (2001). Aberrant methylation during cervical carcinogenesis. *Clin. Cancer Res.* **7,** 584–589.

Waterland, R. A., and Jirtle, R. L. (2003). Transposable elements: Targets for early nutritional effects on epigenetic gene regulation. *Mol. Cell. Biol.* **23,** 5293–5300.

Widschwendter, M., Apostolidou, S., Raum, E., Rothenbacher, D., Fiegl, H., Menon, U., Stegmaier, C., Jacobs, I. J., and Brenner, H. (2008). Epigenotyping in peripheral blood cell DNA and breast cancer risk: A proof of principle study. *PLoS ONE* **16,** e2656.

Wilson, V. L., Smith, R. A., Ma, S., and Cutler, R. G. (1987). Genomic 5-methyldeoxycytidine decreases with age. *J. Biol. Chem.* **262,** 9948–9951.

Yoon, J. H., Smith, L. E., Feng, Z., Tang, M., Lee, C. S., and Pfeifer, G. P. (2001). Methylated CpG dinucleotides are the preferential targets for G-to-T transversion mutations induced by benzo[a] pyrene diol epoxide in mammalian cells: Similarities with the p53 mutation spectrum in smoking-associated lung cancers. *Cancer Res.* **61,** 7110–7117.

Yuasa, Y., Nagasaki, H., Akiyama, Y., Hashimoto, Y., Takizawa, T., Kojima, K., Kawano, T., Sugihara, K., Imai, K., and Nakachi, K. (2009). DNA methylation status is inversely correlated with green tea intake and physical activity in gastric cancer patients. *Int. J. Cancer* **124,** 2677–2682.

Zhang, Y. J., Ahsan, H., Chen, Y., Lunn, R. M., Wang, L. Y., Chen, S. Y., Lee, P. H., Chen, C. J., and Santella, R. M. (2002). High frequency of promoter hypermethylation of RASSF1A and p16 and its relationship to aflatoxin B1-DNA adduct levels in human hepatocellular carcinoma. *Mol. Carcinog.* **35,** 85–92.

Zhao, C. Q., Young, M. R., Diwan, B. A., Coogan, T. P., and Waalkes, M. P. (1997). Association of arsenic-induced malignant transformation with DNA hypomethylation and aberrant gene expression. *Proc. Natl. Acad. Sci. USA* **94,** 10907–10912.

8

Diet, Epigenetic, and Cancer Prevention

Jia Chen and Xinran Xu

Department of Preventive Medicine, Mount Sinai School of Medicine, New York, USA

ABSTRACT

Disruption of the epigenome has been a hallmark of human cancers and has been linked with tumor pathogenesis and progression. Since epigenetic changes can be reversed in principle, studies have been carried out to identify modifiable (such as diet and lifestyle) factors, which possess epigenetic property, in hope for

Advances in Genetics, Vol. 71
0065-2660/10 $35.00
DOI: 10.1016/S0065-2660(10)71009-9

developing epigenetically based prevention/intervention strategies. The goal is to achieve some degree of epigenetic reprogramming, which would maintain normal gene expression status and reverse tumorigenesis through chemoprevention or lifestyle intervention such as diet modification. The ability of dietary compounds to act epigenetically in cancer cells has been studied and evidence continues to surface for constituents in food and dietary supplements to influence the epigenome and ultimately individual's risk of developing cancer. In this chapter, we summarized the existing data, both from animal and human studies, on the capacity of natural food products to influence three key epigenetic processes: DNA methylation, histone modification, and microRNA expression. As discussed in the perspective, while diet-based intervention that targets epigenetic pathways is promising, significant challenges remain in translating these scientific findings into clinical or public health practices in the context of cancer prevention. © 2010, Elsevier Inc.

I. INTRODUCTION

Accumulating evidences have shown that epigenetic abnormalities are associated with multiple human diseases, including cancer. Evidence also continues to surface for constituents in food and dietary supplements to influence the epigenome and ultimately individual's risk of developing cancer. Since epigenetic processes are dynamic, reversible, and susceptible to exogenous factors, they offer a unique opportunity for chemoprevention or diet-based intervention that targets epigenetic pathways.

   The two classic epigenetic processes with profound roles in gene regulation, development, and carcinogenesis are DNA methylation and histone modification. We are also beginning to understand a novel class of posttranscriptional gene regulator, microRNA, which is involved in cell growth, differentiation, apoptosis, and ultimately tumorigenesis (Negrini *et al.*, 2009). All these processes can influence gene expression without altering the sequence of the DNA. It is now evident that in the cancer genome, a large panel of genes harbors aberrant DNA methylation, histone modification, and microRNA expression patterns; disrupted epigenome is a hallmark of cancer cells.

   Of great interest is the fact that effects of aberrant epigenetic changes may be reversed, forming the rationale for epigenetically based prevention/intervention strategies. The goal is to achieve some degree of epigenetic reprogramming, which would maintain normal gene expression status and reverse tumorigenesis through chemoprevention or lifestyle intervention such as diet modification. This chapter will summarize the latest advance in cancer prevention by preventing epigenetic alterations and possibly reversing aberrant epigenetic changes, focusing on dietary compounds. Novel strategies for epigenetically based cancer prevention will also be reviewed.

II. DIETARY COMPOUNDS AFFECTING DNA METHYLATION

Global hypomethylation, accompanied by promoter hypermethylation of cancer-related genes, is a common feature of tumor cells (Esteller, 2008). Global hypomethylation may induce chromosomal instability, reactivate transposons, promote loss of imprinting, and activate proto-oncogenes. Yet, reduced methylation may also protect against C to T transition mutations (Ulrich, 2007). Gene promoter hypermethylation, on the other hand, is associated with the inactivation of genes in virtually all pathways protective of carcinogenesis (e.g., DNA repair, cell cycle control, inflammatory/stress response, detoxification, apoptosis, etc.) (Table 8.1).

Below we summarize existing results on several compounds with methylation-modifying capability and their impact on carcinogenesis.

Table 8.1. Diet Compounds and Their Potential Epigenetic Effects

		Potential epigenetic effects on		
Diet compound	Food source	DNA methylation	Histone modification	microRNA
Folate	Leafy vegetables, fruits, fortified cereal	*		*
B vitamins (B_2, B_6, B_{12})	Meat, nuts, various sources	*		
Methionine	Dairy products, nuts, fish	*		
Choline	Egg, milk, meat sources	*		
Betaine	Spinach, beets, wheat	*		
Phytoestrogen	Soy, legumes, cereal	*	*	
Sulforaphane	Broccoli sprout		*	*
Diallyl sulfide	Garlic		*	
Curcumin	Turmeric		*	
EGCG	Green tea	*	*	
Butyrate	Fermentation of dietary fiber in the digestive tract		*	
Biotin	Egg yolk, animal liver		*	

A. Nutrients in one-carbon metabolism

Among identified compounds that are capable of influencing DNA methylation, nutrients in folate-mediated one-carbon metabolism are of particular interest (Stern et al., 2000). The universal methyl donor in the cellular reactions, S-adenosylmethionine (SAM), is generated in one-carbon pathway. The key nutrients in this pathway are folate, methionine, and several B vitamins (i.e., B_2, B_6, and B_{12}) that are essential cofactors for one-carbon transfer reactions. Other methyl donors, such as choline and betaine, can also affect the SAM status, primarily through choline-mediated one-carbon metabolism, and ultimately impact DNA methylation.

Dietary intake can influence the availability of one-carbon-related cofactors and affects the synthesis of SAM and ultimately DNA methylation. A protective role of nutrients involved in one-carbon metabolism against carcinogenesis is supported by epidemiologic observations of reduced risk of certain cancers with increased folate intakes or biological folate levels (reviewed in Chen et al., 2009). Molecular epidemiologic studies that show effect modification by polymorphisms in one-carbon metabolizing genes add additional support for the causal relationship of folate and cancer, particularly for colorectal and hematopoietic malignancies (Giovannucci, 2002; Sanjoaquin et al., 2005; Ulrich, 2005). Nevertheless, the actual epigenetic link of these associations is far from clear.

Numerous studies have been conducted that demonstrate effects of folate deficiency on DNA methylation in cultured cells. Folate deficiency in the medium usually results in global DNA hypomethylation (Duthie et al., 2000; Jhaveri et al., 2001). Animal studies also support that dietary methyl donors are capable of modulating methylation patterns. They yield important findings: while folate depletion alone is a sufficient perturbing force to diminish the methyl pool, folate-replete diet can restore the DNA methylation status (Miller et al., 1994). For example, rats continuously fed a methyl-deficient diet display hypomethylation of the whole genome as well as promoters in proto-oncogenes such as c-myc, c-fos, and H-ras in the liver (Christman et al., 1993); these abnormalities can be restored after switching back to a methyl-sufficient diet (Pogribny et al., 2009). Interestingly, when rats are fed a methyl-deficient diet for a longer period (18 weeks vs. 9 weeks), switching to methyl-sufficient diet failed to restore completely the methylation status and to prevent the progression of liver carcinogenesis (Pogribny et al., 2009). Besides folate, other methyl donors, such as choline, can also modulate DNA methylation during hepatocarcinogenesis (Asada et al., 2006; Shimizu et al., 2007).

Human data also support the capacity of epigenome modulation by one-carbon nutrients in human population. In two separate studies on healthy postmenopausal women, folate depletion results in lymphocyte DNA hypomethylation, that is reversed following folate repletion (Jacob et al., 1998;

Rampersaud *et al.*, 2000). Folate supplementation in patients with resected colonic adenoma or cancer significantly has reversed a genomic DNA hypomethylation in the normal rectal mucosa (Cravo *et al.*, 1994, 1998). A study on folate and alcohol intake in relation to methylation in colorectal cancer lends strong support that diet is capable of influencing carcinogenesis through epigenetic pathway (van Engeland *et al.*, 2003). For each tested tumor suppressor genes (*p14ARF, p16INK4A, hMLH1, MGMT*, and *RASSF1A*) in this study, the prevalence of promoter hypermethylation is higher in colorectal tumors derived from patients with low folate/high alcohol intake compared to those with high folate/low alcohol intake, although the results do not reach significance due to small sample size. A very recent study to evaluate the effect of choline intakes on cellular methylation in a 12-week controlled choline intervention study suggests that higher than current dietary recommendations (550 mg/day) intake of choline could preserve markers of cellular methylation and attenuate DNA damage among individuals susceptible to folate deficiency (Shin *et al.*, 2010).

Although epidemiologic data generally support the anticarcinogenic property of folate and one-carbon nutrients, issues of dosage and timing of the effect become increasingly important (Kim, 2003). For example, in a study using a mouse model of colorectal cancer (*APC/Min* mice), increasing dietary folate levels significantly reduced the number of ileal polyps in a dose-dependent manner at 3 months, but the association was reversed at 6 months (Song *et al.*, 2000). A more recent study exploring the timing of folic acid exposure in early life reveals that folic acid depletion during pregnancy did not change intestinal tumor incidence; the postweaning depletion, on the other hand, appears to be protective against colorectal neoplasia in female mice (McKay *et al.*, 2008). These intriguing findings suggest that folic acid supplementation may enhance the development and progression of already existing, undiagnosed, premalignant, and malignant lesions.

Findings from the Aspirin-Folate Polyp Prevention Study echo these concerns. This study is a double-blind, placebo-controlled, two-factor, randomized clinical trial of subjects with a recent history of colorectal adenomas. Folic acid supplementation in this trial not only shows no protection against adenoma recurrence, but also confers higher risks of having at least one advanced lesion during 3–5 years follow-up period; folic acid has also been associated with higher risk of having three or more adenomas and of noncolorectal cancers (Cole *et al.*, 2007). It should be mentioned that some subjects reached extremely high doses of folate intake in this study. This trial has raised serious concerns not only about the lack of efficacy but also the potential adverse effects in colorectal cancer prevention and is consistent with a role of folate in promoting the growth of undetected preneoplastic lesions (Ulrich and Potter, 2007). A recent modeling study suggests that folate fortification reduces colorectal cancer rates if started early in life, but can increase rates if begun after age 20 (Luebeck *et al.*, 2008). The precise role of folate in cancer prevention needs to be carefully studied and evaluated.

B. Phytoestrogen

Phytoestrogens, sometimes called "dietary estrogens," are naturally occurring nonsteroidal compounds with pro- or antiestrogenic properties. The major phytoestrogens in the Western diet come from nuts, oilseeds, soy products, cereals, breads, legumes, etc. The major phytoestrogen groups are flavonoids (isoflavones, flavones, and coumestans) and ligands, whereas the former are found in high concentration in soy bean products and the later are mainly found in flax seeds.

Consumption of phytoestrogens have been linked to reduced risks of hormone-sensitive cancers like those of the breast (Trock *et al.*, 2006; Wu *et al.*, 2008) and prostate (Hwang *et al.*, 2009). Timing of exposure is also crucial for the protective effects. For example, consumption of soy foods or exposure to genistein, a potent isoflavone found in soy, during childhood and adolescence in women, and before puberty onset in animals, reduces later mammary cancer risk (Warri *et al.*, 2008). Effects of phytoestrogens differ with respect to different exposure period (Warri *et al.*, 2008). All these findings are consistent with "fetal origins of adult disease" which posits that early developmental exposures involve epigenetic modifications, such as DNA methylation, that influence adult disease susceptibility.

Using cancer cell lines, it has been shown that genistein can partially reverse DNA hypermethylation and reactivate $p16^{INK4a}$, $RAR\beta$, and MGMT (Fang *et al.*, 2005). Genistein can also inhibit the transcription of human telomerase reverse transcriptase (*hTERT*), the catalytic subunit of the human telomerase enzyme, in breast benign (MCF10AT) and cancer (MCF7) cells in a time- and dose-dependent manner, along with downregulation of three major DNA methyltransferases (*DNMT1*, *3a*, and *3b*; Li *et al.*, 2009). In a recent study of 34 healthy premenopausal women randomized to isoflavones daily through one menstrual cycle, isoflavones induced dose-specific changes in $RAR\beta2$ and CCND2 gene methylation performed on intraductal specimens (Qin *et al.*, 2009). This work provides novel insights into methylation effects of dietary isoflavones. However, current evidences are not strong enough to draw any causal relationship or justify certain dietary intervention. More mechanistic studies and well-designed human trials are warranted.

C. Polyphenols

Polyphenols are a class of chemicals with antioxidant activities found in a wide variety of plant foods. Important dietary sources of polyphenols in Western societies are berries, cocoa tea, grape/wine, etc. Green tea polyphenols have attracted a great deal of attention in recent years. Epigallocatechin-3-gallate (EGCG) is the major polyphenol and a potent antioxidant found in green tea. The effects of tea consumption on the risk of human cancer have been investigated in many epidemiological studies, but the results have been inconclusive (Ju *et al.*, 2007). The inconsistency may stem from confounding difficulties in quantifying tea consumption and varied cancer etiology in different populations. Experimental data, however, clearly demonstrate that EGCG has the ability to inhibit DNA methyltransferase activity (Fang *et al.*, 2003). Treatment of cancer cells with this compound results in demethylation of the CpG islands in the promoter regions and the reactivation of methylation-silenced genes in many human malignancies (Fang *et al.*, 2007). For example, promoter demethylation of WIF-1, a key cancer gene, and restoration of WIF-1 expression after EGCG treatment are demonstrated in lung cancer cell lines (Gao *et al.*, 2009). Exposure to EGCG also reduces cellular proliferation and hTERT expression and induces apoptosis in a breast cancer cell line; the downregulation of hTERT gene expression appears to be largely due to epigenetic alterations including promoter methylation and histone acetylation (Berletch *et al.*, 2008).

There are several intervention studies of green tea that have been completed. Early studies have shown the chemopreventive effects of tea in human oral precancerous mucosa lesions (Li *et al.*, 1999), human cervical lesions (Ahn *et al.*, 2003), progression from prostate intraepithelial neoplasia to prostate cancer (Bettuzzi *et al.*, 2006), and colon adenoma recurrence (Shimizu *et al.*, 2008). Although these studies are limited in size, results are promising. Currently, there are more than 30 human trials with tea that are ongoing or are planned based on the NIH Web site.

D. Counteraction of adverse effects of environmental toxicants

Dietary nutrients not only play direct role in maintaining the epigenome that safeguards normal cell differentiation and grow, they also counteract the insults of environmental toxicants on the human epigenome. Such effect is best illustrated by a study of agouti (Avy) mice of which coat color was epigenetically controlled. While maternal exposure to bisphenol-A (BPA), an environmental endocrine disruptor found in plastics, shifts the coat color distribution of mouse offspring toward yellow by decreasing methylation in Agouti gene, maternal dietary supplementation, with either methyl donors like folic acid or the phytoestrogen genistein, negates the DNA hypomethylating effect of BPA and restores the coat color (Dolinoy *et al.*, 2007). This study offers compelling evidence that an altered epigenome induced by environment toxicants can be counteracted by dietary supplements. Emerging epidemiologic data also support this notion. In a randomized, double-blind, placebo-controlled folic acid supplementation trial in an arsenic-exposed population in Bangladesh, folic acid supplementation to participants with low plasma folate enhances arsenic methylation and reduces toxicity, which may ultimately lead to reduction in the risk of arsenic-related health outcomes, including cancer (Gamble *et al.*, 2006).

III. DIETARY COMPOUNDS AFFECTING HISTONE MODIFICATION

Amino acid residues in histone tails can be modified by covalent modification such as acetylation, methylation, phosphorylation, etc. These modifications can affect the structure of chromatin that ultimately regulates gene expression (Fischle *et al.*, 2003; Jenuwein and Allis, 2001). Given the fact that these modifications of histones are potentially reversible, selective agents are being sought that might target abnormal patterns of histone modification, as a means of preventing cancerous transformation and destroying cancer cells.

One hallmark of human cancers is the loss of monoacetylation and trimethylation of histone H4 (Fraga *et al.*, 2005). A particularly active avenue of research involves inhibitors of histone deacetylase (HDAC), which affects histone acetylation status and access of transcription factor to DNA, thereby derepressing epigenetically silenced genes in cancer cells, resulting in cell cycle arrest and/or apoptosis (Myzak and Dashwood, 2006a,b; Xu *et al.*, 2007). Recent interest in HDAC inhibitors has expanded into cancer chemoprevention that is distinct from cancer chemotherapy. Evidence from both animal experiments and human trials have shown that dietary compounds such as sulforaphane (SFN) and diallyl disulfide (DADS) can exert anticancer effects via histone modification. These compounds act as weak ligands for HDAC and exhibit HDAC inhibitory activity (Dashwood *et al.*, 2006; Myzak and Dashwood, 2006a,b). The working hypothesis for these agents is that DNA/chromatin interactions are kept in a constrained state in the presence of HDAC/corepressor complexes, but HDAC inhibitors enable histone acetyltransferase/coactivator complexes to transfer acetyl groups to lysine tails in histones, thereby loosening the interactions with DNA and facilitating transcription factor access and gene activation.

Below we summarize existing results on several compounds with histone-modifying capability.

A. Sulforaphane (SFN)

One of the best-documented histone-modifying compounds is sulforaphane (SFN) and related isothiocyanates. Sulforaphane is found in cruciferous vegetables, such as broccoli and broccoli sprouts. It was first identified as a potent inducer of phase II detoxification enzymes, thus has been proposed as an anticarcinogen (Zhang *et al.*, 1992). New evidence is emerging that SFN can act through other chemoprotective mechanisms (Juge *et al.*, 2007; Myzak and Dashwood, 2006a,b), including inhibition of the HDAC activity (Dashwood *et al.*, 2006; Xu *et al.*, 2007).

The first evidence of SFN inhibiting HDAC activity comes from studies of human colon cancer cells (Myzak *et al.*, 2004) and then in various human prostate lines (Myzak *et al.*, 2006a,b), in which an increase in both global and local histone acetylation status has been observed upon SFN exposure. Findings of HDAC inhibition by SFN has recently been extended to human breast cancer cells (Pledgie-Tracy *et al.*, 2007). *In vivo* studies also corroborate the *in vitro* evidence. SFN reduces the growth of prostate cancer xenografts (Myzak *et al.*, 2007) and suppresses spontaneous intestinal polyps in mice (Myzak *et al.*, 2006a,b) with evidence for altered histone acetylation status and HDAC inhibition.

It has been shown that HDAC is competitively inhibited by the metabolites SFN-*N*-acetylcysteine and SFN-cysteine, rather than the parent compound SFN, and studies have been performed to examine the HDAC inhibition effects of other structurally related isothiocyanates (Dashwood *et al.*, 2006; Myzak *et al.*, 2004). Synthetic isothiocyanate phenylhexyl isothiocyanate provides evidence for HDAC inhibition and chromatin remodeling in human leukemia cells, leading to growth arrest (Ma *et al.*, 2006). In the last report, it was noteworthy that the inhibition of HDAC activity is associated with changes in multiple histone "marks." Specifically, there is a dose-dependent increase in acetylated histones H3 and H4, as well as methylated H3K4, with concomitant loss of the "repressive" histone mark, methylated H3K9.

In a human trial, a single ingestion of 68 g (1 cup) of broccoli sprouts inhibits HDAC activity in circulating peripheral blood mononuclear cells 3–6 h later, with a concomitant induction of histone H3 and H4 acetylation (Myzak *et al.*, 2007). These findings provided the first translational evidence for HDAC inhibition by a natural "whole food," namely broccoli sprouts, and support for an epigenetic mechanism of SFN action at intake levels readily achievable in humans. Currently, there are a handful of NIH-funded trials of sulforaphane and related isothiocyanates in healthy populations as well as in prevention of leukemia and cancers of the prostate, breast, and lung.

B. Diallyl disulfide (DADS)

Diallyl disulfide (DADS) is a major organosulfur compound present in garlic with an antimitotic potential against neoplastic lesions. Consumption of DADS has been linked to low risks of gastrointestinal cancers in epidemiological studies (Fleischauer and Arab, 2001; Fleischauer *et al.*, 2000). Experimental studies have been carried out to determine anticarcinogenic properties of these compounds. Available data reveal that DADS has the ability to modulate numerous biological mechanisms that may influence carcinogenesis and has been shown to inhibit rodent chemically induced carcinogenesis in various organs, including colon. In human tumor colon cell lines, DADS induces a rapid histone hyperacetylation that is transient with a single treatment and is prolonged with a repeated treatment (Druesne *et al.*, 2004). *In vivo* experiments of nontumor colon suggest that histone H4 and H3 acetylation is increased in isolated colonocytes after administration of DADS by intracecal perfusion or gavage (Druesne-Pecollo et al., 2007).

It would be interesting to investigate DADS as a dietary HDAC inhibitor with possible preventive effects against carcinogenesis in normal cells. More studies are needed to determine whether DADS, at physiological dose and with repeated exposure, can modulate gene expression in vivo through histone acetylation modifications, and whether a clear link exists between these effects and a cancer-preventive activity. Currently, there is a NIH-sponsored trial that is underway to investigate the prevention of garlic supplement against stomach cancer.

C. Curcumin

Curcumin (also known as diferuloylmethane) is a principal and active component of turmeric, which is a member of the ginger family. Curcumin has been reported to have both anticancer and anti-inflammatory properties (Aggarwal et al., 2003). A recent study reports that curcumin can inhibit inflammation and restore glucocorticoid efficacy (which is lost under oxidative stress) through upregulation/restoration of HDAC2 activity in the monocytes and macrophages in a concentration-dependent manner (Rahman, 2008). Interestingly, it has been suggested that the anti-inflammatory actions of curcumin propagated through inhibition of histone acetyltransferase (HAT) activity, preventing NF-κB-mediated chromatin acetylation (Kang et al., 2005). Hence, it might be reasonable to propose that in addition to its role as an antioxidant/anti-inflammatory agent, curcumin may also assist in increasing the efficacy of steroids via modulation of HDAC and HAT activity. Currently, epidemiologic data on curcumin are very limited. There are several cancer prevention trials with curcumin, predominantly with against colon cancer but also with pancreatic cancer and multiple myeloma.

D. Dietary fiber—butyrate

Certain fibers, such as resistant starch, can modulate the colorectal lumen in a number of theoretically beneficial ways (Young et al., 2005). Through bacterial fermentation of the fiber there is increase in the luminal concentrations of short chain fatty acids, such as butyrate, along with declining pH (Young et al., 2005). In two observational studies, patients with colorectal cancer were found to have significantly lower fecal butyrate concentrations (Clausen et al., 1991), and in another report, stool pH was inversely associated with colorectal cancer risk (Walker et al., 1986). At high concentrations, butyrate modulates transcriptional regulation in a similar manner as other HDAC inhibitors do, with similar consequences concerning cellular differentiation, cell cycle arrest, apoptosis, invasion, and metastasis. Like other HDAC inhibitors, butyrate is able to induce apoptosis by a mechanism involving the activation of caspases and decreased levels of Bcl-2 (Hinnebusch et al., 2002). However, contradictory to being a HDAC inhibitor, it has recently discovered that more genes are downregulated than upregulated by butyrate; in contrast to global increase in histone acetylation, there appears to be a simultaneous and more localized deacetylation found around certain promoter regions, with resultant transcriptional silencing (Rada-Iglesias et al., 2007). It is likely that butyrates influence on normal and cancer cells is complicated given that butyrate affects different signaling pathways (Scheppach and Weiler, 2004; Scheppach et al., 1995), but is almost certainly mediated in part through important epigenetic mechanisms (Rada-Iglesias et al., 2007). Although therapeutic intervention with butyrate may be far-reaching, an ongoing exposure stemming from the microbial degradation of dietary fiber in the colon could have chemopreventive effects, at least in part by histone deacetylase inhibition (Myzak and Dashwood, 2006a,b).

E. Biotin

A novel class of histone modification, namely, biotinylation, has been of interest recently. Histones H2A, H3, and H4 are modified by covalent binding of biotin to distinct lysine residues (Camporeale et al., 2004; Chew et al., 2006; Kobza et al., 2005). Biotinylated histones have been detected in primary human cells, transformed cell lines, and in other eukaryotes such as Drosophila melanogaster (Camporeale et al., 2006; Gralla et al., 2008; Stanley et al., 2001), and they are catalyzed by holocarboxylase synthetase (HCS) and perhaps biotinidase (BTD).

Studies have shown that biotinylated species of histone H4, that is, K12BioH4 or K8BioH4, are integral components of repeat regions in (peri) centromeric chromatin and mediate silencing of transcriptionally competent chromatin (Camporeale et al., 2007a,b). Biotin deficiency is associated with decreased abundance of biotinylated histones at transposable elements, increasing the transcriptional activity of endogenous retroviruses and genomic instability. A model has been proposed in which biotin homeostasis in human cells is regulated by enrichment of K12BioH4 at promoters driving the expression of biotin transporters (Gralla et al., 2008). Decreased biotinylation of histones caused by long-term biotin deficiency or by HCS knockdown is associated with decreased life span and stress resistance, and aberrant gene-expression patterns in Drosophila (Camporeale et al., 2006). These effects were caused by decreased biotinylation of histones rather than decreased biotinylation of carboxylases, another substrate for biotinylation by HCS (Camporeale et al., 2007a,b).

Biotin is consumed from a wide range of food sources; however, there are few particularly rich dietary sources. Foods with relatively high biotin content include egg yolk, liver, and some vegetables. Biotin is also available from supplements. Both biotin deficiency and supplementation are prevalent in the North American diet. For example, moderate biotin deficiency has been observed in up to 50% of pregnant women (Mock and Stadler, 1997; Mock *et al.*, 1997). About 20% of the US population report taking biotin supplements (Institute of Medicine, 1998), producing super-physiological concentrations of the vitamin in tissues and body fluids (Mock *et al.*, 1995; Zempleni *et al.*, 2001). Biotin deficiency and overdose likely have effects on gene regulation and genome stability that go far beyond the classical coenzymic role of biotin in metabolism. Studies are needed to evaluate its anticarcinogenic or maybe procarcinogenic property in human populations.

IV. DIETARY COMPOUNDS AFFECTING MICRORNA

Aberrant expression of certain microRNAs plays a causal role in tumorigenesis (Negrini *et al.*, 2009). However, dietary influence on microRNAs in the context of cancer is an underexplored area. Strongest evidence so far points to dietary methyl group (folate, B vitamins, methionine, choline, etc.) as potential modulator of microRNAs. Mice fed with methyl-deficient diet display differential microRNA expression in liver compared with those fed with normal diet; these adverse effects appear to be reversible once switched to normal diet (Wang *et al.*, 2009). Upregulation of oncogenic miR-155, miR-221/222, and miR-21 and downregulation of the most abundant liver-specific miR-122 has been reported at early stages of hepatocarcinogenesis (Wang *et al.*, 2009). Folate deficiency has also been shown to result in global increase in microRNA expression in human lymphoblastoid cells in culture; the expression profile returns to normal after normal medium is reintroduced (Marsit *et al.*, 2006). Utilizing samples from a population-based study of head and neck squamous cell carcinomas, overexpression of miR-222 has been observed in individuals with very low folate intake (lowest 1%) compared to those with very high folate intake (highest 1%) (Marsit *et al.*, 2006).

Besides methyl group compounds, one study compared the microRNA expression profiles between environmental cigarette smoke (ECS) and ECS-exposed rats treated with the orally administered chemopreventive agents N-acetylcysteine, oltipraz, indole-3-carbinol, 5,6-benzoflavone, and phenethyl isothiocyanate. The study has shown that some downregulated microRNAs by ECS are protected by chemopreventive agents (Izzotti *et al.*, 2010). Corn oil or fish oil are shown to be capable of protecting the colon from carcinogen-induced microRNA dysregulation (Davidson *et al.*, 2009). Dietary vitamin E has also been shown to affect hepatic microRNA concentrations comparing rats fed diets deficient or sufficient in vitamin E (Gaedicke *et al.*, 2008). Given the importance of microRNA in carcinogenesis and potential modulation by diet, there is an urgent need of systematic studies on the dietary influence on genome-wide profile of microRNA expression in human population.

V. PERSPECTIVES

The ability of dietary compounds to act epigenetically in cancer cells has important implications for cancer prevention given the reversibility of epigenetic processes. In this chapter, we summarized existing data, both from animal and human studies, on the capacity of natural food products to influence three key epigenetic processes, histone modification, DNA methylation, and microRNA expression. However, significant challenges remain in translating these scientific findings into clinical or public health practices in the context of cancer prevention.

Here are several of critical issues to be thoroughly investigated:

1. *Optimal time and duration of prevention.* Given that cancer in general, is a developmental disease, it is critical to consider the issue of "window of susceptibility" that directly influence the timing of an effective cancer prevention strategy. Emerging evidence suggests that diet-induced epigenetic modulation varies during different times in the lifespan. For example, it has been shown that dietary effects may be greatest during embryogenesis and early development (Waterland and Jirtle, 2003). Whether dietary intervention in adult population could result in sufficient and sustaining epigenetic patterns in tissues remains unclear. The opposite effects of folate intervention in polyps prevention trials (as discussed above) also highlight the importance of timing with respect to the developmental stage of the tumor as epigenetic changes appear to occur early in most cancers. In the meantime, there is lack of studies on the long-term effect of dietary pre/intervention. Identifying crucial times and duration for exposure during development and throughout the lifespan is of critical importance in chemoprevention.
2. *Effective dose of intervention and feasibility in public practice.* Most animal and *in vitro* studies presented in this chapter uses exposure levels far exceeding the physiological relevant doses in human population. Studies are needed to investigate the minimum effective dose of these epigenetically active compounds. It is also important to evaluate whether these doses can be achieved in a public health setting.

3. *Nonselective modulation of the epigenome.* Effects of dietary components and chemopreventive agents are likely to influence the entire epigenome in a nonselective manner, thus the consequence may not be unidirectional depending upon specific pathways or target organs. For example, while HDAC inhibitors may have anticarcinogenic activities, inhibiting HDAC activity, by oxidative stress, may enhance inflammatory gene expression leading to a chronic inflammatory response in disorders such as chronic obstructive pulmonary disease (Rahman *et al.*, 2004).

4. *Biomarkers for cancer prevention.* Because cancer is a developmental disease and epigenetic changes occur early in tumor development, the most desirable prevention strategy would be lifelong, nontoxic dietary intervention. This poses daunting challenges to evaluate the efficacy of most prevention trials. Should cancer outcome be used as the end point, trials would require massive size and long duration and become prohibitively expensive. It is crucial to identify epigenetic marks to use for early cancer detection and to monitor response to preventive or therapeutic interventions. The ability of dietary compounds to act epigenetically in cancer cells has important implications for cancer prevention given the reversibility of epigenetic processes.

Acknowledgments

This work was supported by grants from the National Cancer Institutes (R01 CA109753; 3R01CA109753-04S1).

References

Aggarwal, B. B., Kumar, A., and Bharti, A. C. (2003). Anticancer potential of curcumin: Preclinical and clinical studies. *Anticancer Res.* **23,** 363–398.

Ahn, W. S., Yoo, J., Huh, S. W., Kim, C. K., Lee, J. M., Namkoong, S. E., Bae, S. M., and Lee, I. P. (2003). Protective effects of green tea extracts (polyphenon E and EGCG) on human cervical lesions. *Eur. J. Cancer Prev.* **12,** 383–390.

Asada, K., Kotake, Y., Asada, R., Saunders, D., Broyles, R. H., Towner, R. A., Fukui, H., and Floyd, R. A. (2006). LINE-1 hypomethylation in a choline-deficiency-induced liver cancer in rats: dependence on feeding period. *J. Biomed. Biotechnol.* **2006,** 17142.

Berletch, J. B., Liu, C., Love, W. K., Andrews, L. G., Katiyar, S. K., and Tollefsbol, T. O. (2008). Epigenetic and genetic mechanisms contribute to telomerase inhibition by EGCG. *J. Cell. Biochem.* **103,** 509–519.

Bettuzzi, S., Brausi, M., Rizzi, F., Castagnetti, G., Peracchia, G., and Corti, A. (2006). Chemoprevention of human prostate cancer by oral administration of green tea catechins in volunteers with high-grade prostate intraepithelial neoplasia: a preliminary report from a one-year proof-of-principle study. *Cancer Res.* **66,** 1234–1240.

Camporeale, G., Shubert, E. E., Sarath, G., Cerny, R., and Zempleni, J. (2004). K8 and K12 are biotinylated in human histone H4. *Eur. J. Biochem.* **271,** 2257–2263.

Camporeale, G., Giordano, E., Rendina, R., Zempleni, J., and Eissenberg, J. C. (2006). Drosophila melanogaster holocarboxylase synthetase is a chromosomal protein required for normal histone biotinylation, gene transcription patterns, lifespan, and heat tolerance. *J. Nutr.* **136**, 2735–2742.

Camporeale, G., Oommen, A. M., Griffin, J. B., Sarath, G., and Zempleni, J. (2007a). K12-biotinylated histone H4 marks heterochromatin in human lymphoblastoma cells. *J. Nutr. Biochem.* **18**, 760–768.

Camporeale, G., Zempleni, J., and Eissenberg, J. C. (2007b). Susceptibility to heat stress and aberrant gene expression patterns in holocarboxylase synthetase-deficient drosophila melanogaster are caused by decreased biotinylation of histones, not of carboxylases. *J. Nutr.* **137**, 885–889.

Chen, J., Xu, X., Liu, A., and Ulrich, C. M. (2009). Folate and cancer: epidemiological perspective. *In* "Folate in Health and Disease" (Lynn B. Bailey, ed.), 2nd edn, pp. 190–230. CRC Press.

Chew, Y. C., Camporeale, G., Kothapalli, N., Sarath, G., and Zempleni, J. (2006). Lysine residues in N-terminal and C-terminal regions of human histone H2A are targets for biotinylation by biotinidase. *J. Nutr. Biochem.* **17**, 225–233.

Christman, J. K., Sheikhnejad, G., Dizik, M., Abileah, S., and Wainfan, E. (1993). Reversibility of changes in nucleic acid methylation and gene expression induced in rat liver by severe dietary methyl deficiency. *Carcinogenesis* **14**, 551–557.

Clausen, M. R., Bonnen, H., and Mortensen, P. B. (1991). Colonic fermentation of dietary fibre to short chain fatty acids in patients with adenomatous polyps and colonic cancer. *Gut* **32**, 923–928.

Cole, B. F., Baron, J. A., Sandler, R. S., Haile, R. W., Ahnen, D. J., Bresalier, R. S., McKeown-Eyssen, G., Summers, R. W., Rothstein, R. I., Burke, C. A., Snover, D. C., Church, T. R., et al. (2007). Folic acid for the prevention of colorectal adenomas: A randomized clinical trial. *JAMA* **297**, 2351–2359.

Cravo, M., Fidalgo, P., Pereira, A. D., Gouveia-Oliveira, A., Chaves, P., Selhub, J., Mason, J. B., Mira, F. C., and Leitao, C. N. (1994). DNA methylation as an intermediate biomarker in colorectal cancer: modulation by folic acid supplementation. *Eur. J. Cancer Prev.* **3**, 473–479.

Cravo, M. L., Pinto, A. G., Chaves, P., Cruz, J. A., Lage, P., Nobre, Leitao C., and Costa Mira, F. (1998). Effect of folate supplementation on DNA methylation of rectal mucosa in patients with colonic adenomas: correlation with nutrient intake. *Clin. Nutr.* **17**, 45–49.

Dashwood, R. H., Myzak, M. C., and Ho, E. (2006). Dietary HDAC inhibitors: time to rethink weak ligands in cancer chemoprevention? *Carcinogenesis* **27**, 344–349.

Davidson, L. A., Wang, N., Shah, M. S., Lupton, J. R., Ivanov, I., and Chapkin, R. S. (2009). n-3 Polyunsaturated fatty acids modulate carcinogen-directed non-coding microRNA signatures in rat colon. *Carcinogenesis* **30**, 2077–2084.

Dolinoy, D. C., Huang, D., and Jirtle, R. L. (2007). Maternal nutrient supplementation counteracts bisphenol A-induced DNA hypomethylation in early development. *Proc. Natl. Acad. Sci.* **104**, 13056–13061.

Druesne, N., Pagniez, A., Mayeur, C., Thomas, M., Cherbuy, C., Duee, P. H., Martel, P., and Chaumontet, C. (2004). Repetitive treatments of colon HT-29 cells with diallyl disulfide induce a prolonged hyperacetylation of histone H3 K14. *Ann. NY Acad. Sci.* **1030**, 612–621.

Druesne-Pecollo, N., Chaumontet, C., Pagniez, A., Vaugelade, P., Bruneau, A., Thomas, M., Cherbuy, C., Duee, P. H., and Martel, P. (2007). In vivo treatment by diallyl disulfide increases histone acetylation in rat colonocytes. *Biochem. Biophys. Res. Commun.* **354**, 140–147.

Duthie, S. J., Narayanan, S., Blum, S., Pirie, L., and Brand, G. M. (2000). Folate deficiency in vitro induces uracil misincorporation and DNA hypomethylation and inhibits DNA excision repair in immortalized normal human colon epithelial cells. *Nutr. Cancer* **37**, 245–251.

Esteller, M. (2008). Epigenetics in cancer. *N. Engl. J. Med.* **358**, 1148–1159.

Fang, M. Z., Wang, Y., Ai, N., Hou, Z., Sun, Y., Lu, H., Welsh, W., and Yang, C. S. (2003). Tea polyphenol (-)-epigallocatechin-3-gallate inhibits dna methyltransferase and reactivates methylation-silenced genes in cancer cell lines. *Cancer Res.* **63**, 7563–7570.

Fang, M. Z., Chen, D., Sun, Y., Jin, Z., Christman, J. K., and Yang, C. S. (2005). Reversal of hypermethylation and reactivation of p16ink4a, rarβ, and mgmt genes by genistein and other isoflavones from soy. *Clin. Cancer Res.* **11,** 7033–7041.

Fang, M., Chen, D., and Yang, C. S. (2007). Dietary polyphenols may affect DNA methylation. *J. Nutr.* **137,** 223S–228S.

Fischle, W., Wang, Y., and Allis, C. D. (2003). Histone and chromatin cross-talk. *Curr. Opin. Cell Biol.* **15,** 172–183.

Fleischauer, A. T., and Arab, L. (2001). Garlic and cancer: a critical review of the epidemiologic literature. *J. Nutr.* **131,** 1032S–1040S.

Fleischauer, A. T., Poole, C., and Arab, L. (2000). Garlic consumption and cancer prevention: Meta-analyses of colorectal and stomach cancers. *Am. J. Clin. Nutr.* **72,** 1047–1052.

Fraga, M. F., Ballestar, E., Villar-Garea, A., Boix-Chornet, M., Espada, J., Schotta, G., Bonaldi, T., Haydon, C., Ropero, S., Petrie, K., Iyer, N. G., and Perez-Rosado, A. (2005). Loss of acetylation at Lys16 and trimethylation at Lys20 of histone H4 is a common hallmark of human cancer. *Nat. Genet.* **37,** 391–400.

Gaedicke, S., Zhang, X., Schmelzer, C., Lou, Y., Doering, F., Frank, J., and Rimbach, G. (2008). Vitamin E dependent microRNA regulation in rat liver. *FEBS Lett.* **582,** 3542–3546.

Gamble, M. V., Liu, X., Ahsan, H., Pilsner, J. R., Ilievski, V., Slavkovich, V., Parvez, F., Chen, Y., Levy, D., Factor-Litvak, P., and Graziano, J. H. (2006). Folate and arsenic metabolism: A double-blind, placebo-controlled folic acid-supplementation trial in Bangladesh. *Am. J. Clin. Nutr.* **84,** 1093–1101.

Gao, Z., Xu, Z., Hung, M. S., Lin, Y. C., Wang, T., Gong, M., Zhi, X., Jablon, D. M., and You, L. (2009). Promoter demethylation of WIF-1 by epigallocatechin-3-gallate in lung cancer cells. *Anticancer Res.* **29,** 2025–2030.

Giovannucci, E. (2002). Epidemiologic studies of folate and colorectal neoplasia: a review. *J. Nutr.* **132,** 2350S–2355S.

Gralla, M., Camporeale, G., and Zempleni, J. (2008). Holocarboxylase synthetase regulates expression of biotin transporters by chromatin remodeling events at the SMVT locus. *J. Nutr. Biochem.* **19,** 400–408.

Hinnebusch, B. F., Meng, S., Wu, J. T., Archer, S. Y., and Hodin, R. A. (2002). The effects of short-chain fatty acids on human colon cancer cell phenotype are associated with histone hyperacetylation. *J. Nutr.* **132,** 1012–1017.

Hwang, Y. W., Kim, S. Y., Jee, S. H., Kim, Y. N., and Nam, C. M. (2009). Soy food consumption and risk of prostate cancer: a meta-analysis of observational studies. *Nutr. Cancer* **61,** 598–606.

Institute of Medicine (1998). Dietary reference intakes for folate, thiamin, riboflavin, niacin, vitamin B12, panthothenic acid, biotin, and choline 1st edn, National Academies Press, Washington, DC.

Izzotti, A., Calin, G. A., Steele, V. E., Cartiglia, C., Longobardi, M., Croce, C. M., and De Flora, S. (2010). Chemoprevention of cigarette smoke–induced alterations of micrornaRNA expression in rat lungs. *Cancer Prev. Res.* **3,** 62–72.

Jacob, R. A., Gretz, D. M., Taylor, P. C., James, S. J., Pogribny, I. P., Miller, B. J., Henning, S. M., and Swendseid, M. E. (1998). Moderate folate depletion increases plasma homocysteine and decreases lymphocyte DNA methylation in postmenopausal women. *J. Nutr.* **128,** 1204–1212.

Jenuwein, T., and Allis, C. D. (2001). Translating the histone code. *Science* **293,** 1074–1080.

Jhaveri, M. S., Wagner, C., and Trepel, J. B. (2001). Impact of extracellular folate levels on global gene expression. *Mol. Pharmacol.* **60,** 1288–1295.

Ju, J., Lu, G., Lambert, J. D., and Yang, C. S. (2007). Inhibition of carcinogenesis by tea constituents. *Semin. Cancer Biol.* **17,** 395–402.

Juge, N., Mithen, R. F., and Traka, M. (2007). Molecular basis for chemoprevention by sulforaphane: A comprehensive review. *Cell. Mol. Life Sci.* **64,** 1105–1127.

Kang, J., Chen, J., Shi, Y., Jia, J., and Zhang, Y. (2005). Curcumin-induced histone hypoacetylation: The role of reactive oxygen species. *Biochem. Pharmacol.* **69,** 1205–1213.

Kim, Y. (2003). Role of folate in colon cancer development and progression. *J. Nutr.* **133,** 3731S–3739S.

Kobza, K., Camporeale, G., Rueckert, B., Kueh, A., Griffin, J. B., Sarath, G., and Zempleni, J. (2005). K4, K9 and K18 in human histone H3 are targets for biotinylation by biotinidase. *FEBS J.* **272,** 4249–4259.

Li, N., Sun, Z., Han, C., and Chen, J. (1999). The chemopreventive effects of tea on human oral precancerous mucosa lesions. *Proc. Soc. Exp. Biol. Med.* **220,** 218–224.

Li, Y., Liu, L., Andrews, L. G., and Tollefsbol, T. O. (2009). Genistein depletes telomerase activity through cross-talk between genetic and epigenetic mechanisms. *Int. J. Cancer* **125,** 286–296.

Luebeck, E. G., Moolgavkar, S. H., Liu, A. Y., Boynton, A., and Ulrich, C. M. (2008). Does folic acid supplementation prevent or promote colorectal cancer? Results from model-based predictions, Cancer epidemiol biomarkers prev.

Ma, X., Fang, Y., Beklemisheva, A., Dai, W., Feng, J., Ahmed, T., Liu, D., and Chiao, J. W. (2006). Phenylhexyl isothiocyanate inhibits histone deacetylases and remodels chromatins to induce growth arrest in human leukemia cells. *Int. J. Oncol.* **28,** 1287–1293.

Marsit, C. J., Eddy, K., and Kelsey, K. T. (2006). MicroRNA responses to cellular stress. *Cancer Res.* **66,** 10843–10848.

McKay, J. A., Williams, E. A., and Mathers, J. C. (2008). Gender-specific modulation of tumorigenesis by folic acid supply in the Apc mouse during early neonatal life. *Br. J. Nutr.* **99,** 550–558.

Miller, J. W., Nadeau, M. R., Smith, J., Smith, D., and Selhub, J. (1994). Folate-deficiency-induced homocysteinaemia in rats: Disruption of S-adenosylmethionine's co-ordinate regulation of homocysteine metabolism. *Biochem. J.* **298**(Pt 2), 415–419.

Mock, D. M., and Stadler, D. D. (1997). Conflicting indicators of biotin status from a cross-sectional study of normal pregnancy. *J. Am. Coll. Nutr.* **16,** 252–257.

Mock, D. M., Lankford, G. L., and Mock, N. I. (1995). Biotin accounts for only half of the total avidin-binding substances in human serum. *J. Nutr.* **125,** 941–946.

Mock, D. M., Stadler, D. D., Stratton, S. L., and Mock, N. I. (1997). Biotin status assessed longitudinally in pregnant women. *J. Nutr.* **127,** 710–716.

Myzak, M. C., and Dashwood, R. H. (2006a). Histone deacetylases as targets for dietary cancer preventive agents: Lessons learned with butyrate, diallyl disulfide, and sulforaphane. *Curr. Drug Targets* **7,** 443–452.

Myzak, M. C., and Dashwood, R. H. (2006b). Chemoprotection by sulforaphane: Keep one eye beyond Keap1. *Cancer Lett.* **233,** 208–218.

Myzak, M. C., Karplus, P. A., Chung, F. L., and Dashwood, R. H. (2004). A novel mechanism of chemoprotection by sulforaphane: Inhibition of histone deacetylase. *Cancer Res.* **64,** 5767–5774.

Myzak, M. C., Dashwood, W. M., Orner, G. A., Ho, E., and Dashwood, R. H. (2006a). Sulforaphane inhibits histone deacetylase in vivo and suppresses tumorigenesis in Apc-minus mice. *FASEB J.* **20,** 506–508.

Myzak, M. C., Hardin, K., Wang, R., Dashwood, R. H., and Ho, E. (2006b). Sulforaphane inhibits histone deacetylase activity in BPH-1, LnCaP and PC-3 prostate epithelial cells. *Carcinogenesis* **27,** 811–819.

Myzak, M. C., Tong, P., Dashwood, W. M., Dashwood, R. H., and Ho, E. (2007). Sulforaphane retards the growth of human PC-3 xenografts and inhibits HDAC activity in human subjects. *Exp. Biol. Med. (Maywood)* **232,** 227–234.

Negrini, M., Nicoloso, M. S., and Calin, G. A. (2009). MicroRNAs and cancer–new paradigms in molecular oncology. *Curr. Opin. Cell Biol.* **21,** 470–479.

Pledgie-Tracy, A., Sobolewski, M. D., and Davidson, N. E. (2007). Sulforaphane induces cell type-specific apoptosis in human breast cancer cell lines. *Mol. Cancer Ther.* **6,** 1013–1021.

Pogribny, I. P., Shpyleva, S. I., Muskhelishvili, L., Bagnyukova, T. V., James, S. J., and Beland, F. A. (2009). Role of DNA damage and alterations in cytosine DNA methylation in rat liver carcinogenesis induced by a methyl-deficient diet. *Mutat. Res.* **669**, 56–62.

Qin, W., Zhu, W., Shi, H., Hewett, J. E., Ruhlen, R. L., MacDonald, R. S., Rottinghaus, G. E., Chen, Y. C., and Sauter, E. R. (2009). Soy isoflavones have an antiestrogenic effect and alter mammary promoter hypermethylation in healthy premenopausal women. *Nutr. Cancer* **61**, 238–244.

Rada-Iglesias, A., Enroth, S., Ameur, A., Koch, C. M., Clelland, G. K., Respuela-Alonso, P., Wilcox, S., Dovey, O. M., Ellis, P. D., Langford, C. F., Dunham, I., Komorowski, J., *et al.* (2007). Butyrate mediates decrease of histone acetylation centered on transcription start sites and down-regulation of associated genes. *Genome Res.* **17**, 708–719.

Rahman, I. (2008). Dietary polyphenols mediated regulation of oxidative stress and chromatin remodeling in inflammation. *Nutr. Rev.* **66**(Suppl 1), S42–S45.

Rahman, I., Marwick, J., and Kirkham, P. (2004). Redox modulation of chromatin remodeling: Impact on histone acetylation and deacetylation, NF-kappaB and pro-inflammatory gene expression. *Biochem. Pharmacol.* **68**, 1255–1267.

Rampersaud, G. C., Kauwell, G. P., Hutson, A. D., Cerda, J. J., and Bailey, L. B. (2000). Genomic DNA methylation decreases in response to moderate folate depletion in elderly women. *Am. J. Clin. Nutr.* **72**, 998–1003.

Sanjoaquin, M. A., Allen, N., Couto, E., Roddam, A. W., and Key, T. J. (2005). Folate intake and colorectal cancer risk: A meta-analytical approach. *Int. J. Cancer* **113**, 825–828.

Scheppach, W., and Weiler, F. (2004). The butyrate story: Old wine in new bottles? *Curr. Opin. Clin. Nutr. Metab. Care* **7**, 563–567.

Scheppach, W., Bartram, H. P., and Richter, F. (1995). Role of short-chain fatty acids in the prevention of colorectal cancer. *Eur. J. Cancer* **31A**, 1077–1080.

Shimizu, K., Onishi, M., Sugata, E., Sokuza, Y., Mori, C., Nishikawa, T., Honoki, K., and Tsujiuchi, T. (2007). Disturbance of DNA methylation patterns in the early phase of hepatocarcinogenesis induced by a choline-deficient L-amino acid-defined diet in rats. *Cancer Sci.* **98**, 1318–1322.

Shimizu, M., Fukutomi, Y., Ninomiya, M., Nagura, K., Kato, T., Araki, H., Suganuma, M., Fujiki, H., and Moriwaki, H. (2008). Green tea extracts for the prevention of metachronous colorectal adenomas: a pilot study. *Cancer Epidemiol. Biomarkers Prev.* **17**, 3020–3025.

Shin, W., Yan, J., Abratte, C. M., Vermeylen, F., and Caudill, M. A. (2010). Choline intake exceeding current dietary recommendations preserves markers of cellular methylation in a genetic subgroup of folate-compromised men. *J. Nutr.*

Song, J., Medline, A., Mason, J. B., Gallinger, S., and Kim, Y. (2000). Effects of dietary folate on intestinal tumorigenesis in the apcmin mouse. *Cancer Res.* **60**, 5434–5440.

Stanley, J. S., Griffin, J. B., and Zempleni, J. (2001). Biotinylation of histones in human cells. Effects of cell proliferation. *Eur. J. Biochem.* **268**, 5424–5429.

Stern, L. L., Mason, J. B., Selhub, J., and Choi, S. W. (2000). Genomic DNA hypomethylation, a characteristic of most cancers, is present in peripheral leukocytes of individuals who are homozygous for the C677T polymorphism in the methylenetetrahydrofolate reductase gene. *Cancer Epidemiol. Biomarkers Prev.* **9**, 849–853.

Trock, B. J., Hilakivi-Clarke, L., and Clarke, R. (2006). Meta-analysis of soy intake and breast cancer risk. *J. Natl Cancer Inst.* **98**, 459–471.

Ulrich, C. M. (2005). Nutrigenetics in cancer research–folate metabolism and colorectal cancer. *J. Nutr.* **135**, 2698–2702.

Ulrich, C. M. (2007). Folate and cancer prevention: A closer look at a complex picture. *Am. J. Clin. Nutr.* **86**, 271–273.

Ulrich, C. M., and Potter, J. D. (2007). Folate and cancer–timing is everything. *JAMA* **297**, 2408–2409.

van Engeland, M., Weijenberg, M. P., Roemen, G. M., Brink, M., de Bruine, A. P., Goldbohm, R. A., van den Brandt, P. A., Baylin, S. B., de Goeij, A. F., and Herman, J. G. (2003). Effects of dietary folate and alcohol intake on promoter methylation in sporadic colorectal cancer: The Netherlands cohort study on diet and cancer. *Cancer Res.* **63**, 3133–3137.

Walker, A. R., Walker, B. F., and Walker, A. J. (1986). Faecal pH, dietary fibre intake, and proneness to colon cancer in four South African populations. *Br. J. Cancer* **53**, 489–495.

Wang, B., Majumder, S., Nuovo, G., Kutay, H., Volinia, S., Patel, T., Schmittgen, T. D., Croce, C., Ghoshal, K., and Jacob, S. T. (2009). Role of microRNA-155 at early stages of hepatocarcinogenesis induced by choline-deficient and amino acid-defined diet in C57BL/6 mice. *Hepatology* **50**, 1152–1161.

Warri, A., Saarinen, N. M., Makela, S., and Hilakivi-Clarke, L. (2008). The role of early life genistein exposures in modifying breast cancer risk. *Br. J. Cancer* **98**, 1485–1493.

Waterland, R. A., and Jirtle, R. L. (2003). Transposable elements: targets for early nutritional effects on epigenetic gene regulation. *Mol. Cell. Biol.* **23**, 5293–5300.

Wu, A. H., Yu, M. C., Tseng, C. C., and Pike, M. C. (2008). Epidemiology of soy exposures and breast cancer risk. *Br. J. Cancer* **98**, 9–14.

Xu, W. S., Parmigiani, R. B., and Marks, P. A. (2007). Histone deacetylase inhibitors: molecular mechanisms of action. *Oncogene* **26**, 5541–5552.

Young, G. P., Hu, Y., Le Leu, R. K., and Nyskohus, L. (2005). Dietary fibre and colorectal cancer: A model for environment–gene interactions. *Mol. Nutr. Food Res.* **49**, 571–584.

Zempleni, J., Helm, R. M., and Mock, D. M. (2001). In Vivo biotin supplementation at a pharmacologic dose decreases proliferation rates of human peripheral blood mononuclear cells and cytokine release. *J. Nutr.* **131**, 1479–1484.

Zhang, Y., Talalay, P., Cho, C. G., and Posner, G. H. (1992). A major inducer of anticarcinogenic protective enzymes from broccoli: Isolation and elucidation of structure. *Proc. Natl. Acad. Sci. USA* **89**, 2399–2403.

Epigenetic Databases

9 Epigenetic Databases and Computational Methodologies in the Analysis of Epigenetic Datasets

Maté Ongenaert

OncoMethylome Sciences, Liege, Belgium

Advances in Genetics, Vol. 71
0065-2660/10 $35.00
DOI: 10.1016/S0065-2660(10)71010-5

ABSTRACT

Epigenetics research is one of the emerging research fields in biomedical research. During the last few decades, a collection of useful tools (both to design the experiments and to analyze the results) and databases are developed. This review chapter discusses basic tools which are used to detect CpG islands and the Transcription Start Site (TSS) and discusses experimental design and analysis, mainly of DNA-methylation experiments.

During the last years, an enormous amount of experimental data had been generated and published. Therefore, we describe some epigenetic databases, with a special focus on DNA methylation and cancer. Some general cancer databases are discussed as well, as they might reveal the link between the results from epigenetic experiments and their biological influence on the development or progression of cancer.

Next, some novel computational approaches in epigenetics are discussed, for instance used to predict the methylation state of a promoter in certain circumstances.

To show a possible data analysis strategy of an epigenetic dataset in cancer research, there is a showcase where a DNA-methylation dataset, generated on colorectal cancer samples, is analyzed. This demonstrates how a DNA-methylation dataset might look like and the different steps in a possible analysis strategy and how to interpret the results. © 2010, Elsevier Inc.

I. INTRODUCTION

As epigenetics literally is the layer above the genetic information, adapted detection methodologies are developed. Accordingly, the (computational) analysis strategies must be able to deal with the information obtained. In addition, in the last decade, more and more high-throughput methodologies have been developed and the analysis methodologies must be able to deal with an immense amount of data generated and be tailored to detect the biologically relevant information in these datasets.

Epigenetic changes are described in multiple species and in a variety of events, such as during embryonic development, host defense mechanisms, and protection from viral transcripts. This clearly demonstrates that epigenetics is crucial in the organism's function. However, changes in the epigenetic code are also described during the initiation, development, and progression of cancer. In addition to genetic and chromosomal changes, epigenetic alterations seem to play a key role in the deregulation of tumor suppression mechanisms, cell-cycle control, apoptosis initiation, and cellular communication. DNA methylation (the addition of a methyl group on a cytosine residue) is the most described epigenetic alteration, in addition to histone tail modifications (such as methylation and acetylation). DNA methylation in the promoter region of genes may affect the transcription of the gene.

The detection of DNA methylation and other epigenetic alterations opens perspectives to be used in order to detect cancer development in an early stage, classify (stage, grade, aggressiveness) the tumor, and propose treatments based on the epigenetic profile.

This chapter describes some basic tools often used in epigenetic research. Some databases where raw and/or processed data from epigenetic experiments is stored in are discussed. In the next part of this chapter, some computational methodologies, specifically created or adapted to deal with epigenetic data on a genome-wide scale, are taken a detailed look at.

II. BASIC EPIGENETIC TOOLS

The specific research field of epigenetics needs specific tools to design the experiments and to analyze the generated datasets. Some of these tools are entirely novel in order to answer equally novel research questions, while other tools are rather modified from existing tools used in biotechnology or genetic research.

A. CpG detection

The so-called CpG islands (CGIs) play a prominent role in DNA-methylation research. As the cytosine residue in a CG dinucleotide can become methylated, the various detection methodologies are designed in order to distinguish methylated and unmethylated cytosine residues.

In the human genome, the CG dinucleotide is very uncommon and occurs much less frequent than one would expect, based on the nucleotide frequencies. The CG dinucleotide is chemically unstable as the cytosine residue is prone to deamination to adenine. However, some regions are relatively rich in CG dinucleotides. Such regions (frequently occurring in the promoter regions of genes) are called CGIs. The largest part (70%) of the CpG dinucleotides outside CGIs is methylated, while cytosine residues within CGIs generally are unmethylated. The methylated cytosine accounts for 0.75–1% of all DNA bases (Bestor, 2000).

A CGI is identified based on four parameters (default or most common values/settings in most detection methodologies given):

– The minimal length (common: 200 bp): the region where the other parameters are true.

– The window width (common: 100–1 bp shift/step): a rather technical setting used in the algorithms, defining the sliding window width to calculate the remaining parameters in. Imagine a sequence of 1000 bp to detect CGIs in. The CG content and O/E ratio will be calculated in the first window (e.g., 100 bp) if it meets the criteria set. Next, the window slides with the selected shift/step and again the calculations are made and matched with the criteria.
– The CG content (common value: 55%): the minimal number of CG content, this is the number of Cs and Gs in relation with the length of the DNA fragment analyzed (the window).
– The O/E ratio (observed/expected; common value: 0.60): the number of CG dinucleotides that is observed in the fragment/window, divided by expected number of CG dinucleotides (calculated on the frequency of occurrence of C and G).

Tools using these parameters for the identification of CGIs in DNA sequences include:

– CpGReport and NewCpGReport: part of the EMBOSS suite (Rice *et al.*, 2000), based on the given values for the different parameters, the given sequences are checked for CGIs
– Cpgplot: visually plots the CG content, O/E ratio, and the location of identified CGIs

There exist other methodologies and algorithms with other criteria and methodologies to detect CGIs, such as CpGcluster (Hackenberg *et al.*, 2006) and CpG_MI (Su *et al.*, 2010). The definition what you consider a CGI may be of importance, as you might for instance only design primers within the CGI regions.

B. Transcription start site and translation start site

DNA methylation often occurs in the CGIs in the promoter region of genes and promoter hypermethylation may cause transcriptional silencing. In order to identify the promoter region, the Transcription Start Site (TSS) has either to be experimentally determined or to be predicted. The region upstream of this TSS is annotated as the promoter region, it is not transcribed and plays a key role in regulating the transcription. A region of approximately 1000 bp around the TSS generally is considered biologically important. The major part of the experimental designs focuses on the promoter region (and CGIs within this promoter region), relatively close to the TSS.

As CG dinucleotides in the first exon and intron sometimes show to have equally important biological consequences, the location of the Translational Start Site (and the start codon and CDS—coding sequence) is crucial as well. The TSS can either be experimentally validated or predicted. Some genes have multiple (alternative) transcripts and thus multiple Transcriptional Start Sites as well. There are a number of databases that contain TSS annotation: DBTSS, EPD, and Ensembl.

1. DBTSS (DataBase of human Transcriptional Start Sites)

DBTSS (Wakaguri *et al.*, 2008) is a database which contains precise positional information for TSSs of eukaryotic mRNAs. Using the newly developed oligo-capping method, full-length cDNAs were obtained and used to extend existing cDNA libraries toward the 5′ end. In recent updates of the database, the information obtained by massive sequencing methodologies (from a total of 33 different cell types or cultured conditions) was added.

The promoter region is annotated with known single-nucleotide polymorphisms (SNPs) and the sequence region around a selected TSS can easily be explored and retrieved.

Application/main features	Location of TSS, extraction of cDNA sequences relative to TSS (up- and downstream; retrieve promoter sequence section) annotated with other the location of SNPs (inspection of the promoter region window)
Usage	Web-interface, entire dataset download
Data sources and coverage	327 809 148 experimentally sequenced cDNA tags from 31 culture conditions, 17,879 RefSeq genes covered. Mapping of cDNAs and annotation is automated
Advantages	Excellent coverage (most of RefSeq genes covered), easy sequence retrieval around TSS
Disadvantages	One single gene at a time (no batch downloads)

2. EPD (European Promoter Database)

EPD (Cavin *et al.*, 1998) is an annotated, nonredundant collection of eukaryotic POL II promoters, for which the TSS has been determined experimentally. Access to promoter sequences is provided by pointers to positions in nucleotide sequence entries. All information is either directly extracted from scientific literature or, starting from release 73, compiled by a new *in silico* primer-extension method.

Application/main features	Location of TSS as a pointer to EMBL sequences
Usage	Web-interface, entire dataset download
Data sources and coverage	4806 Promoters in a variety of species, including 1871 promoters in human. Promoter selection is curated (from literature sources or the primer-extension methodology) and a promoter must comply with strict including rules: such as being recognized and experimentally defined, or homologous and sufficiently similar to an experimentally defined promoter
Advantages	Manual revision of promoter entries, complying with well-defined rules. Several eukaryotic species covered
Disadvantages	One single gene at a time (no batch downloads), limited coverage (less than 2000 human promoters)

3. Ensembl

Ensembl (Flicek *et al.*, 2010) is a joint project between EMBL—EBI and the Wellcome Trust Sanger Institute to develop a software system which produces and maintains automatic annotation on selected eukaryotic genomes. These genomes include human, mouse, rat, and the annotation includes promoter regions, transcription, and translational start sites and positions of exons and introns, gathered from different sources (and computational predictions). These are all mapped on the assembly of the genomes.

Ensembl has a much extended querying system (including BioMart which is particularly suited for providing "data mining" like searches of complex descriptive or sequence data, see Fig. 9.1). Access to Ensembl data is possible via the Ensembl.org Website or by using programmatic access as Ensembl provides an application programming interface (API). The data and data structures and the database schemes behind can be consulted, downloaded, and locally installed and modified or extended if desired.

Application/main features	Broad usage genome browser and automated annotation system. Includes sequences and annotations (TSS, exon–intron structure, functions, and pathways, . . .) in a selection of genomes. Information is stored in database schemes and can thus be queried in complex ways
Usage	Web-interface (single searches and in bulk using BioMart), programmatic access (including distributed annotation system, DAS), download of the complete database schemes and the contained data, download of complete flat datafiles

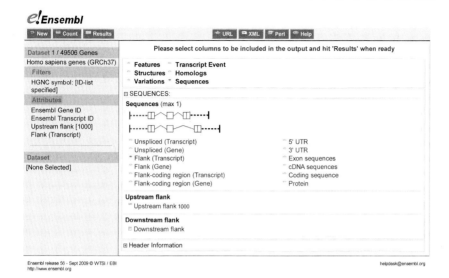

Figure 9.1. Ensemble query using BioMart. In this example, we are going to retrieve the promoter sequence (1000 upstream the TSS) for all transcripts of the BRCA1 gene.

Data sources and coverage	Automated pipelines annotate transcripts from EMBL, DBTSS, and NCBI sources in more than 50 species. These automated annotations are combined with manually curated information sources (Havana project)
Advantages	Ultimate coverage in a wide variety of species, excellent data-retrieval options (BioMart and custom queries), integration of both annotation and sequence features
Disadvantages	Rather steep learning curve, advanced retrieval and querying options might require training

C. Genome browsers, annotation databases, and mapping tools

In order to map sequences to the (human) genome, design primers, find the location of amplicons used in biomarker research, or to visualize results of analysis, soon or later a so-called genome browser will be needed in order to visualize the sequence, the annotation, and the experimental results.

These genome browsers provide an annotated assembly of the genome and allow scrolling and zooming trough the entire genome as it was a road map. While navigating throughout the chromosomes, the desired information can be displayed. This information is commonly displayed in the so-called "information tracks": tracks on the dynamic figures with information, selected from a variety of possibilities such as conservation across genomes, location of exons and introns, gene annotation, CpG percentage, location of SNPs. Depending on the zoom level, there are less or more details visible. By making use of DAS (Dowell *et al.*, 2001), the annotation within different databases can be exchanged. This system allows showing information from various data sources to be shown within one interface such as a track of a genome browser.

Most genome browsers allow to query and extract the sequence and annotation data in the underlying data structures. One of the genome browsers was discussed before (Ensembl), a frequently used other genome browser is the UCSC genome browser.

1. UCSC genome browser

The UCSC genome browser (Rhead *et al.*, 2010) mainly focuses on the "genome browser" application and the information is displayed in several tracks. Each element on such a track is clickable and leads to more detailed information of the selected item (see Fig. 9.2).

Other useful tools of the UCSC genome browser (in the perspective of epigenetic research) include:

— BLAT (BLAST-like alignment tool; Kent, 2002) is designed to quickly find sequences of 95% and greater similarity of length 25 bases or more. It may miss more divergent or shorter sequence alignments. It will find perfect sequence matches of 25 bases, and sometimes find them down to 20 bases. BLAT can thus be used to find primers and amplicons in the genome as these hybridize perfectly to the target sequence in the genome. BLAT does this mapping job extremely fast, as it keeps an indexed version of the entire genome in memory. This index consists of all nonoverlapping 11-mers in the genome.
— In *silico* PCR: finds given primers and the according possible amplicons in the entire genome. Based on the entry of the primer pair and some parameters describing the primer matching and amplicon properties, this tool can be used to find the amplicon of the given primer pair or to check for a specific amplicons while designing and selecting primers.

Application/main features   Broad usage genome browser and automated annota-
                                   tion system for the human genome. Browse through
                                   the genome and see annotation in the so-called
                                   annotation tracks (within UCSC or user-submitted)

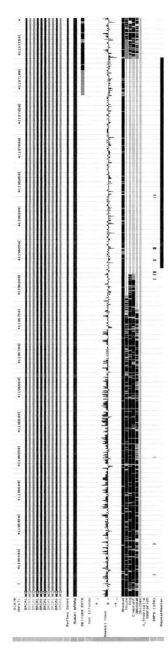

Figure 9.2. UCSC genome browser, showing a region and the selected tracks with annotations of the BRCA1 gene.

Usage	Web-interface, download of complete flat datafiles (whole genome or by chromosome)
Data sources and coverage	Automated pipelines, information is stored in so-called annotation tracks (half of the annotation tracks are computed at UCSC from publicly available sequence data, the remaining tracks are provided by collaborators)
Advantages	User-friendly genome browser with a large user-selectable annotation tracks and seamless zooming in and out
Disadvantages	Limited to human genome, rather limited querying options

D. Bisulfite conversion

Treatment of DNA with bisulfite converts cytosine residues into uracil (deamination reaction). However, the 5-methylcytosine residues remain unaffected. Thus, bisulfite treatment introduces specific changes in the DNA sequence that depend on the methylation state of individual cytosine residues. This single-nucleotide resolution makes the bisulfite treatment one of the first steps in the determination and detection of the methylation state of individual cytosine residues.

It is thus no surprise that the *in silico* conversion of a DNA fragment is an everyday task for a researcher in the epigenetics world. Using a word or text processing application may not be sufficient and time intensive while still inaccurate and inconvenient. One of the Web-based tools that facilitate bisulfite conversion, in addition to other frequently executed tasks such as reversing a sequence and retrieve the complementary strand, is BiQ Analyzer—Conversions (BiConverter; Bock *et al.*, 2005).

E. Experimental design and analysis

The experimental procedures in epigenetics research are in most cases modified from standard molecular techniques such as PCR, sequencing, or restriction analysis. As the fifth base (the methylated cytosine) becomes one of the key players, the methods must have the discriminating power between a regular cytosine and a 5-methylcytosine.

In order to design methylation-specific PCR (MSP) primers and bisulfite sequencing primers or to select suitable restriction enzymes, various computational tools can be used in order to prepare these epigenetics experiments.

1. Primer design (MSP/bisulfite sequencing)

After bisulfite treatment, there can be designed primers that are able to discriminate between a methylated cytosine residue (which remains a C) and a non-methylated cytosine (converted into a T). This adds additional complexity and difficulties to design PCR primers.

The design of MSP primers (and/or the design of bisulfite sequencing primers) can be performed with the following tools:

- MethPrimer (Li and Dahiya, 2002) is based on the primer design algorithms of primer3 (Rozen and Skaletsky, 2000). In addition, it is able to detect CGIs and CpG-specific parameters are added (such as the number of CGs in the primer, the presence of a CG near the 3′or 5′end).
- MethMarker facilitates the design and optimization of gene-specific DNA-methylation assays. Beyond its use as an epigenetic primer-design tool, it provides extensive support for epigenetic biomarker optimization. It can design assays for bisulfite single-nucleotide primer extension (SNuPE), bisulfite pyrosequencing, MSP, MethyLight, and MeDIP-qPCR.
- MethBLAST (Pattyn et al., 2006) is not an experiment design tool as the ones above, but uses a BLAST search on the bisulfite converted version (methylated and unmethylated) of the database sequences. It can thus be used to find amplicons in the converted genomes or to check the specificity of generated MSP or bisulfite sequencing primers (it has the possibility to enter both a forward and reverse primer at the same time).

2. Methylation-sensitive restriction analysis (COBRA)

Combined bisulfite restriction analysis (COBRA; Eads and Laird, 2002) is a DNA-methylation analysis strategy, making use of methylation-sensitive restriction enzymes. The length of the fragments can then be used to determine the methylation state of the residues in the recognition sequences of the used enzymes.

- MethMarker (discussed earlier) contains a list of methylation-sensitive restriction enzymes and can be used to set up a COBRA experiment
- Methyl-typing (Yang et al., 2009) facilitates the search for suited COBRA restriction enzymes and the locations they cover by covering the entire flow from input, retrieval of the promoter sequence, identification of CGIs, and the search for methylation-sensitive enzymes in the selected islands.
- The restriction enzyme catalogue of New England Biolabs (NEB) and the NEB database REBASE (Roberts et al., 2010) contains a section to indicate the methylation sensitivity of each enzyme and helps selecting restriction enzymes that can be used together with the same buffer solution

3. Experiment analysis

Once the experiments are finished, the resulting files have to be analyzed in order to determine the methylation state (and the degree of methylation) of each investigated CpG. As more and more high-throughput methodologies are used, there is a strong need for fast and accurate analysis methodologies able to start with raw data file and process the data all the way to summarizing reports or graphical representations.

- MethMarker (discussed earlier) is able to process data from COBRA, SNuPE, MSP, MethyLight, and pyrosequencing experiments. It indicates the methylation state of each covered cytosine residue and can then be used to discriminate between classes (e.g., cancer and normal) using logistic regression models
- BiQ Analyzer (Bock *et al.*, 2005) can process the raw bisulfite sequencing files, which are aligned to a genomic sequence in the next step. It incorporates data quality control (such as indicating low conversion efficiencies of the bisulfite treatment). It prepares the data for analysis with statistical software packages and gives a graphical representation of the data using the so-called *lollipop-diagrams*.

4. Overview of tools

The discussed software tools are a limited set of useful tools that can be used in epigenetic research. The selection is made to give representative applications that can be used to design and/or analyze the mainly used experimental techniques (MSP, bisulfite sequencing, COBRA). Below is the summary of the main features and a comparative Table 9.1 summarizes the application areas of the tools.

In most cases, a combination of tools will be needed in order to design the experiment and analyze the results. For instance, a promoter region of interest can be designed primers for using MethPrimer. In order to check whether the primers might amplify other regions in the genome, MethBLAST can be used. Analysis (classification normals vs. cancer and which CpGs discriminate best between these classes) could be done in MethMarker. A graphic representation of the methylation ratio of the investigated CpGs can be made using BiQ Analyzer.

BiConverter

Application/main features	Inverting and converting DNA sequences (bisulfite conversion—fully methylated or unmethylated, inverting, complement, reverse complement)
Usage	Web-interface

Table 9.1. Overview of Features and Possible Applications for the Described Tools (Divided in Features for Design and Analysis)

| Tool Name | Design | | | | | Analysis | | Usage & complexity |
	MSP	Bisulfite sequencing	COBRA	Bisulfite SNuPE	Others	Preparation for analysis	Data analysis features	Web (W) or Stand-alone (S)
BiConverter	++	++	++	++	+/-	NA	NA	W, easy
MethPrimer	+++	+++	-	-	-	NA	NA	W, easy-intermediate
MethMarker	+++	+++	+++	+++	+++	+++	+++	S, intermediate-expert
MethBLAST	++	++	+	+	+	NA	NA	W, easy-intermediate
BiQ Analyzer	-	+++	-	-	-	+++	NA	S, intermediate-expert
Methyl-typing	+/-	+/-	+++	-	-	++	NA	W, intermediate

Application where tool is most appropriate is highlighted in bold.
+++, very well suited; ++, well suited; +, can be applied; -, not suited; NA, feature not available.

| Advantages | User-friendly bisulfite and reverse/complement conversion tool |
| Disadvantages | Limited use (only conversions) |

MethPrimer

Application/main features	Designing bisulfite conversion-based methylation PCR Primers. Currently, it can design primers for two types of bisulfite PCR: MSP and bisulfite-sequencing PCR (BSP) or bisulfite-restriction PCR. Uses primer3 in the background, several options for both primers (size, CpGs, Tm difference) and amplicon (length, Tm, number, and position of CpGs)
Usage	Web-interface
Advantages	User-friendly MSP/bisulfite sequencing primer design tool
Disadvantages	Does not check primer specificity in the entire genome

MethMarker

Application/main features	*Design*: designing experiments for COBRA, bisulfite SNuPE, bisulfite pyrosequencing, MSP, MethyLight, and MeDIP-qPCR *Analysis*: optimize (choose which CpG sites discriminate the sample classes) and validate DNA-methylation biomarkers that provide robust classification (using logistic regression models, training, and test sets can be defined and the models be both trained and evaluated)
Usage	Java program (stand-alone)
Advantages	Design and analysis of a large number of detection technologies, strong biomarker analysis tools including classification and training/evaluation, graphical representations, and graphics of analysis/classification
Disadvantages	Might require some training, limited choice in analysis strategies (only logistic regression)

MethBLAST

| Application/main features | BLAST versus *in silico* bisulfite modified DNA databases (methylated forward and reverse and unmethylated forward and reverse) can be used to locate primers or amplicons used in MSP, COBRA, bisulfite-PCR-SCCP (BiPS), and methylation-sensitive single-nucleotide primer extension (Ms-SNuPE) |

Usage	Web-interface
Advantages	Fast, easy to use, and understandable (regular BLAST-interface)
Disadvantages	Rather limited use, makes use of nr-databases (nonredundant) instead of blasting entire genomes

BiQ Analyzer

Application/main features	Starting from bisulfite sequencing reads: align these reads on the genomic sequence of interest, perform quality control, and calculate methylation ratios for each cytosine residue that can be used in data processing programs to classify samples, etc. Generated so-called lollipop-graphics, graphically representing the methylation ration for each CG sites
Usage	Stand-alone java program
Advantages	Quality-control steps and manual curation steps improve and fasten bisulfite sequencing mapping and analysis, good reporting and graphical outputs
Disadvantages	No data analysis options

Methyl-typing

Application/main features	Starting from a list of genes, the promoter region is downloaded (using DBTSS), CGIs are identified and locations of methylation-sensitive restriction enzymes are searched and listed. This automated flow facilitates the choice of enzymes to perform COBRA analysis. Location of MSP or bisulfite sequencing primers can be visualized as well
Usage	Web-interface
Advantages	All steps in the process automated (bisulfite conversion, CGI identification, listing of possible restriction enzymes to use)
Disadvantages	Rather limited use (only design of COBRA experiments), no experiment analysis

III. EPIGENETIC DATABASES

Who can imagine biomedical research without databases in all different research areas and disciplines? Each year, new databases arise, whether they are very specific or of general use. Year after year, the number of databases increases as can be seen in the *Database issue* of the journal Nucleic Acids Research. Also in epigenetic research, there exist a number of very useful databases, ranging from broad and generic databases to focused databases within a specific area or application.

A. Chromatin

1. ChromDB

ChromDB (Gendler *et al.*, 2008) displays chromatin-associated proteins, including RNAi-associated proteins, for a broad range of organisms. The primary focus is to display sets of highly curated plant genes predicted to encode proteins associated with chromatin remodeling. Model animal and fungal proteins are included in the database to facilitate a complete, comparative analysis of the chromatin proteome and to make the database applicable to all chromatin researchers and educators.

2. CREMOFAC

CREMOFAC (Shipra *et al.*, 2006) is a dedicated Web-database for chromatin-remodeling factors. The database harbors factors from different organisms reported in literature and facilitates a comprehensive search for them. In addition, it also provides in-depth information for the factors reported in the three widely studied mammals namely, human, mouse, and rat. Further, information on pathways and phylogenetic relationships has also been covered.

B. DNA methylation

1. Methdb

Methdb (Grunau *et al.*, 2001) is a broad and general methylation database. It contains data from all kinds of experiments in different research settings (diseases including cancer and environmental influences). It is sample based while the resolution (which cytosine residues were methylated and which not) depends on the used technology. This can range from single base-pair resolution after bisulfite sequencing, but can be a whole genome CG methylation degree content after an HPLC experiment. The data is to be submitted in a certain format and is guided using data entry forms.

Data can be searched and accessed based on tissue, sex, locus/gene, phenotype, method, and environment. As the data has to be user-submitted, the level of detail depends on the original submission. No overviews or data summaries are available in the database itself, although the data can be accessed through the DAS distribution protocol in order to access the data in combination with, for example, sequence or expression data.

The experiments provide data from a broad range of research areas, but overall the coverage (available data in the database vs. all available data described in literature) is low due to the user-submission strategy, indicating the need for a standardized submission system, encouraged by the journal publishers (such as the MIAME standard for microarray data Brazma *et al.*, 2001).

Application/main features	General purpose methylation database: list results of methylation experiments of different kinds (ranging from overall genome methylation degree to site-specific measurements). It lists information about the tissue, the phenotype of the sample, the gene, the methodology and methylation ratios up to the detail given by the submitter (depending on the detection technology used)
Usage	Web-interface, entire dataset download, programmatic access (DAS)
Data sources and coverage	21,256 Datapoints in different species contains information of 208 gene loci. Data is user-submitted (single entry or in batch)
Advantages	Covers several species and a large range of methodologies and sample types
Disadvantages	Limited to user-submission: rather limited coverage; limited genome coverage

2. MethprimerDB

MethprimerDB (Pattyn *et al.*, 2006) does not contain data from methylation experiments but contains the (published) primers, used to generate the experimental data. This database can be useful to use published primers as a positive control or to check in which regions the published primers are. The data is linked with MethDB.

Application/main features	Enlists primers/probes to test methylation for a number of methodologies (MSP, Bisulfite-PCR-SSCP, Ms-SNuPE, COBRA, bisulfite sequencing)
Usage	Web-interface, entire dataset download
Data sources and coverage	259 Datapoints in human, rat, and mouse. User-submitted data, often from publications

Advantages	Lists primer sequences for MSP/bisulfite sequencing, often from literature, which is information that is sometimes difficult to summarize as hidden in the articles full text or tables
Disadvantages	Limited to user-submission: rather limited coverage

C. DNA methylation and cancer

In the oncology research area, the last decade DNA methylation arose as one of the most described mechanisms. DNA-methylation detection can be used for screening and early detection purposes, to predict progression and aggressiveness or to aid in the treatment choices made. Therefore, some methylation databases are entirely focused toward the oncology area.

This could lead to better understanding which biological processes are affected in which cancer types, whether the DNA methylation signature changes while progressing and help stratifying patients based on their methylation pattern.

1. PubMeth

PubMeth (Ongenaert *et al.*, 2008a) is a cancer methylation database that includes genes that are reported to be methylated in various cancer types. A query can be based either on genes (to check in which cancer types the genes are reported as being methylated) or on cancer types (which genes are reported to be methylated in the cancer (sub) types of interest).

PubMeth is based on text-mining of Medline/PubMed abstracts, combined with (computer assisted) manual revision and annotation of preselected abstracts. The text-mining approach results in increased speed and selectivity (as for instance many different aliases of a gene are searched at once), while the manual screening significantly raises the specificity and quality of the database (Fig. 9.3).

The summarized overview of the results is very useful in case multiple genes or cancer types are searched at the same time and help researchers identify publicly described biomarkers without having to launch several time-consuming and incomplete (and thus inaccurate) PubMed searches and to review the articles returned by such searches. Often, only summarizing data is available in the abstract while in the full text article, there is more information, often covered in tables and figures. This "hidden" data is covered in PubMeth due to the manual expert revision and annotation.

Application/main features	Database of genes, reported as methylated in cancer in human samples contains the number of samples (cancer, normals, and cell lines) and the methylation ratio.

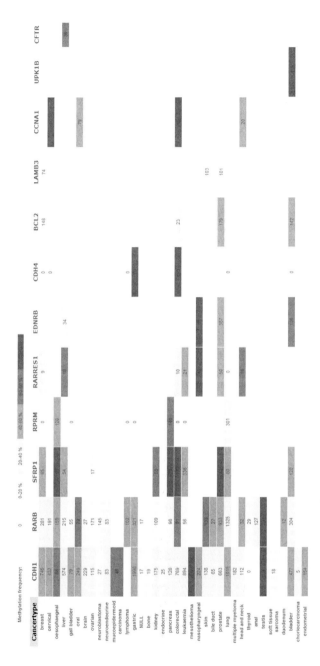

Figure 9.3. Results of a gene-centric query in PubMeth. The cancer types are sorted based on relevance, the numbers within the cells represent the number of samples, while the color gives an indication about the methylation frequency.

	Searching can start either from a list of genes or by selecting cancer (sub) types. A graphical summarizing overview of the results is given
Usage	Web-interface, entire dataset download
Data sources and coverage	3941 Datapoints, 440 genes, about 1000 literature reference. Information gathered by automated text-mining, manually reviewed and curated
Advantages	Summarization of which genes are methylated in which cancer (sub) types are vice versa. User-friendly, multiple genes/cancer types can be searched and visualized at the same time, good quality due to manual curation
Disadvantages	Limited coverage due to initial text-mining of abstracts

2. MethyCancer

Methycancer (He *et al.*, 2008) hosts both highly integrated data of DNA methylation, cancer-related gene, mutation, and cancer information from public resources, and the CGI clones derived from large-scale sequencing. Interconnections between different data types were analyzed and presented. The graphical *MethyView* shows DNA methylation in the context of genomic and genetic data, facilitating the research in cancer to understand both genetic and epigenetic mechanisms. Both can cause dramatic changes in gene expression or the regulation thereof during the development and progression of cancer. The graphical representation actually is a genome browser (comparable with the UCSC genome browser) with epigenetics-related annotation tracks. From within the graphical view, switching to very detailed information is very intuitive (Fig. 9.4).

Application/main features	Genome-like browser application, with in the information tracks information from several sources (CGI clones mapped on the genome, public methylation data (from HEP, MethDB, and others), cancer and gene/mutation information, correlation among DNA methylation, gene expression, and cancer. Searches can start from genes, cancer types, or methylation features.
Usage	Web-interface, entire dataset download
Data sources and coverage	64,681 Distinct genomic loci, termed MethyLoci. 485 Annotated cancer genes, of which 323 are supported by experimentally validated methylation data and 114 matched with CGI predictions, 6615 candidate cancer

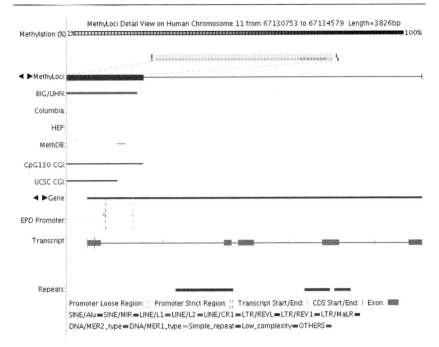

Figure 9.4. The graphical genome browser like "MethyView" representation of data in the Methy-Cancer database. Several tracks containing genetics and epigenetics-related information are available.

	genes, of which 3698 and 1900 are supported by experimental methylation data and CGI predictions. Data-integration (automated) from various sources
Advantages	Integration of several data sources (methylation, cancer) in one user-friendly genome browser like view. Relation with expression/methylation in a certain cancer type can be assessed
Disadvantages	Graphical view does not allow extracting information the graphic contains

3. MeInfoText

MeInfoText (Fang *et al.*, 2008) presents comprehensive association information about gene methylation and cancer, the profile of gene methylation among human cancer types and the gene methylation profile of a specific cancer type, based on association mining from literature.

Application/main features	Searches start with gene and/or cancer type. Based on cooccurrence with epigenetics and cancer/methylation-related keywords, literature references are found back. Based on the number of references retrieved, some measures are calculated (confidence and support). A table with main cancer types and the number of returned references, in combination with the gene and methylation-related keywords, is given
Usage	Web-interface
Data sources and coverage	Automated annotation of literature, precomputed measures
Advantages	Ability to search in literature abstracts starting from either gene name and cancer type, fast as precomputed, good coverage
Disadvantages	Only one gene summary at a time, no manual annotation/curation, only based on cooccurrence of gene name and cancer/methylation-related keywords, measures not always representative

IV. CANCER DATABASES

In addition to epigenetic databases in cancer, there also exist cancer-related databases that are not focused on epigenetics but might be useful to understand the underlying biological mechanisms of cancer development and progression. Epigenetic effects and genetic effects are often related, have an influence on each other, or have an additive or adverse effect. In this part, some genetic databases in the cancer area are briefly introduced.

A. Oncomine

Oncomine (Rhodes *et al.*, 2007) is a bioinformatics initiative aimed at collecting, standardizing, analyzing, and delivering cancer transcriptome data to the biomedical research community. The analysis has identified the genes, pathways, and networks deregulated across 18,000 cancer gene expression microarrays, spanning the majority of cancer types and subtypes.

The information and the analysis results are presented in an intuitive user-interface including a lot of summarizing tables, figures, and graphics. The search options and filters are very fast and easy to use. A biologist with a minimal knowledge of analysis strategies is able to see expression-related results in the cell lines and patient samples in the cancer types of choice within minutes.

The analysis options are very extended, ranging from checking in which cancer types a certain gene in over- or underexpressed in to an analysis which genes are most underexpressed in high-grade versus low-grade cancer samples.

As epigenetics (and specifically DNA methylation in the promoter region) is often related with transcriptional silencing, Oncomine can be used to check whether methylation and expression are related with each other. The database and the precomputed analysis of the large number of microarray studies covered make Oncomine a very useful resource for all researchers in the cancer area.

Application/main features	Expression results of a large selection of microarray experiments in the cancer field. Results from experiments are analyzed and can be accessed by an advanced search functionality with a large variety of filters to apply (on experiment, cancer type, dataset, etc.)
Usage	Web-interface
Data sources and coverage	484 Datasets, 38,306 samples, over 18,000 microarrays, processed results
Advantages	Offers lots of search options and filtering options, processed results ready to explore, easy and intuitive navigation, large collection of datasets in different cancer types and large collection of covered cancer cell lines
Disadvantages	Requires some expertise and training, results cannot be exported in the free edition

B. The Roche Cancer Genome Database

The Roche Cancer Genome Database (RCGDB) (Kuntzer *et al.*, 2010) integrates different kinds of genetic data. The database is designed to integrate the disparate cancer genome data like single-nucleotide variants, single-nucleotide polymorphisms, and chromosomal aberrations (CGH and FISH).

Taking the Knudson two-hit hypothesis (Knudson, 1971) in mind, it is useful to combine methylation data, SNP, CGH, and other genetic polymorphisms and defects into account. In many cases, one copy of the gene is affected by a genetic variation or chromosomal aberration, while the other paternal copy is targeted by an epigenetic event such as DNA methylation of the promoter region.

Application/main features	Integrates various sources of mutation data, including SNPs, nucleotide variants, chromosomal aberrations
Usage	Web-interface

Data sources and coverage	Integration of various data sources
Advantages	Standardized interface for different genetic data, user-friendly interface
Disadvantages	Can only assess one gene at a time

V. COMPUTATIONAL APPROACHES IN EPIGENETICS

A. Promoter region analysis and epigenome prediction

A substantial amount of bioinformatics research has been focused on the prediction of epigenetic information (such as methylation state) from characteristics of the genome sequence (the sequence itself, the CG content, presence of certain motifs, etc.).

These computational predictions can substitute for experimental data to some degree and help speed up the research that studies novel epigenetic mechanisms or covers species not well known. In addition, the prediction algorithms give additional insight in the epigenetic mechanisms by building models from training data and applying the found rules on a test dataset. A number of algorithms, trained and tested on epigenetic data, are discussed here. Most of them rely on artificial intelligence (AI) and machine learning in order to find a so-called classifier that can discriminate between, for instance, methylated versus not-methylated.

1. ConDist

ConDist (Hackenberg *et al.*, 2009) is a tool to statistically analyze quantitative gene and promoter properties. The software includes approximately 200 quantitative features of gene and promoter regions for seven commonly studied species (including human, mouse, rat, and fruit fly).

All the features in the annotation database are based on quantitative gene and promoter properties. These features are grouped in six categories: base composition, physical properties of DNA and chromatin, evolution, general gene/protein properties, overlap with genomic elements, and gene expression. The publication describing ConDist uses the annotation features to predict differentially methylated promoters and the difference between tissue-specific and housekeeping genes.

The tool statistically analyzes the differences between the provided datasets for the chosen features and generates reports, including results after randomization in order to estimate the statistical power and false positive and negative rate.

2. Epigenome prediction

Epigenome prediction is the prediction of an epigenetic state (e.g., methylation) by using sequence features. Based on an experimental dataset, the algorithms are trained. The algorithms can then be applied on any dataset. The selected sequence features can be primary sequence properties (such as the CG content, presence of a sequence motif) or can be much more complex (such as location in the 3D structure). There exist a number of studies that describe epigenome prediction. A selection of them is discussed here.

- By training support vector machines on epigenetic data for CGIs on human chromosomes 21 and 22, informative DNA attributes were identified that correlate with open versus compact chromatin structures. These DNA attributes are used to predict the epigenetic states of all CGIs genome-wide. This way, "bona fide" CGIs were predicted. The predictions show to be applicable in different tissues and cell types (Bock *et al.*, 2007). The predictions and calculated scores are publicly available as an annotation layer to be used with the USCS genome browser.
- Das *et al.* (2006) describe a computational pattern recognition method that is used to predict the methylation landscape of human brain DNA. The method can be applied both to CGIs and to non-CGI regions. It computes the methylation propensity for an 800-bp region centered on a CpG dinucleotide based on specific sequence features within the region.
- One of the first studies describing an epigenome prediction methodology is the one from Feltus *et al.* (2003). By overexpressing one of the main maintenance DNA methyltransferases (DNMT1), they found out that the methylation state of majority of CGIs was not affected by the overexpression of DNMT1. A subset of methylation-prone CGIs that were consistently hypermethylated in multiple DNMT1 overexpressing clones was identified. DNA pattern recognition algorithms and supervised learning techniques were used to derive a classification function based on the frequency of seven novel sequence patterns that was capable of discriminating methylation-prone from methylation-resistant CGIs with 82% accuracy.
- A similar pattern recognition methodology, in combination with a genome-wide sequence alignment strategy and data analysis of reexpression experiments in cancer cell lines of different origins (Hoque *et al.*, 2008) was used to find methylation biomarkers in a very broad range of cancer types. These showed a very good validation potential showing the power of the computational strategies to enrich toward cancer—specifically methylated sequences.

VI. APPLICATIONS AND DATA ANALYSIS IN CANCER EPIGENETICS

Cancer is probably the research areas where most epigenetic datasets is generated in. Epigenetics has shown to be able to discriminate between normal and cancer cells and between different stages within one cancer type. Data analysis strategies can reveal the functional mechanisms behind the observations.

Epigenetic biomarkers can be used to improve the (early stage) diagnosis. In addition, the treatment strategy could be adapted to the epigenetic signature, observed in the patient. In the last decade, the use of genome-wide or high-throughput techniques has risen. In order to analyze these enormous amounts of generated data, fast and accurate data analysis strategies must be available.

A. Reexpression analysis

Many DNA-methylation studies start by using (cancer) cell lines. One genome-wide way of identifying possibly methylated genes in cell lines is by analyzing expression levels before and after treatment with DAC (decitabine-5-Aza-2′-deoxycytidine). This is a cytosine analogue that cannot be methylated. After cell divisions after the treatment, DAC is built in the genome instead of cytosine and the methylation will be lost.

Thus, a gene that is reexpressed after the DAC-treatment is possibly methylated. The loss of DNA methylation might cause a higher expression of the gene. Another compound often used in "pharmacological unmasking" is TSA (Trichostatin A): it inhibits the class I and II mammalian histone deacetylase (HDAC) families of enzymes and thus interferes with the removal of acetyl groups from the histones.

As expression levels can be determined by using commercially available microarrays (such as from Agilent or Affymetrix), a reexpression analysis can reveal genome-wide reexpression data. There are studies where a twofold over-expression after DAC-treatment is taken as a threshold while there are specific algorithms developed to deal with reexpression data, such as relaxation ranking (Ongenaert et al., 2008b). An example of an advanced bioinformatics analysis of reexpression datasets is given in Schuebel et al. (2007) where methylation biomarkers are discovered based on their behavior after treatment with varying concentrations of DAC and TSA.

B. CpG array analysis

A direct measurement on methylation on a genome-wide scale can be achieved by the so-called CpG arrays. This type of microarrays does not measure the expression levels by measuring RNA copies, but directly measures DNA methylation of the spotted probes.

An example of such a CpG array is the HumanMethylation27 BeadChip (Illumina). Upon treatment with bisulfite, unmethylated cytosine bases are converted to uracil, while methylated cytosine bases remain unchanged. The assay interrogates these chemically differentiated loci using two site-specific probes, one designed for the methylated locus (M bead type) and another for the unmethylated locus (U bead type). Single-base extension of the probes incorporates a labeled ddNTP, which is subsequently stained with a fluorescence reagent. The level of methylation for the interrogated locus can be determined by calculating the ratio of the fluorescent signals from the methylated versus unmethylated sites.

For each investigated CpG site, the ratio ranges from 0 (unmethylated) to 1 (completely methylated). This ratio (the beta-value) can be calculated by the Illumina software package BeadStudio. The package methylumi provides convenient mechanisms for loading the results of the Illumina methylation platform into R/Bioconductor (Gentleman et al., 2004). Classes based on common Bioconductor classes for encapsulating the data and to facilitate data manipulation are at the core of the package. It contains methods and functions to perform quality control, normalization, and plotting. The package can be used to assess the quality and make further analysis possible as the data is in a format, ready to be used by other R/BioConductor packages.

Further in this chapter, a showcase describes how to deal with CpG array data and how this platform can be used to identify possible methylation biomarkers, differentially methylated in colorectal cancer (CRC) versus normal, adjacent tissue.

C. ChIP-Chip/ChIP-Seq data analysis

Specific DNA sites in direct physical interaction with transcription factors and other proteins can be isolated by chromatin immunoprecipitation (ChIP). ChIP produces a library of target DNA sites that a given factor was bound to *in vivo*. Afterward, the captured sequences can be sequenced (ChIP-Seq) or analyzed using a microarray (ChIP-Chip).

For epigenetics research, in the ChIP step, antibodies designed to bind on a member of the Polycomb (Pc) complex (such as SUZ12 or EZH2) can be used. It is shown that sequences, bound by members of the Pc complex are often prone to DNA methylation (Bennett et al., 2009; Vire et al., 2006). Other possibilities are antibodies for histone modifications (such as H3K27 methylation) or directly for a methylated cytosine (called MeDIP; Mohn et al., 2009).

The methylation state of the captured DNA sequences can be determined by bisulfite treatment, followed by sequencing or by using a CpG array as described before. Advantage of a sequencing methodology is that it will reveal methylation in all covered sites with the ultimate resolution (a single cytosine). Currently the disadvantage is the price of next-generation sequencing. The price-aspect is in favor of the Chip platforms. The Chip platforms also don't require an extended and customized analysis strategy as the suppliers provide analysis and annotation features in the software. Using a sequencing approach, the sequencing reads have to be mapped on the genome, counted, and the cytosine residues are to be detected and annotated.

The mapping of sequencing reads resulting from bisulfite treated sequences is more complex than mapping regular next-generation sequencing reads. Therefore, specific mapping tools has been developed, such as BSMAP (Xi and Li, 2009) able to map high-throughput bisulfite reads at whole genome level with feasible memory and CPU usage.

D. Showcase: Analysis of a DNA-methylation dataset in CRC

1. CpG array analysis

As a showcase how to process and analyze a DNA-methylation dataset in the cancer area, this part briefly demonstrates a possible analysis strategy. The experiment data referred to, are publicly available, and so are all analysis tools used. The experimental dataset is available in GEO (Gene Expression Omnibus; Barrett et al., 2009) under the record GSE17648. The data consist of the so-called beta-values from the Illumina Infinium HumanMethylation27 Bead-Chip with results for 44 samples (2 pairs of CRC tissue and adjacent normal mucosa tissue). The beta-value is generated using the Infinium platform and reports on the intensity ratio between methylated- and nonmethylated-specific assays for each covered CpG site on the chip. Since assay performance can vary between loci, this ratio does not consistently report percent methylation.

The dataset in GEO is already processed and normalized but starting from raw machine data, these so-called beta-values can easily be obtained by using R/Bioconductor and the package methylumi. There also exists more general packages to deal with all kinds of data (methylation, expression) from the Illumina bead platforms, such as the beadarray package (Dunning et al., 2007).

After reading in the data (as available in GEO) as a CSV file (comma separated values) into R (by using the read.csv or read.table commands), the data is available for processing and analyzing using commonly used microarray analysis strategies, frequently used in the analysis of expression microarrays. This analysis can start by more or less straightforward data control and exploration steps, such as plotting XY graphs (as shown in Fig. 9.5) or histograms. These basic, initial analysis and representation steps allow getting insight in the data and give a first indication of the variation and heterogeneity within the dataset as well as the potential important signals or artifacts.

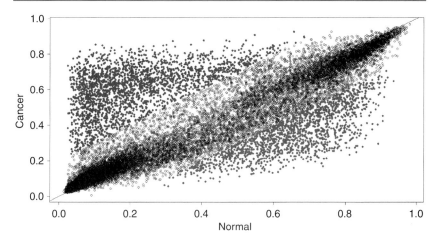

Figure 9.5. Beta-values of the sample GSM440333 (normal adjacent mucosa) and the colorectal cancer sample of the same patient plotted versus each other. There clearly are CpG probes with a large difference in methylation ratio between the two samples. The dark grey data points represent probes with more than 0.20 difference between cancer and normal (higher methylation degree in cancer), while the light grey data points represent probes that are methylated in a higher ratio in the normal sample compared to the cancer sample. (For interpretation of the references to color in this figure legend, the reader is referred to the Web version of this chapter.)

2. Analysis of top-ranking genes

A more advanced analysis would be to find significantly different CpG probes by using a ranking strategy in combination with a false discovery rate (FDR) analysis in order to estimate the significance and power of the results. This kind of analysis can be performed by the R/BioConductor package RankProd (Hong *et al.*, 2006). Principle is that after permutation of the data, the ranking is performed for each iteration. If in almost all random permutations the signal is lost, there is a high change there is a genuine signal within the dataset and the difference between the classes (in this case: cancer vs. normal) observed is real and not by chance. The limit is set at 5%: if in more than 5% of the permutation iterations there is at least an equally good signal as there was in the not-permutated dataset, the probe will not be retained.

By defining two classes (in this case: the normal adjacent tissue samples and the CRC samples) and choosing 100 permutations for estimating the FDR, RankProd will start the analysis. The RankProd analysis does not expect a certain distribution of data and therefore can also be used on the beta-values representing the rate of methylated versus unmethylated probes.

```
#load the library
library(RankProd)
#define the classes: 0 -- normal (first 22 samples);
1 -- colorectal cancer
cl<-c(rep(0,22),rep(1,22))
#start RankProd analysis with two classes and non-log
converted data (as the beta-values are used)
RP.out<-RP(data[,1:44],cl, logged=FALSE, rand=123)
#plot the results with cutoff for the PFP 0.05
plotRP(RP.out, cutoff = 0.05)
#output the results (beta-values higher in cancer than
normal, false discovery rate < 0.05)
topGene(RP.out, cutoff = 0.05, method = "pfp")
toplist<-topGene$Table1
#the toplist table contains as first column the row index
number of the selected probes of the "data" matrix, and can
directly be used for further analysis functions (such as
illustrated by the heatmap)
```

The RankProd analysis very clearly shows the difference in beta-values between normal adjacent samples and the CRC samples. Figure 9.6 highlights a large number of CpG probes that significantly show differences in methylation-rate (beta-value) in CRC samples versus normal adjacent tissue.

The top-listed probe that significantly has a higher methylation rate in cancer versus normal is cg26189983, designed for the gene TNFRSF1B (a tumor necrosis factor receptor). Next is cg06627364, followed by cg05037688 which is located in EGFL7. EGFL7 is described in literature as potentially methylated in cancer cells (Saito et al., 2009). In colon cancer patients, deregulated expression is associated with pathological tumor features of poor prognosis (Diaz et al., 2008). These are strong indications that the promoter region of the EGFL7 gene can actually be methylated in colon cancer and might actually have an influence on the transcriptional levels.

The list with a significant difference in beta-value can be further analyzed in a number of ways:

- Literature research: manual or semiautomated searches can reveal information about the top-ranking genes in the cancer type or application of interest. Example of a Web-based tool that identifies cooccurring biomedical concepts associated with the top-ranking genes is CoPub (Frijters et al., 2008)
- Gene ontology (GO) terms enrichment: GO provides a controlled vocabulary of terms for describing gene product characteristics and gene product annotation data. If in the top-ranking genes certain GO terms are overrepresented, this can lead to insight in which biological mechanisms

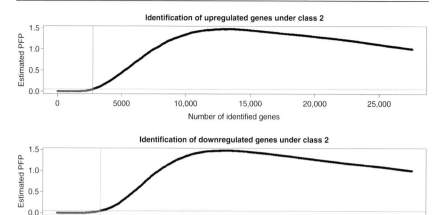

Figure 9.6. Result of the RankProd analysis with class 1 normal adjacent tissue and class 2 the colorectal cancer samples. The upper graph shows the PFP percentage (predicted false positives) of probes that are methylated at a higher ratio in the cancer samples in comparison with the normal adjacent tissue. The cut-off (red probes) is set at 0.05 (5% false positives). (For interpretation of the references to color in this figure legend, the reader is referred to the Web version of this chapter.)

might be involved. There exist a number of tools that can be used to identify overrepresented GO terms, such as GoStat (Beissbarth and Speed, 2004)
- Pathway overrepresentation might be used as well to identify affected pathways. This can for instance be done using Ingenuity Pathway Analysis software
- Expression profiles of the top-ranking genes can give further insight in which tissues, diseases, cancer types, … is up- or downregulated. Such expression profiles are in the previously discussed OncoMine for oncology-related datasets. Microarray submission databases such as NCBI GEO (Gene Expression Omnibus; Barrett *et al.*, 2009) and EBI ArrayExpress (Rocca-Serra *et al.*, 2003) can also be used to see the expression profile of the gene in the various datasets. ArrayExpress gives an overview of the experiments where the gene in question is differentially expressed.

Taking all probes into account that have higher beta-values in the cancer samples in comparison with the normal tissue samples with a maximum of 5% false positives (according to the RankProd analysis), the heatmap in Fig. 9.7 is obtained.

```
#load gplots package, including heatmap.2
library(gplots)
```

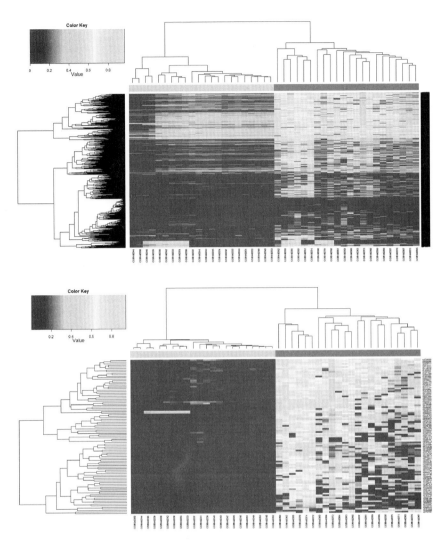

Figure 9.7. Top heatmap of the beta-values. Represented in the columns: the samples (red: cancer samples; green: normal tissue samples). In the rows: selected CpG probes (about 3000 probes) using RankProd, higher beta-value in cancer samples and a false discovery rate of less than 5%. Bottom detailed heatmap based on the top-listed 100 probes, ranked based on the false discovery rate as calculated by RankProd. (For interpretation of the references to color in this figure legend, the reader is referred to the Web version of this chapter.)

```
#draw the heatmap, all samples, genes in toplist after
RankProd analysis. Colors of samples: green for normals,
red for cancer samples. 100 variations of colors to
represent beta-values, according to topo color-scheme
```

```
heatmap.2(data[toplist[,1],],col="topo".colors(100),
scale  =  "none",key  =  TRUE,  symkey=FALSE,  density.
info="none", trace = "none", cexRow = 0.5, ColSideColors =
c(rep("green",22),rep("red",22)))
```

As expected, the beta-values of the selected probes can be used to distinguish CRC samples from normal adjacent tissue samples. The hierarchical clustering algorithm (hclust) nicely clusters the different samples classes. There is not a single misclassified sample, although there is a relatively high degree of variability in the cancer samples. This variability seems to be less present in the normal tissue samples. However, some samples (the four samples on the left in the heatmap) overall show lower beta-values, compared to the remaining normal samples.

This example analysis shows that clustering, classification, and diverse other machine learning techniques can be applied to epigenetic datasets much like they are used in the analysis and graphical representation of expression microarrays. This kind of analysis might extract relevant biological mechanisms and can be used to design clinical trial tests on a larger number of samples as possible epigenetic biomarkers are discovered.

3. Promoter sequence analysis

After identifying interesting genes that could serve as DNA-methylation biomarkers, the next step for further validation will involve taking a detailed look into the (promoter) sequence and the properties of the CGIs. Using DBTSS, the EGFL7 promoter sequence of 2000 bp around the TSS is retrieved (-1000 to $+1000$, the chosen TSS (NM_201446) at position 0). According to newcpgreport, this sequence contains a CGI of 124 bp length (at positions -163 to -40) just upstream of the TSS.

According to the Match-algorithm, using the Transfac database of transcription factors (Wingender et al., 1996) there is a binding site for v-Myb around position -115 and a binding site for Elk-1 around position -94. Both are thus located within the CGI in the promoter region. The binding of V-Myb has been described to be influenced by the methylation state of the target sequence (Klempnauer, 1993).

Based on the promoter sequence and the region of interest (the CGI), MSP- or bisulfite-sequencing primers can be designed. Using MethPrimer, one of the generated MSP primer pairs for the methylated state is the following: TTAGATTTTGATGGTTTAGGGGAC–AAAAAACGACTTTTTATAC TCCGTC. The amplicon starts at position -226 and ends at position -20 (amplicon length 207 bp). These primers are blasted using the bisulfite converted genome by methBLAST. The only perfect hits are indeed on chromosome 9. The next BLAST hits for one of the primers already have at least four mismatching residues.

4. Conclusion

This showcase demonstrates the enormous potential in analyzing high-throughput epigenetic datasets using robust methodologies that were initially developed to answer genetic questions. In this example, using a public dataset and freely available software and analysis packages, within less than 2 h, potential methylation biomarkers in colon cancer were found. Using various datasets it becomes feasible (without too high costs) to perform a combinatory analysis making use of for instance gene expression data, DNA-methylation data, and micro-RNA array data.

In this combinatory analysis, the expression of micro-RNAs could be correlated with gene expression data, and DNA-methylation data, revealing interactions between all these different mechanisms. The number of biological questions that can be answered drastically increases by being able to combine the different data sources. Only limits are the computational power, the capacities of the bioinformatician in charges and the bright mind of the researchers to come up with novel research questions to be answered . . .

VII. CONCLUSIONS

The amount of data in the epigenetics field has been drastically increased the last decades. In order to design the experiments and to analyze the generated data, specified tools and databases are designed, ranging from broadly applicable tools to tools and databases, tailored for a particular application.

In the cancer field, there is a strong interest in DNA-methylation and other epigenetic changes during the initiation, development, and progression of cancer. There exist some databases that contain the generated data in this field. However, the coverage is relatively low, mainly due to the fact that for epigenetic data there are neither guidelines to follow nor data repositories to submit data to when submitting a manuscript to a journal. The different experimental methodologies their genomic coverage and resolution are within a broad range.

With the availability of sequencing techniques, it now becomes feasible to determine the epigenetic state in a sample, at the highest resolution possible: the single base pair. In order to be able to analyze such huge amount of data, to store the results and to be able to compare sample sets with each other, a huge effort is still to be made. A central depository system with submission guidelines (comparable with the MIAME guidelines for microarray experiments) and homogenized analysis procedures would be desirable. Such a central data storage system would contain thousands of samples from different tissues, diseases, treatments, etc., all with a high resolution and genome coverage. This would dramatically speed up epigenetic research and give better insight in the biological mechanisms behind, clearing the road for even more advanced applications, such as the use of compounds that influence the epigenetic state in order to fight cancer.

Acknowledgments

The author would like to thank Wim Van Criekinge and Leander Van Neste for their feedback. The comments and recommendations of an anonymous reviewer helped drastically improve the manuscript and are highly appreciated.

References

Barrett, T., Troup, D. B., Wilhite, S. E., Ledoux, P., Rudnev, D., Evangelista, C., Kim, I. F., Soboleva, A., Tomashevsky, M., Marshall, K. A., Phillippy, K. H., Sherman, P. M., *et al.* (2009). NCBI GEO: archive for high-throughput functional genomic data. *Nucleic Acids Res.* 37, D885–D890.

Beissbarth, T., and Speed, T. P. (2004). GOstat: find statistically overrepresented Gene Ontologies within a group of genes. *Bioinformatics* 20, 1464–1465.

Bennett, L. B., Schnabel, J. L., Kelchen, J. M., Taylor, K. H., Guo, J., Arthur, G. L., Papageorgio, C. N., Shi, H., and Caldwell, C. W. (2009). DNA hypermethylation accompanied by transcriptional repression in follicular lymphoma. *Genes Chromosom. Cancer* 48, 828–841.

Bestor, T. H. (2000). The DNA methyltransferases of mammals. *Hum. Mol. Genet.* 9, 2395–2402.

Bock, C., Reither, S., Mikeska, T., Paulsen, M., Walter, J., and Lengauer, T. (2005). BiQ Analyzer: visualization and quality control for DNA methylation data from bisulfite sequencing. *Bioinformatics* 21, 4067–4068.

Bock, C., Walter, J., Paulsen, M., and Lengauer, T. (2007). CpG island mapping by epigenome prediction. *PLoS Comput. Biol.* 3, e110.

Brazma, A., Hingamp, P., Quackenbush, J., Sherlock, G., Spellman, P., Stoeckert, C., Aach, J., Ansorge, W., Ball, C. A., Causton, H. C., Gaasterland, T., Glenisson, P., *et al.* (2001). Minimum information about a microarray experiment (MIAME)-toward standards for microarray data. *Nat. Genet.* 29, 365–371.

Cavin, P. R., Junier, T., and Bucher, P. (1998). The Eukaryotic Promoter Database EPD. *Nucleic Acids Res.* 26, 353–357.

Das, R., Dimitrova, N., Xuan, Z., Rollins, R. A., Haghighi, F., Edwards, J. R., Ju, J., Bestor, T. H., and Zhang, M. Q. (2006). Computational prediction of methylation status in human genomic sequences. *Proc. Natl. Acad. Sci. USA* 103, 10713–10716.

Diaz, R., Silva, J., Garcia, J. M., Lorenzo, Y., Garcia, V., Pena, C., Rodriguez, R., Munoz, C., Garcia, F., Bonilla, F., and Dominguez, G. (2008). Deregulated expression of miR-106a predicts survival in human colon cancer patients. *Genes Chromosom. Cancer* 47, 794–802.

Dowell, R. D., Jokerst, R. M., Day, A., Eddy, S. R., and Stein, L. (2001). The distributed annotation system. *BMC Bioinform.* 2, 7.

Dunning, M. J., Smith, M. L., Ritchie, M. E., and Tavare, S. (2007). Beadarray: R classes and methods for Illumina bead-based data. *Bioinformatics* 23, 2183–2184.

Eads, C. A., and Laird, P. W. (2002). Combined bisulfite restriction analysis (COBRA). *Methods Mol. Biol.* 200, 71–85.

Fang, Y. C., Huang, H. C., and Juan, H. F. (2008). MeInfoText: associated gene methylation and cancer information from text mining. *BMC Bioinform.* 9, 22.

Feltus, F. A., Lee, E. K., Costello, J. F., Plass, C., and Vertino, P. M. (2003). Predicting aberrant CpG island methylation. *Proc. Natl. Acad. Sci. USA* **100,** 12253–12258.

Flicek, P., Aken, B. L., Ballester, B., Beal, K., Bragin, E., Brent, S., Chen, Y., Clapham, P., Coates, G., Fairley, S., Fitzgerald, S., Fernandez-Banet, J., *et al.* (2010). Ensembl's 10th year. *Nucleic Acids Res.* **38,** D557–D562.

Frijters, R., Heupers, B., van, B. P., Bouwhuis, M., van, S. R., de, V. J., Polman, J., and Alkema, W. (2008). CoPub: a literature-based keyword enrichment tool for microarray data analysis. *Nucleic Acids Res.* **36,** W406–W410.

Gendler, K., Paulsen, T., and Napoli, C. (2008). ChromDB: the chromatin database. *Nucleic Acids Res.* **36,** D298–D302.

Gentleman, R. C., Carey, V. J., Bates, D. M., Bolstad, B., Dettling, M., Dudoit, S., Ellis, B., Gautier, L., Ge, Y., Gentry, J., Hornik, K., Hothorn, T., *et al.* (2004). Bioconductor: open software development for computational biology and bioinformatics. *Genome Biol.* **5,** R80.

Grunau, C., Renault, E., Rosenthal, A., and Roizes, G. (2001). MethDB–a public database for DNA methylation data. *Nucleic Acids Res.* **29,** 270–274.

Hackenberg, M., Previti, C., Luque-Escamilla, P. L., Carpena, P., Martinez-Aroza, J., and Oliver, J. L. (2006). CpGcluster: a distance-based algorithm for CpG-island detection. *BMC Bioinform.* **7,** 446.

Hackenberg, M., Lasso, G., and Matthiesen, R. (2009). ContDist: a tool for the analysis of quantitative gene and promoter properties. *BMC Bioinform.* **10,** 7.

He, X., Chang, S., Zhang, J., Zhao, Q., Xiang, H., Kusonmano, K., Yang, L., Sun, Z. S., Yang, H., and Wang, J. (2008). MethyCancer: the database of human DNA methylation and cancer. *Nucleic Acids Res.* **36,** D836–D841.

Hong, F., Breitling, R., McEntee, C. W., Wittner, B. S., Nemhauser, J. L., and Chory, J. (2006). RankProd: a bioconductor package for detecting differentially expressed genes in meta-analysis. *Bioinformatics* **22,** 2825–2827.

Hoque, M. O., Kim, M. S., Ostrow, K. L., Liu, J., Wisman, G. B., Park, H. L., Poeta, M. L., Jeronimo, C., Henrique, R., Lendvai, A., Schuuring, E., Begum, S., *et al.* (2008). Genome-wide promoter analysis uncovers portions of the cancer methylome. *Cancer Res.* **68,** 2661–2670.

Kent, W. J. (2002). BLAT–the BLAST-like alignment tool. *Genome Res.* **12,** 656–664.

Klempnauer, K. H. (1993). Methylation-sensitive DNA binding by v-myb and c-myb proteins. *Oncogene* **8,** 111–115.

Knudson, A. G. (1971). Mutation and cancer: statistical study of retinoblastoma. *Proc. Natl. Acad. Sci. USA* **68,** 820–823.

Kuntzer, J., Eggle, D., Lenhof, H. P., Burtscher, H., and Klostermann, S. (2010). The Roche Cancer Genome Database (RCGDB). *Hum. Mutat* **4,** 407–413.

Li, L. C., and Dahiya, R. (2002). MethPrimer: designing primers for methylation PCRs. *Bioinformatics* **18,** 1427–1431.

Mohn, F., Weber, M., Schubeler, D., and Roloff, T. C. (2009). Methylated DNA immunoprecipitation (MeDIP). *Methods Mol. Biol.* **507,** 55–64.

Ongenaert, M., Van, N. L., De, M. T., Menschaert, G., Bekaert, S., and Van, C. W. (2008a). PubMeth: a cancer methylation database combining text-mining and expert annotation. *Nucleic Acids Res.* **36,** D842–D846.

Ongenaert, M., Wisman, G. B., Volders, H. H., Koning, A. J., Zee, A. G., Van, C. W., and Schuuring, E. (2008b). Discovery of DNA methylation markers in cervical cancer using relaxation ranking. *BMC Med. Genomics* **1,** 57.

Pattyn, F., Hoebeeck, J., Robbrecht, P., Michels, E., De, P. A., Bottu, G., Coornaert, D., Herzog, R., Speleman, F., and Vandesompele, J. (2006). methBLAST and methPrimerDB: web-tools for PCR based methylation analysis. *BMC Bioinform.* **7,** 496.

Rhead, B., Karolchik, D., Kuhn, R. M., Hinrichs, A. S., Zweig, A. S., Fujita, P. A., Diekhans, M., Smith, K. E., Rosenbloom, K. R., Raney, B. J., Pohl, A., Pheasant, M., *et al.* (2010). The UCSC Genome Browser database: update 2010. *Nucleic Acids Res.* **38,** D613–D619.

Rhodes, D. R., Kalyana-Sundaram, S., Mahavisno, V., Varambally, R., Yu, J., Briggs, B. B., Barrette, T. R., Anstet, M. J., Kincead-Beal, C., Kulkarni, P., Varambally, S., Ghosh, D., *et al.* (2007). Oncomine 3.0: genes, pathways, and networks in a collection of 18,000 cancer gene expression profiles. *Neoplasia* **9,** 166–180.

Rice, P., Longden, I., and Bleasby, A. (2000). EMBOSS: the European Molecular Biology Open Software Suite. *Trends Genet.* **16,** 276–277.

Roberts, R. J., Vincze, T., Posfai, J., and Macelis, D. (2010). REBASE–a database for DNA restriction and modification: enzymes, genes and genomes. *Nucleic Acids Res.* **38,** D234–D236.

Rocca-Serra, P., Brazma, A., Parkinson, H., Sarkans, U., Shojatalab, M., Contrino, S., Vilo, J., Abeygunawardena, N., Mukherjee, G., Holloway, E., Kapushesky, M., Kemmeren, P., *et al.* (2003). ArrayExpress: a public database of gene expression data at EBI. *C. R. Biol.* **326,** 1075–1078.

Rozen, S., and Skaletsky, H. (2000). Primer3 on the WWW for general users and for biologist programmers. *Methods Mol. Biol.* **132,** 365–386.

Saito, Y., Friedman, J. M., Chihara, Y., Egger, G., Chuang, J. C., and Liang, G. (2009). Epigenetic therapy upregulates the tumor suppressor microRNA-126 and its host gene EGFL7 in human cancer cells. *Biochem. Biophys. Res. Commun.* **379,** 726–731.

Schuebel, K. E., Chen, W., Cope, L., Glockner, S. C., Suzuki, H., Yi, J. M., Chan, T. A., Van, N. L., Van, C. W., van den, B. S., van, E. M., Ting, A. H., *et al.* (2007). Comparing the DNA hypermethylome with gene mutations in human colorectal cancer. *PLoS Genet.* **3,** 1709–1723.

Shipra, A., Chetan, K., and Rao, M. R. (2006). CREMOFAC–a database of chromatin remodeling factors. *Bioinformatics* **22,** 2940–2944.

Su, J., Zhang, Y., Lv, J., Liu, H., Tang, X., Wang, F., Qi, Y., Feng, Y., and Li, X. (2010). CpG_MI: a novel approach for identifying functional CpG islands in mammalian genomes. *Nucleic Acids Res.* **38,** e6.

Vire, E., Brenner, C., Deplus, R., Blanchon, L., Fraga, M., Didelot, C., Morey, L., Van, E. A., Bernard, D., Vanderwinden, J. M., Bollen, M., Esteller, M., *et al.* (2006). The Polycomb group protein EZH2 directly controls DNA methylation. *Nature* **439,** 871–874.

Wakaguri, H., Yamashita, R., Suzuki, Y., Sugano, S., and Nakai, K. (2008). DBTSS: database of transcription start sites, progress report 2008. *Nucleic Acids Res.* **36,** D97–D101.

Wingender, E., Dietze, P., Karas, H., and Knuppel, R. (1996). TRANSFAC: a database on transcription factors and their DNA binding sites. *Nucleic Acids Res.* **24,** 238–241.

Xi, Y., and Li, W. (2009). BSMAP: whole genome bisulfite sequence MAPping program. *BMC Bioinform.* **10,** 232.

Yang, C. H., Chuang, L. Y., Cheng, Y. H., Gu, D. L., Chen, C. H., and Chang, H. W. (2009). Methyl-Typing: An improved and visualized COBRA software for epigenomic studies. *FEBS Lett* **4,** 739–744.

Index